CAMBRIDGE TRACTS IN MATHEMATICS

---

78. *Finite groups and finite geometries*

p

T. TSUZUKU
Professor of Mathematics, Hokkaido University, Japan

Finite groups and finite geometries

TRANSLATED BY
A. SEVENSTER AND T. OKUYAMA

CAMBRIDGE UNIVERSITY PRESS
CAMBRIDGE
LONDON NEW YORK NEW ROCHELLE
MELBOURNE SYDNEY

Published by the Press Syndicate of the University of Cambridge
The Pitt Building, Trumpington Street, Cambridge CB2 1RP
32 East 57th Street, New York, NY 10022, USA
296 Beaconsfield Parade, Middle Park, Melbourne 3206, Australia

Originally published in Japanese as *Yugengun to Yugenkika*
by Iwanami Shoten, Tokyo, Japan 1976 and

First published in English by the Cambridge University Press 1982 as
Finite groups and finite geometries
English translation Cambridge University Press 1982

Printed in Great Britain at the University Press, Cambridge

British Library Cataloguing in Publication Data

Tsuzuku, Tosiro
Finite groups and finite geometries
(Cambridge tracts in mathematics; 78)

1. Finite groups 2. Geometry
I. Title II. Series
512′.2 QA171 80-41377

ISBN 0 521 22242 7

Dedicated to the memory of

TADASI NAKAYAMA
1912–1964

Contents

Preface

Concepts such as geometric symmetry, symmetric transformations and so on, that are introduced in elementary school mathematics, form the basis for the interrelation between groups and geometries. Symmetric transformation is a special type of motion, i.e. a transformation of Euclidean space that preserves distances, and the group of all motions – the group of automorphisms of the Euclidean space – characterizes Euclidean geometry. This is the basic idea of the so-called Erlangen Program of Klein – the group theoretical observation – according to which the greater part of geometry, especially classical geometry, can be considered in these terms. Recently, this approach has pervaded many parts of mathematics so that for the study of sets endowed with some structure, the group of its automorphisms often plays an important role.

The study of automorphism groups is also a powerful tool in the study of finite geometries. Given a finite geometry, the group of automorphisms is a finite group. Interesting geometric structures often give rise to interesting finite groups. Conversely, from the group theoretical point of view, there are many interesting problems in the areas related to finite geometries, such as the construction of new finite geometries and the determination of the finite geometry corresponding to a given finite group.

The aim of this book is to give a self-contained introduction to some fundamental results about relations between finite groups and finite geometries from the group theoretical point of view. In order to guide the reader to an understanding of the recent problem of the characterization of projective transformation groups as permutation groups, I explain the fundamental results in related areas (elementary to intermediate level difficulty) necessary to gain this understanding. A few recent results are included in the final part of the book.

Apart from high-school mathematics, the reader is expected to have a modest knowledge of linear algebra – reviewed briefly in chapter 1. The theory of finite groups and finite geometries is developed from basic principles to the point where problems mentioned above may be understood.

Since 1960, research on finite groups has mushroomed and diversified rapidly. Therefore, even though this book is restricted to areas related with finite geometries, it covers only a small part of this topic and offers far from a full picture. It is my hope that this book will arouse the reader's interest in this subject.

I was encouraged to write this book by Professor Nagayoshi Iwahori of Tokyo University, who, together with Professor Hiroshi Nagao of Osaka University was kind enough to give many useful suggestions with regard to content and organization of this book. My colleagues at Hokkaido University, Professor Hiroshi Kimura and Dr Shiro Iwasaki, and also Mr Tomoyuki Wada of Otaru University of Commerce corrected the manuscript and the proof-sheets and pointed out many mistakes and imperfections. Mr Hideo Arai of Iwanami Shoten was very helpful and patient while I was preparing the manuscript for this book. I want to express my heartfelt feelings of gratitude for their kindness.

March 1976 TOSIRO TSUZUKU

Preface to the English edition

This book is a complete translation of the Japanese text of my *Yugengun to Yugenkika* published by Iwanami Shoten (Tokyo, 1976).

I owe the translation to Dr Arjen Sevenster and Dr Tetsuro Okuyama. I would like to express my heartfelt thanks to Dr Sevenster who has demonstrated how to combine scrupulous respect for the text with its effective adaptation to the spirit of the English language, and to Dr Okuyama who has checked the text during translation and made many useful suggestions for improvement.

Hokkaido University, Japan

September 1980 TOSIRO TSUZUKU

1

Introduction

1.1 Notation and preliminaries

In this section we introduce some notation and definitions that will be used in this book.

$\mathbb{N}$, $\mathbb{Z}$, $\mathbb{Q}$, $\mathbb{R}$ and $\mathbb{C}$ will denote the sets of natural numbers, integers, rational numbers, real numbers and complex numbers respectively.

If m and n are elements of $\mathbb{Z}$ and if m is a divisor of n, we write $m|n$ or $n \equiv 0 \pmod{m}$. If m is not a divisor of n, we write $m \nmid n$ or $n \not\equiv 0 \pmod{m}$. If p is a prime number and if e is a natural number or zero such that $p^e|n$, $p^{e+1} \nmid n$, we write $p^e \mathrel{\top} n$ and we denote this p^e by n_p or $n(p)$. If π is a subset of the set of prime numbers, we define n_π for n an *element of* $\mathbb{Z}$ by $n_\pi = \prod_{p \in \pi} n_p$ and n_π is called the π-component of n. If $n = n_\pi$ or $n = -n_\pi$, n is called a π-number.

If n is an element of $\mathbb{Z}$ greater than 1, then n can be written as a product of powers of primes:

$$n = \prod_{p \text{ prime}} n_p = \prod_{\substack{p \text{ prime} \\ e_p \geq 1}} p^{e_p}.$$

This is called the prime factorization of n and this prime factorization is unique. If n is an element of $\mathbb{Z}$, $|n|$ will denote its absolute value. If m and n are elements of $\mathbb{Z}$ their greatest common divisor (i.e. the greatest positive integer that divides both m and n) is denoted by (m, n). For any pair of integers m and n, (m, n) exists and is uniquely determined. If $|m| = \prod p^{e_p}$ and $|n| = \prod p^{e'_p}$, then $(m, n) = \prod p^{e''_p}$, where e''_p is the smallest of e_p and e'_p (or equal to, say, e_p when e_p and e'_p are equal). If e'''_p denotes the greatest of e_p and e'_p (or, say, e_p when e_p and e'_p are equal), then $\prod p^{e'''_p}$ is the least common multiple of m and n, i.e. the smallest positive integer that is divided by both m and n.

If A is a set and if a is an element of A, we write $a \in A$ (or $A \ni a$).

If B is a subset of A, we write $B \subseteq A$. The corresponding negations are denoted by $a \notin A$ and $B \nsubseteq A$ respectively. If $B \subseteq A$ but $B \neq A$, we write $B \subsetneqq A$. If A_i are sets for all i of some index set I, $\bigcup_{i \in I} A_i$ and $\bigcap_{i \in I} A_i$ denote their union and intersection respectively. The symbol $\emptyset$ stands for the empty set. The set of all x that satisfy condition ... is denoted by $\{x | ...\}$. The set consisting of the elements $a, b, \ldots,$ is denoted by $\{a, b, \ldots\}$. The direct product of the two sets A and B, written as $A \times B$, is defined by

$$A \times B = \{(a, b) | a \in A, b \in B\}.$$

Similarly, the direct product of sets $A_1, \ldots, A_n$ is defined by

$$A_1 \times \ldots \times A_n = \{(a_1, \ldots, a_n) | a_i \in A_i, 1 \leq i \leq n\}.$$

The number of elements of A is denoted by $|A|$ and the set of all subsets of A is denoted by 2^A.

'For all (or any) elements a of A' is written in abbreviated form as '$\forall a \in A$' and, 'for some element a of A' is written as '$\exists a \in A$'. In the same way '$\forall a, b \in A$' means 'for all elements $a, b \in A$'. If $a, b, \ldots$ are all different, the notation '$a, b, \ldots (\neq)$' will be used.

If P and Q are statements, '$P \Rightarrow Q$' is used to denote the statement: 'If P is valid, then Q is valid'. The negation of this statement is denoted by '$P \not\Rightarrow Q$'. Instead of '$P \Rightarrow Q$ and $Q \Rightarrow P$' we write '$P \Leftrightarrow Q$'. Further '$P \rightarrow Q$' is used for 'P is valid, hence Q is valid'.

If f is a map from A into B we write $f: A \rightarrow B$ or $A \xrightarrow{f} B$. (We use the same symbol '$\rightarrow$' as a logical symbol, but this cannot cause confusion.) The collection of all maps from A into B is denoted by $M(A, B)$.

If $f: A \rightarrow B$ is a map and if a is an element of A, then the image of a under f is denoted by a^f or $f(a)$. For $S \subseteq A$ we define $f(S)(= S^f) = \{f(a) | a \in S\}$ and for $T \subseteq B$ we define $f^{-1}(T) = \{a \in A | f(a) \in T\}$. We call $f(S)$ the image of S under f and we call $f^{-1}(T)$ the inverse image of T under f. If $f(A) = B$, then f is called surjective; if $|f^{-1}(b)| \leq 1$ ($\forall b \in B$) then f is called injective. If $f: A \rightarrow B$ and $g: B \rightarrow C$ are maps then the composition $f \cdot g: A \rightarrow C$ is defined by $(f \cdot g)(a) = g(f(a))$ for $a \in A$. The map $f: A \times A \rightarrow \{0, 1\}$, defined by

$$f(a, b) = \begin{cases} 0 & \text{if } a \neq b \\ 1 & \text{if } a = b, \end{cases}$$

will be denoted by δ. Instead of $\delta(a, b)$ we will sometimes write $\delta_{a,b}$.

If for $i \in I$, A_i are subsets of A such that

(1) $A = \bigcup_{i \in I} A_i$

(2) $A_i \cap A_j = \varnothing$ if $i \neq j$,

then A is called the direct sum of $\{A_i | i \in I\}$ (as sets). In this case we write $A = \sum_{i \in I} A_i$ (or, if $I = \{1, 2, \ldots\}$, $A = A_1 + A_2 + \ldots$). The summation $A = \sum_{i \in I} A_i$ is also called a decomposition of A (as a set) into disjoint subsets.

A subset E of $A \times A$ is called a relation; instead of $(a, b) \in E$ we usually write $a \underset{E}{\sim} b$ or simply $a \sim b$ and instead of $(a, b) \notin E$, we write $a \nsim b$. We often denote the relation by $\sim$ rather than by E. An equivalence relation is a relation $\sim$ satisfying:

(i) $a \sim a \quad \forall a \in A$;

(ii) $a \sim b \Rightarrow b \sim a$;

(iii) $a \sim b$ and $b \sim c \Rightarrow a \sim c$.

Putting $C_a = \{x \in A | a \sim x\}$, we have by (ii) and (iii) that if $C_a \neq C_b$, then $C_a \cap C_b = \varnothing$. Therefore, if we pick all the different subsets from the collection $\{C_a | a \in A\}$ and if we denote this collection of subsets by $\{A_i | i \in I\}$, then we have a decomposition of A into disjoint subsets. The A_i are called equivalence classes under $\sim$. If conversely $A = \sum_{i \in I} A_i$ is a decomposition of A into disjoint subsets, then the relation $E = \bigcup_{i \in I} A_i \times A_i \subset A \times A$ is an equivalence relation and the decomposition into disjoint subsets defined by E is just the original decomposition $A = \sum_{i \in I} A_i$. Hence, there is a one-to-one correspondence between decompositions of A into disjoint subsets and equivalence relations on A.

A map $f: B \times A \to A$ is called an operation of B on A. If $(b, a) \in B \times A$, then the element $f(b, a)$ of A is denoted by a^b, ba, $b \cdot a$ and so on and called the image of a by the operation of b. An operation of A on A is called a law of composition on A. If certain laws of composition of A or operations of other sets on A are given, then A is said to carry an algebraic structure or to be an algebraic system.

We assume that the reader is acquainted with the basic concepts and theorems in linear algebra over $\mathbb{C}$. For instance:

(1) Denoting the determinant of a square matrix X by $\det X$,

we have the following result: if A is a (n, m)-matrix, if B is a (m, n)-matrix and if $\det AB \neq 0$ then $n \leq m$.

(2) If $X = (x_{ij})$ is a (n, n)-matrix, then the trace of X is defined by trace $X = \sum_{i=1}^{n} x_{ii}$. If A and P are both (n, n)-matrices and if P is non-singular (i.e. $\det P \neq 0$), then trace A = trace $P^{-1}AP$.

(3) Let A be an (n, n)-matrix. If there exists a non-zero $(n, 1)$-matrix B (i.e. a non-zero column vector) and a complex number α such that $AB = \alpha B$, then α is called an eigenvalue of A, and B is called an eigenvector of A corresponding to the eigenvalue α. An eigenvalue α is a root of the equation of the nth degree with complex coefficients $\det(xE - A) = 0$, where E is the (n, n)-identity matrix, i.e. the matrix which has ones on the diagonal and zeros elsewhere. The nth degree polynomial $\det(xE - A)$ is called the characteristic polynomial of A. The coefficient of x^{n-1} in $\det(xE - A)$ is equal to $-$trace A and the constant term is equal to $(-1)^n \det A$.

(4) The collection of all polynomials with rational coefficients $\mathbb{Q}[X]$ has many properties that correspond with properties of $\mathbb{Z}$. The role of prime numbers in $\mathbb{Z}$ is played by the irreducible polynomials of $\mathbb{Q}[X]$ (i.e. polynomials that cannot be written as the product of two non-constant polynomials). Since every polynomial can be written as a product of irreducible polynomials (unique to within order and constant factors), the greatest common divisor and the least common multiple are defined and unique to within constant factors.

(5) The nth degree polynomial with complex coefficients

$$x^n + \lambda_1 x^{n-1} + \ldots + \lambda_n = 0$$

has n complex roots.

(6) ...

We finish this section by stating and proving some properties of cyclotomic polynomials, algebraic numbers etc. that we shall have occasion to use.

For two integers m and n, if $(m, n) = 1$ we say that m is prime to n or that m and n are relatively prime.

The Euler-function φ assigns to every positive integer n the number of positive integers less than n that are relatively prime to n. If $n = p_1^{e_1} \dots p_r^{e_r}$ is the prime-factorization of n, then

$$\varphi(n) = \prod_{i=1}^{r} (p_i^{e_i} - p_i^{e_i - 1}).$$

We want to include a proof of this assertion, because it is fundamental.

For $a \in \mathbb{Z}$ the residue class modulo n is defined by $(a)_n = \{x \in \mathbb{Z} | x \equiv a \pmod{n}\}$. The collection of all residue classes modulo n is denoted by $\mathbb{Z}_n$. For $a, b \in \mathbb{Z}$, dividing both a and b by n we get $a = t_1 n + s_1$ and $b = t_2 n + s_2 (0 \le s_1, s_2 < n)$. Then $a \equiv b \pmod{n} \Leftrightarrow s_1 = s_2$. The relation $a \sim b$ defined by $a \equiv b \pmod{n}$ is an equivalence relation and the equivalence class determined by a is just $(a)_n$. The following facts are easy to check:

(i) $\forall a \in \mathbb{Z}$ there is an i such that $0 \le i < n$ and $(a)_n = (i)_n$.

(ii) $(a)_n \neq (b)_n \Rightarrow (a)_n \cap (b)_n = \varnothing$.

(iii) $\mathbb{Z}_n = \{(0)_n, (1)_n, \dots, (n-1)_n\}$.

If $(a, n) = 1$, then $(a)_n$ is called a prime residue class. Since $(a, n) = (b, n)$ whenever $a \equiv b \pmod{n}$, this definition is independent from the choice of the representant a of $(a)_n$. If we represent the collection of all prime residue classes modulo n by $\mathbb{Z}_n^*$, we have by (iii) $\varphi(n) = |\mathbb{Z}_n^*|$. If $n = n_1 n_2$ and if $(n_1, n_2) = 1$, then:

(iv) $(a)_{n_1} \cap (b)_{n_2} \neq \varnothing, \forall a, b \in \mathbb{Z}$. Hence there exists a $c \in \mathbb{Z}$ such that $(c)_{n_1} = (a)_{n_1}$ and $(c)_{n_2} = (b)_{n_2}$.

(v) $(a)_{n_1} \cap (a)_{n_2} = (a)_n$.

(vi) $(a)_{n_1}$ and $(a)_{n_2}$ are prime residue classes $\Leftrightarrow (a)_n$ is a prime residue class.

Proof

(iv) $(n_1, n_2) = 1 \rightarrow tn_1 + sn_2 = 1 (\exists t, s \in \mathbb{Z}) \rightarrow (a-b)tn_1 + (a-b)sn_2 = a - b \rightarrow a + (b-a)tn_1 = b + (a-b)sn_2 \in (a)_{n_1} \cap (b)_{n_2}$.

(v) $a + tn_1 = a + sn_2 \rightarrow tn_1 = sn_2 \rightarrow n_1 | s \rightarrow s = rn_1 \rightarrow a + sn_2 = a + rn_1 n_2 \in (a)_n$. Therefore $(a)_{n_1} \cap (a)_{n_2} \subseteq (a)_n$ and $(a)_{n_1} \cap (a)_{n_2} \supseteq (a)_n$ is evident.

(vi) Evident.

By (vi) there is a map $f: \mathbb{Z}_n^* \rightarrow \mathbb{Z}_{n_1}^* \times \mathbb{Z}_{n_2}^*$ defined by $f((a)_n) = ((a)_{n_1}, (a)_{n_2})$. This map is injective by (v) and surjective by (iv).

Therefore $|\mathbb{Z}_n^*| = |\mathbb{Z}_{n_1}^*| \times |\mathbb{Z}_{n_2}^*|$, i.e. $\varphi(n) = \varphi(n_1) \cdot \varphi(n_2)$ and hence $\varphi(n) = \varphi(p_1^{e_1}) \ldots \varphi(p_r^{e_r})$. The numbers less than p^r that are not relatively prime to p^r are $p, 2p, \ldots, p^{r-1} \cdot p$, hence $\varphi(p^r) = p^r - p^{r-1}$ and we have

$$\varphi(n) = \prod_{i=1}^{r} (p_i^{e_i} - p_i^{e_i - 1}). \qquad \blacksquare$$

The complex roots of the equation $X^n - 1 = 0$ are called the nth roots of unity. These roots correspond with the n points of the unit circle in the Gaussian plane that starting from the point corresponding with 1 divide the unit circle in n equal arcs. $\omega = e^{2\pi i/n} = \cos 2\pi/n + i \sin 2\pi/n$ is one of the nth roots of unity and since $\omega, \omega^2, \ldots, \omega^n = 1$ are all different nth roots of unity, we conclude that every nth root of unity is included in this set. An nth root of unity with the property that its nth power is the first power to become 1 is called a primitive nth root of unity. Since $\omega^m = 1 \Leftrightarrow m \equiv 0 \pmod{n}$ we conclude:

$$\omega^{md} = 1 \Leftrightarrow md \equiv 0 \pmod{n} \Leftrightarrow d \equiv 0 \pmod{n/(n, m)}.$$

Hence

$$\omega^m \text{ is a primitive } n\text{th root of unity} \Leftrightarrow (n, m) = 1.$$

Therefore the number of primitive nth roots of unity is equal to $\varphi(n)$. If d is an integer such that $d \mid n$ and $1 \leq d < n$, then $X^d - 1$ is a factor of $X^n - 1$. Hence, the least common multiple $f(X)$ (with leading coefficient 1) of $\{X^d - 1 \mid d \mid n \text{ and } 1 \leq d < n\}$ is also a factor of $X^n - 1$; i.e. $X^n - 1 = f(X) \cdot \Phi_n(X)$ for some polynomial $\Phi_n(X)$. The polynomial $\Phi_n(X)$ is called the nth cyclotomic polynomial; it has integer coefficients and its leading coefficient is equal to 1. It is easy to verify that $\Phi_n(X)$ can be written as:

$$\Phi_n(X) = \prod_{\substack{\omega_i \text{ primitive } n\text{th} \\ \text{root of } 1}} (X - \omega_i).$$

A complex number α is called algebraic if it satisfies an equation with rational coefficients:

$$a_0 x^r + \ldots + a_r = 0 \quad (a_0, \ldots, a_r \in \mathbb{Q})$$

If α satisfies an equation with $a_0 = 1, a_1, \ldots, a_r \in \mathbb{Z}$, then α is called an algebraic integer. Of course, rational numbers are algebraic and integers are algebraic integers. We have the following lemma.

Lemma 1.1.1 *If the rational number α is an algebraic integer, then α is an integer.*

Proof Suppose $\alpha \notin \mathbb{Z}$, then α can be written as: $\alpha = m/n$ with $(n, m) = 1$ and $n > 1$. Since α is an algebraic integer by assumption there is a polynomial $f(X) = X^r + a_1 X^{r-1} + \ldots + a_r (a_1, \ldots, a_r \in \mathbb{Z})$ such that $f(\alpha) = 0$. Hence

$$m^r + a_1 m^{r-1} n + \ldots + a_{r-1} m n^{n-1} + a_r n^r = 0,$$

but this contradicts the assumption $(n, m) = 1$. ∎

Lemma 1.1.2 *Let $y_1, \ldots, y_N, z \in \mathbb{C}$ such that at least one of the $y_1, \ldots, y_N$ is not equal to zero. If there exist rational numbers $a_{ij} (1 \leq i, j \leq N)$ such that*

$$zy_i = \sum_{j=1}^{N} a_{ij} y_j \quad (1 \leq i \leq N), \tag{1.1.1}$$

then z is algebraic. If all the a_{ij} are integers, then z is an algebraic integer.

Proof The equations (1.1.1) say that z is an eigenvalue of the matrix $A = (a_{ij})$, hence z is a root of the characteristic polynomial

$$f(x) = \det(xE - A) = x^N + \alpha_1 x^{N-1} + \ldots + \alpha_{N-1} x + \alpha_N.$$

Since the coefficients α_i are linear combinations of monomials of the a_{ij} with coefficients ± 1, the proof is finished. ∎

Lemma 1.1.3 *The products and sums of algebraic integers are algebraic integers.*

Proof Let α and β be algebraic integers, satisfying the equations $x^n + \lambda_1 x^{n-1} + \ldots + \lambda_n = 0 \, (\lambda_1, \ldots, \lambda_n \in \mathbb{Z})$ and $x^m + \mu_1 x^{m-1} + \ldots + \mu_m = 0 \, (\mu_1, \ldots, \mu_m \in \mathbb{Z})$ respectively. Putting $y_{i,j} = \alpha^i \beta^j (0 \leq i \leq n-1,$

$0 \leq j \leq m-1$), we have:

$$(\alpha+\beta)y_{ij} = \begin{cases} y_{i+1,j} + y_{i,j+1} & (i \leq n-2, j \leq m-2) \\ -\sum_{k=1}^{n} \lambda_k y_{n-k,j} + y_{n-1,j+1} & (i = n-1, j \leq m-2) \\ y_{i+1,m-1} - \sum_{k=1}^{m} \mu_k y_{i,m-k} & (i \leq n-2, j = m-1) \\ -\sum_{k=1}^{n} \lambda_k y_{n-k,m-1} - \sum_{k=1}^{m} \mu_k y_{n-1,m-k} & (i=n-1, j=m-1). \end{cases}$$

Therefore $\alpha+\beta$ is an algebraic integer by lemma 1.1.2. The proof for $\alpha \cdot \beta$ is similar. ■

If we expand the product $(X-x_1)\dots(X-x_n)$, we get

$$(X-x_1)\dots(X-x_n) = X^n - E_1 X^{n-1} + \dots + (-1)^n E_n$$

where

$$E_1 = \sum_{i=1}^{n} x_i, \; E_2 = \sum_{i<j} x_i x_j, \dots, E_r = \sum_{i_1 < \dots < i_r} x_{i_1} \dots x_{i_r}, \dots,$$

$$E_n = x_1 x_2 \dots x_n.$$

$E_1, \dots, E_n$ are invariant under all permutations of $x_1, \dots, x_n$ and are called the elementary symmetric polynomials. Generally, a polynomial with complex coefficients $f(x_1, \dots, x_n)$ that is invariant under all permutations of $x_1, \dots, x_n$ is called a symmetric polynomial.

Lemma 1.1.4 *For every symmetric polynomial $f(x_1, \dots, x_n)$ there exists a polynomial $g(y_1, \dots, y_n)$ such that $f(x_1, \dots, x_n) = g(E_1, \dots, E_n)$ and such that the coefficients of $g(y_1, \dots, y_n)$ are polynomial expressions with integer coefficients of the coefficients of $f(x_1, \dots, x_n)$.*

Proof The degree of a monomial $\lambda x_1^{r_1} \dots x_n^{r_n} (\lambda \in \mathbb{C}, \lambda \neq 0)$ is defined to be $\sum_{i=1}^{n} r_i$. Let $f_r(x_1, \dots, x_n)$ represent the sum of all monomials with degree r occurring in $f(x_1, \dots, x_r)$. Then f can be represented as $f = \sum_r f_r(x_1, \dots, x_n)$ and each $f_r(x_1, \dots, x_n)$ is symmetric. Hence, it suffices to prove the lemma for the case

$f(x_1, \ldots, x_n) = f_r(x_1, \ldots, x_n)$ (in this case f is called a homogeneous polynomial of degree r). Let $M = M(x_1, \ldots, x_n)$ be one of the monomials occurring in $f(x_1, \ldots, x_n)$. Then all monomials obtained from M by a permutation of $(x_1, \ldots, x_n)$ occur in $f(x_1, \ldots, x_n)$ and their sum $f_0(x_1, \ldots, x_n)$ is symmetric. Therefore it suffices to prove the lemma for the case $f(x_1, \ldots, x_n) = f_0(x_1, \ldots, x_n)$. Among the monomials occurring in f there is a monomial M that can be written as

$$M(x_1, \ldots, x_n) = \lambda_0 (x_1 \ldots x_l)^a (x_{l+1} \ldots x_m)^b \ldots (x_{p+1} \ldots x_{p+q})^t$$
$$\lambda_0 \in \mathbb{C}, a > b > \ldots > t.$$

Now we order the triples (r, a, l) in the following way: $(r, a, l) > (r_1, a_1, l_1)$ if (i) $r > r_1$ or (ii) $r = r_1$ and $a > a_1$ or (iii) $r = r_1$ and $a = a_1$ and $l > l_1$ and we proceed by induction. If $r = 1$, then $f(x_1, \ldots, x_n) = \lambda_0 E_1$ which verifies our assertion, so suppose $r > 1$. If all the $x_1, \ldots, x_n$ occur in $M(x_1, \ldots, x_n)$, then $f(x_1, \ldots, x_n) = E_n \cdot f_1(x_1, \ldots, x_n)$ for some f_1 and we can apply the induction hypothesis to f_1. So we may assume $p + q < n$. If $a = 1$ then $f(x_1, \ldots, x_n) = \lambda_0 E_r$, hence we may take $a > 1$. Let

$$M_2(x_1, \ldots, x_n) = \lambda_0 (x_1 \ldots x_l)^{a-1} (x_{l+1} \ldots x_m)^b \ldots (x_{p+1} \ldots x_{p+q})^t$$

and let $f_2(x_1, \ldots, x_n)$ be the sum of all the monomials obtained from M_2 by permutations of $x_1, \ldots, x_n$. Then:

$$E_l \cdot f_2 = (\sum x_1 \ldots x_l) \sum \lambda_0 (x_1 \ldots x_l)^{a-1} (x_{l+1} \ldots x_m)^b \ldots (x_{p+1} \ldots x_{p+q})^t$$
$$= f(x_1 \ldots x_n) + f_3(x_1 \ldots x_n) \quad \text{for some } f_3(x_1 \ldots x_n).$$

Since the theorem is true for f_2 and f_3 by the induction hypothesis, it is also true for f. ■

Lemma 1.1.5 (Gauss) *Let f be a polynomial with integer coefficients. If f is reducible over $\mathbb{Q}$, then f is reducible over $\mathbb{Z}$, i.e. if there exist non-constant polynomials $\tilde{g}$ and $\tilde{h}$ with rational coefficients, such that $f = \tilde{g} \cdot \tilde{h}$, then there exist non-constant polynomials g and h with integer coefficients such that $f = gh$.*

Proof We may assume that the greatest common divisor of the coefficients of f is equal to 1 (polynomials over $\mathbb{Z}$ with this property are called primitive). Writing the coefficients of $\tilde{g}$ and $\tilde{h}$ as fractions and multiplying both sides of the equality $f = \tilde{g} \cdot \tilde{h}$ by the product of the denominators of the coefficients of $\tilde{g}$ and $\tilde{h}$, we arrive at an equality: $af(x) = bg(x) \cdot h(x)$, where a and b are integers and g and h are primitive polynomials. It suffices to prove that $|a| = |b|$, i.e. that $g(x)h(x)$ is primitive. Assume $g(x)h(x)$ is not primitive, and let p be a prime divisor of the greatest common divisor of the coefficients of $g(x)h(x)$. Putting $g(x) = a_m x^m + a_{m-1} x^{m-1} + \ldots + a_0$, there exists an i_0 such that $a_{i_0} \not\equiv 0 \pmod p$, $a_i \equiv 0 \pmod p$, $\forall i < i_0$. Similarly, putting $h(x) = b_n x^n + b_{n-1} x^{n-1} + \ldots + b_0$, there exists a j_0 such that $b_{j_0} \not\equiv 0 \pmod p$, $a_j \equiv 0 \pmod p$, $\forall j < j_0$. The coefficient of $x^{i_0 + j_0}$ in $g(x) \cdot h(x)$ is equal to:

$$\sum_{i+j=i_0+j_0} a_i b_j = a_{i_0} b_{j_0} + \sum_{\substack{i<i_0 \text{ or } j<j_0 \\ i+j=i_0+j_0}} a_i b_j \not\equiv 0 \pmod p. \quad \text{Contradiction.}$$

■

1.2 Groups

In this section, we introduce some basic definitions and theorems about groups.

1.2.1 Groups, subgroups and cosets

Let G be a set and let f be a law of composition on G. Instead of $f(a, b)$ we will write ab and we will call ab the product of a and b. The set G together with this operation is called a group if the following conditions are satisfied:

(i) if a, b and c are arbitrary elements of G, then $(ab)c = a(bc)$ (associativity);

(ii) there exists an element $e \in G$ such that $ea = ae = a$ for all elements $a \in G$ (existence of identity); and

(iii) for every element $a \in G$ there exists an element $b \in G$ such that $ab = ba = e$ (existence of inverse).

The element e of condition (ii) is uniquely determined. (For suppose $e' \in G$ also satisfies condition (ii), then $e' = ee' = e$.) This uniquely determined element e is called the identity (or unit) of G and is denoted by 1. The element b of condition (iii) is uniquely determined by a. (For, suppose $b' \in G$ also satisfied condition (iii), then $b' = b'e = b'(ab) = (b'a)b = eb = b$.) This uniquely determined element b is called the inverse of a and is denoted by a^{-1}. The number of elements $|G|$ of G is called the order of G; groups of finite order are called finite groups, those of non-finite order infinite groups. For any three elements a_1, a_2 and $a_3 \in G$ we have $(a_1a_2)a_3 = a_1(a_2a_3)$ by (i) and we may denote this element simply by $a_1a_2a_3$. For n elements $a_1, \ldots, a_n$ of $G (n \geq 4)$ we define the product $a_1 \ldots a_n$ inductively by $a_1 \ldots a_n = (a_1 \ldots a_{n-1})a_n$. If m is an integer such that $1 \leq m < n$, then

$$(a_1 \ldots a_m)(a_{m+1} \ldots a_n) = a_1 \ldots a_n,$$

as is easily verified. In other words: if n elements of G are given in a certain order, then their product in this order is independent of the way brackets are inserted. This is expressed by saying that products obey the general associative law. As is clear from the foregoing, to prove the general associative law, we only need to assume condition (i). So we have proved the following theorem.

Theorem 1.2.1 *If an operation on G satisfies condition* (i) (*i.e. if it is associative*), *then it also satisfies the general associative law.*

The product of n identical factors a is denoted by a^n. The relation $(a^{-1})^n = (a^n)^{-1}$ is easy to check and this element is denoted by a^{-n}. For subsets S and T of G the product ST of S and T is defined by

$$ST = \{xy \mid x \in S, y \in T\}.$$

If $S_1, \ldots, S_n (n \geq 3)$ are subsets of G, their product is inductively defined by $S_1 \ldots S_n = (S_1 \ldots S_{n-1})S_n$. If for example S has only one element a, then ST will be written as aT. Further, for any subset S of G, S^{-1} is defined by

$$S^{-1} = \{a^{-1} \mid a \in S\}.$$

If $a, b \in G$, then $(ab)^{-1} = b^{-1}a^{-1}$ and if S and T are subsets of G, then $(ST)^{-1} = T^{-1}S^{-1}$.

If in addition to the conditions (i), (ii) and (iii) the following condition is also satisfied:

(iv) for all elements $a, b \in G$, $ab = ba$ (commutativity),

then G is called an Abelian or commutative group. The operation of an Abelian group is often written additively, i.e. we write $a + b$ instead of ab. The element $a + b$ is then called the sum of a and b and G is called an additive group. The identity element of an additive group is called its zero element. To distinguish groups that are not additively written from additive groups, the former are often called multiplicative groups.

Let H be a non-empty subset of G that is closed under the operation defined on G, i.e. if $a, b \in H$, then $ab \in H$ or alternatively $HH \subseteq H$. If moreover H together with this operation is a group, then H is called a subgroup of G. Notation: $G \geq H$ or $H \leq G$. If $G \geq H$ and $G \neq H$, then H is called a proper subgroup of G. Notation: $G \gneq H$ or $G > H$. G itself and the subset of G consisting of the identity only are evidently subgroups of G. These are called the trivial subgroups of G. If H is a subgroup of G then $HH \subseteq H$ and $H^{-1} \subseteq H$ since for any element $a \in H, a^{-1} \in H$. Conversely, if H is a non-empty subset of G satisfying $HH \subseteq H$ and $H^{-1} \subseteq H$, then it is easy to verify that H is a subgroup of G. So we have proved:

Theorem 1.2.2 *Let H be a non-empty subset of a group G. H is a subgroup of G if and only if $HH \subseteq H$ and $H^{-1} \subseteq H$.*

For a subset S of G the intersection $\bigcap_{G \geq H \text{ and } H \supseteq S} H$ is a subgroup of G. It is the smallest subgroup of G that contains the subset S. It is called the subgroup of G generated by S and denoted by $\langle S \rangle$. S is called a system of generators of $\langle S \rangle$. If S has only one element a, we write $\langle S \rangle = \langle a \rangle$. The subgroup $\langle a \rangle$ is called the cyclic subgroup generated by a and a is called a generator of $\langle a \rangle$. It is easy to verify that $\langle a \rangle = \{a^n \mid n \in \mathbb{Z}\}$. The order of $\langle a \rangle$ is called the order of a and is denoted by $|a|$.

Example $\mathbb{Z}$, $\mathbb{Q}$, $\mathbb{R}$ and $\mathbb{C}$ are groups under addition (or the usual addition). Consider the set $\mathbb{Z}_n$, defined in §1.1. If $\alpha = (a)_n$ and $\beta = (b)_n$ are two elements of $\mathbb{Z}_n$, then we define their sum by: $\alpha + \beta = (a + b)_n$. It is easy to check that this definition is independent of the choice of the representants of α and β and that $\mathbb{Z}_n$ becomes an additive group under the sum defined above. The subsets of $\mathbb{Q}$, $\mathbb{R}$ and $\mathbb{C}$ obtained by omitting 0 are denoted by $\mathbb{Q}^\#$, $\mathbb{R}^\#$ and $\mathbb{C}^\#$ respectively and easily seen to be commutative groups under the usual multiplication. $\mathbb{Z}^\# = \mathbb{Z} - \{0\}$ is closed under multiplication (i.e. if $a, b \in \mathbb{Z}^\#$, then $ab \in \mathbb{Z}^\#$), associativity holds and there exists an identity element in $\mathbb{Z}^\#$, but in general there are no inverses in $\mathbb{Z}^\#$. Therefore $\mathbb{Z}^\#$ is not a group under multiplication. Consider $\mathbb{Z}_n^*$ ($\mathbb{Z}_n^*$ is not $\mathbb{Z}_n - \{0\}$, but the set of all prime residue classes modulo n as defined in §1.1.) If $\alpha = (a)_n$ and $\beta = (b)_n$ are two elements of $\mathbb{Z}_n^*$, then their product is defined by $\alpha\beta = (ab)_n$. This definition is independent of the choice of the representants of α and β, and $\mathbb{Z}_n^*$ together with this multiplication becomes a group. $\mathbb{Z}_n$ and $\mathbb{Z}_n^*$ are finite groups, all the other groups mentioned above are infinite. The order of $\mathbb{Z}_n$ and $\mathbb{Z}_n^*$ is given by $|\mathbb{Z}_n| = n$ and $|\mathbb{Z}_n^*| = \varphi(n)$. Especially, if p is prime, $|\mathbb{Z}_p| = p$ and $|\mathbb{Z}_p^*| = p - 1$. The groups $\mathbb{Z}$ and $\mathbb{Z}_n$ are cyclic, the groups $\mathbb{Q}$, $\mathbb{R}$, $\mathbb{C}$, $\mathbb{Q}^\#$, $\mathbb{R}^\#$ and $\mathbb{C}^\#$ are not, they do not even have finite systems of generators. $\mathbb{Q}$, $\mathbb{R}$ and $\mathbb{C}$ have no other subgroups of finite order than the subgroup consisting of the identity element ($=0$) only. $\mathbb{Q}^\#$, $\mathbb{R}^\#$ and $\mathbb{C}^\#$ have subgroups of finite order other than that consisting of the identity element ($=1$) only. (For example, $\{-1, 1\}$ is a subgroup of order 2.)

We want to show that all subgroups of finite order of $\mathbb{C}^\#$ are cyclic. Let G be a subgroup of finite order of $\mathbb{C}^\#$ and let a be an element of G such that $|a| = m = m_1 m_2$ for certain m_1 and m_2 with $(m_1, m_2) = 1$. (Here, $|a|$ denotes the order, not the absolute value of a.) Since $(m_1, m_2) = 1$, there are t_1 and $t_2 \in \mathbb{Z}$ such that $t_1 m_1 + t_2 m_2 = 1$. Hence $a = a^{t_1 m_1 + t_2 m_2} = a^{t_1 m_1} a^{t_2 m_2}$. Putting $a_1 = a^{t_2 m_2}$ and $a_2 = a^{t_1 m_1}$ we have $a = a_1 a_2 = a_2 a_1$ and $|a_1| = m_1$ and $|a_2| = m_2$. Now, let $m = p_1^{e_1} \dots p_r^{e_r}$ be the prime factorization of m, then by repeating the above argument, we find $a_1, \dots, a_r \in G$ such that $a = a_1 \dots a_r$ and $|a_i| = p_i^{e_i}$ $(i = 1, \dots, r)$. If $a, b \in G$ such that $|a| = m$ and $|b| = n$ then there are prime numbers $p_1, \dots, p_t$ such that $m = p_1^{e_1} \dots p_t^{e_t}, n = p_1^{f_1} \dots p_t^{f_t}$ with $e_i \geq 0$, $f_i \geq 0$. Hence there are elements $a_1 \dots a_t, b_1, \dots, b_t \in G$ such that $a = a_1 \dots a_t, b = b_1 \dots b_t$, $|a_i| = p_i^{e_i}$ and $|b_i| = p_i^{f_i}$ $(i = 1, \dots, t)$. (If $e_i = 0$ or $f_i = 0$, then we put $a_i = 1$ or $b_i = 1$ respectively.) Define c_i by $c_i = a_i$ if $|a_i| \geq |b_i|$ and $c_i = b_i$ if $|b_i| \geq |a_i|$, and consider the product $c = c_1 \dots c_t \in G$. Then

$|c|$ equals the least common multiple of m and n. Since the order of G is finite, we conclude by repeating this procedure a finite number of times, that there exists an element $d \in G$ such that $|a| \,\big|\, |d|$ for all $a \in G$. Therefore, $a^{|d|} = 1$ for all $a \in G$. Since $\langle d \rangle$ is a subgroup of G we have $|d| \leq |G|$. Suppose $|d| < |G|$, then the equation $X^{|d|} - 1 = 0$ would have more roots than $|d|$ in $\mathbb{C}$. Contradiction, hence $|d| = |G|$, i.e. $\langle d \rangle = G$.

Using a similar method we can prove that $\mathbb{Z}_p^*$ (p prime) is cyclic. To do this we need the following lemma, where as usual $\mathbb{Z}[X]$ denotes the collection of all polynomials with integer coefficients.

Lemma 1.2.3 *Let $f(X) \in \mathbb{Z}[X]$ and $a \in \mathbb{Z}$. Then:*

(1) $f(a) \equiv 0 \pmod{p} \Rightarrow f(X) = (X - a)g(X) + m$ for some $g(X) \in \mathbb{Z}[X]$ and some $m \in \mathbb{Z}$ with $m \equiv 0 \pmod{p}$.

(2) If $(a_1)_p, \ldots, (a_r)_p$ are different elements of $\mathbb{Z}_p$ and if $f(a_i) \equiv 0 \pmod{p} (i = 1, \ldots, r)$, then $f(X)$ can be written as

$$f(X) = (X - a_1) \ldots (X - a_r)h(X) + u(X)$$

for some $h(X), u(X) \in \mathbb{Z}[X]$ such that $\deg u(X) < r$ *and all coefficients of $u(X)$ are divisible by p.*

Proof (1) Dividing $f(X)$ by $X - a$, we get $f(X) = (X - a)g(X) + m$ for some polynomial $g(X) \in \mathbb{Z}[X]$ and some $m \in \mathbb{Z}$. (Since deg $(X - a) = 1$, either the degree of the rest is zero, or the rest is zero.) Since $f(a) = m$, we conclude $m \equiv 0 \pmod{p}$.

(2) By induction on r. If $r = 1$, the assertion is true by (1). Let us assume that the assertion is true for $r - 1$, then

$$f(X) = (X - a_1) \ldots (X - a_{r-1})g_{r-1}(X) + u_{r-1}(X)$$

for some $g_{r-1}(X)$, $u_{r-1}(X) \in \mathbb{Z}[X]$ such that deg $u_{r-1}(X) < r - 1$ and all coefficients of $u_{r-1}(X)$ are divisible by p. Hence: $f(a_r) = (a_r - a_1) \ldots (a_r - a_{r-1})g_{r-1}(a_r) + u_{r-1}(a_r)$, where $f(a_r) \equiv 0 \pmod{p}$ and $u_{r-1}(a_r) \equiv 0 \pmod{p}$. Therefore $(a_r - a_1) \ldots (a_r - a_{r-1})g_{r-1}(a_r) \equiv 0 \pmod{p}$ and since by assumption $a_r - a_i \not\equiv 0 \pmod{p}$ $(i = 1, \ldots, r - 1)$, we conclude that $g_{r-1}(a_r) \equiv 0 \pmod{p}$. So we can write $g_{r-1}(X) = (X - a_r)g_r(X) + m_r$ for some $g_r(X) \in \mathbb{Z}[X]$ and $m_r \in \mathbb{Z}$ such that $m_r \equiv 0 \pmod{p}$. Substituting this in the expression for $f(X)$ above, yields the desired result. ∎

The element $(a)_p \in \mathbb{Z}_p$ is called a root modulo p of $f(X) \in \mathbb{Z}[X]$ if $f(a) \equiv 0 \pmod{p}$. This definition is independent of the choice of the representant of $(a)_p$. The number of roots modulo p of $f(X)$ does not exceed the degree of $f(X)$ by (2). Now, if t is the least common multiple of the orders of all elements of $\mathbb{Z}_p^*$, then there exists an element $\alpha \in \mathbb{Z}_p^*$ such that $|\alpha| = t$. (This is proved in exactly the same way as for $\mathbb{C}^*$. In fact, the only thing we used about $\mathbb{C}^*$ is that it is a finite Abelian group.) Every element $\beta \in \mathbb{Z}_p^*$ satisfies $|\beta| \mid t$, hence $\beta^t = 1$. Since $|\mathbb{Z}_p^*| = p - 1$ we conclude that the equation $X^t - 1 = 0$ has $p - 1$ different roots modulo p and therefore $t = p - 1$, i.e. $\mathbb{Z}_p^* = \langle \beta \rangle$. We sum up our results in the following theorem:

Theorem 1.2.4 (1) *The finite subgroups of* $\mathbb{C}^*$ *are all cyclic.*
(2) *For any prime* p *the group* $\mathbb{Z}_p^*$ *is cyclic of order* $p - 1$.

Let H and K be subgroups of a group G, then the subsets Ha, aK and HaK of G are called a left coset of H, a right coset of K and a double coset of H and K respectively. The element a is called a representant of these cosets. The collection of all (distinct) left cosets of H is denoted by $H\backslash G$. A subset of G obtained by choosing one element from each left coset of H is called a left representative system of H. G/K and $H\backslash G/K$ represent the collection of all right cosets of K and the collection of all double cosets of H and K. Right representative systems of K and double representative systems of H and K are defined similarly. If $x \in Ha$, then $x = ya$ for some $y \in H$. Hence: $Hx = Hya = Ha$. Therefore, if $a, b \in G$ such that $Ha \neq Hb$, then $Ha \cap Hb = \varnothing$, i.e. we have a decomposition of G into disjoint subsets

$$G = \sum_{Ha \in H\backslash G} Ha,$$

called the decomposition into left cosets of H. Similarly, we have a decomposition into right cosets of K

$$G = \sum_{aK \in G/K} aK,$$

and a decomposition into double cosets of H and K

$$G = \sum_{HaK \in H\backslash G/K} HaK.$$

For $a, b \in G$ we have

$$aH = bH \Leftrightarrow Ha^{-1} = (aH)^{-1} = (bH)^{-1} = Hb^{-1}.$$

Therefore, if $\{a_\nu | \nu \in I\}$ is a right representative system of H, then $\{a_\nu^{-1} | \nu \in I\}$ is a left representative system of H, hence $|G \backslash H| = |H/G|$. This number is called the index of H in G, and it is denoted by $|G:H|$. From $ax = ay \Leftrightarrow x = y$ we conclude $|H| = |aH|$. Hence:

Theorem 1.2.5 (Lagrange) *If G is a finite group, then:*

(1) $|G| = |G:H||H|$ *for any subgroup H of G.*

(2) *Especially,* $|a| \, \big| \, |G|$ *for any element* $a \in G$, *hence* $a^{|G|} = 1$ *for all* $a \in G$.

1.2.2 Conjugacy, normal subgroups and quotient groups

If a and b are elements of a group G and if S and T are subsets of G, then we define

$$b^a = a^{-1}ba, \quad S^a = \{x^a | x \in S\}, \quad S^T = \bigcup_{x \in T} S^x.$$

The subsets S_1 and S_2 of G are called G-conjugate if there exists an element $a \in G$ such that $S_1^a = S_2$. Notation: $S_1 \underset{G}{\sim} S_2$ or $S_1 \sim S_2$. This relation $\sim$ is obviously an equivalence relation on 2^G. An equivalence class under $\sim$ is called a G-conjugacy class or simply conjugacy class. Since G can be regarded as a subset of 2^G, we can consider the restriction of the relation $\sim$ to G, giving the relation defined by $a \sim b \Leftrightarrow c^{-1}ac = b$ for some $c \in G$. The equivalence classes of G under this relation are called conjugacy classes and the number of conjugacy classes of G is called the class number of G.

For any subset S of G we define

$$\mathcal{N}_G(S) = \{a \in G | S^a = S\} \text{ and } \mathcal{C}_G(S) = \{a \in G | x^a = x \quad \forall x \in S\}.$$

$\mathcal{N}_G(S)$ and $\mathcal{C}_G(S)$ are subgroups of G called the normalizer and the centralizer of S (in G) respectively. Note that $\mathcal{C}_G(S)$ is the set of all elements that commute with every element of S. If $S = \{a\}$, then $\mathcal{N}_G(S) = \mathcal{C}_G(S)$ and this subgroup is denoted by $\mathcal{C}_G(a)$. Since $S^x = S^y \Leftrightarrow xy^{-1} \in \mathcal{N}_G(S)$, the number of subsets of G conjugate to S

is $|G:\mathcal{N}_G(S)|$. Especially, the number of elements in the conjugacy class that contains a is $|G:\mathcal{C}_G(a)|$.

If $H \leq G$ and $\mathcal{N}_G(H) = G$, then H is called a normal subgroup of G. Notation: $H \trianglelefteq G$. The subgroups $\{1\}$ and G are obviously normal subgroups; they are called the trivial normal subgroups of G. Groups whose only normal subgroups are the trivial subgroups are called simple groups. Since every subgroup of an Abelian group is normal, a simple Abelian group has to be a group of prime order. If $H, K \leq G$ and if $K \leq \mathcal{N}_G(H)$, then we say that K normalizes H. Notation: $H \rtimes K$. If $H \trianglelefteq G$ and if aH and bH are two elements of G/H, then: $aH \cdot bH = abb^{-1}HbH = abHH = abH$. Therefore the product of two elements of G/H is again an element of G/H, i.e. this defines a multiplication on G/H. It is easy to verify that G/H with this multiplication becomes a group: the quotient group of G with respect to H. The identity element of G/H is H and the inverse of aH is $a^{-1}H$. The subgroup $\mathcal{C}_G(G)$ is called the centre of G and is denoted by $Z(G)$. It follows immediately from the definitions that $Z(G) \trianglelefteq G$, hence we can consider the quotient group $G/Z(G)$. Then $Z(G/Z(G))$ is a subgroup of $G/Z(G)$ and the union (in G) of all elements of this subgroup forms a subgroup $Z^2(G)$ of G containing $Z(G)$, such that

$$Z(G/Z(G)) = Z^2(G)/Z(G).$$

Continuing in this way, we can define a sequence of normal subgroups

$$Z(G) = Z^{(1)}(G) \leq Z^{(2)}(G) \ldots \leq Z^{(i-1)}(G) \leq Z^{(i)}(G) \ldots$$

satisfying

$$Z(G/Z^{(i-1)}(G)) = Z^{(i)}(G)/Z^{(i-1)}(G) \quad (i \geq 2).$$

$Z^i(G)$ is called the ith centre of G and the sequence $Z^{(1)}(G), Z^{(2)}(G), \ldots$ is called the upper central series.

If S is a subset of G, then $\bigcap_{S \subset H \text{ and } H \trianglelefteq G} H$ is the smallest normal subgroup of G that contains S. This normal subgroup is called the normal subgroup of G generated by S and it is denoted by $\langle\langle S \rangle\rangle$. S is called a system of generators. The relation $\langle\langle S \rangle\rangle = \langle S^G \rangle$ is easy to check.

For $a, b \in G$ the element $a^{-1}b^{-1}ab = a^{-1}a^b$ is called the commutator of a and b and denoted by $[a, b]$. For $a, b, c \in G$ we put: $[a, b, c] = [[a, b], c]$. We have the following fundamental relations:

Theorem 1.2.6 *For x, y and $z \in G$ the following relations hold:*
(1) $[xy, z] = [x, z]^y[y, z], \quad [x, yz] = [x, z][x, y]^z$
(2) $[x, y^{-1}, z]^y[y, z^{-1}, x]^z[z, x^{-1}, y]^x = 1.$

Proof (1)

$$[xy, z] = y^{-1}x^{-1}z^{-1}xyz = y^{-1}(x^{-1}z^{-1}xz)y(y^{-1}z^{-1}yz)$$
$$= [x, z]^y \cdot [y, z].$$

The proof of the second relation is similar.
(2)
$$[x, y^{-1}, z]^y = y^{-1}[x^{-1}yxy^{-1}, z]y = y^{-1}(yx^{-1}y^{-1}xz^{-1}x^{-1}yxy^{-1}z)y$$
$$= (x^{-1}y^{-1}xz^{-1}x^{-1})(yxy^{-1}zy) = (xzx^{-1}yx)^{-1}(yxy^{-1}zy).$$
In the same way,
$$[y, z^{-1}, x]^z = (yxy^{-1}zy)^{-1}(zyz^{-1}xz)$$
and
$$[z, x^{-1}, y]^x = (zyz^{-1}xz)^{-1}(xzx^{-1}yx).$$
The desired identity follows at once from these three expressions. ∎

Let H and K be subsets of G, then the commutator subgroup of H and K is defined by

$$[H, K] = \langle \{[a, b] \mid a \in H, b \in K\} \rangle.$$

Especially, $[G, G]$ is called the commutator subgroup of G and is also denoted by $D(G)$. From $[a, b]^c = [a^c, b^c]$ we conclude: $G \trianglerighteq D(G)$. The ith commutator subgroup $D^{(i)}(G)$ is defined inductively as follows: $D^1(G) = D(G)$ and $D^{(i)}(G) = D(D^{(i-1)}(G))$. Since $[G, G]$ is the subgroup of G generated by the subset $\{[a, b] = a^{-1}b^{-1}ab \mid a, b \in G\}$, we have:

Theorem 1.2.7 *Suppose $G \trianglerighteq H$, then:*

$$G/H \textit{ is commutative} \Leftrightarrow H \geq [G, G].$$

If G_1 and G_2 are groups and if $f:G_1 \to G_2$ is a map satisfying the condition

$$f(ab)=f(a)f(b) \quad \forall a, b\in G_1,$$

then f is called a homomorphism. Notation: $G_1 \overset{f}{\precsim} G_2$ or simply $G_1 \precsim G_2$. The relations $f(1)=1$ and $f(a^{-1})=(f(a))^{-1}$ $(a\in G_1)$ are easy to check. If f is a surjective homomorphism, i.e. if $f(G_1)=G_2$, then G_2 is said to be homomorphic to G_1. If f is a surjective and one-to-one homomorphism then f is called an isomorphism from G_1 onto G_2, and we say that G_2 is isomorphic to G_1 or that G_1 and G_2 are isomorphic. Notation: $G_1 \overset{f}{\simeq} G_2$ or simply $G_1 \simeq G_2$. If $G_1 \overset{f}{\precsim} G_2$ is a homomorphism, then the inverse image of the identity of G_2, i.e. the set

$$f^{-1}(1)=\{a\in G \mid f(a)=1\}$$

is called the kernel of f and it is denoted by Ker f. From $f(a)=1 \Rightarrow f(b^{-1}ab)=f(b^{-1})f(a)f(b)=f(b)^{-1}f(b)=1$, it follows that Ker f is a normal subgroup of G_1. It is also easy to verify that $f(G_1)$ is a subgroup of G_2.

Let N be a normal subgroup of G. It is easy to check that the map f from G to the quotient group G/N defined by

$$f(a)=aN \quad a\in G$$

is a surjective homomorphism, called the natural homomorphism from G onto G/N. The natural homomorphism f satisfies Ker $f=N$.

Now we have the following fundamental result:

Theorem 1.2.8 (Homomorphism theorem) *If f is a homomorphism from G_1 into G_2 then:*

$$G_1/\mathrm{Ker}\, f \simeq f(G_1).$$

Proof Define $\tilde{f}:G_1/\mathrm{Ker}\ f \to f(G_1)$ by

$$\tilde{f}(a\ \mathrm{Ker}\, f)=f(a) \quad \text{for } a\in G_1.$$

If $b\in \mathrm{Ker}\, f$, then $f(ab)=f(a)f(b)=f(a)$, hence the definition of $\tilde{f}(a\ \mathrm{Ker}\, f)$ is independent from the choice of the representant a.

It is easy to check that $\tilde{f}$ is an isomorphism between $G_1/\mathrm{Ker}\, f$ and $f(G_1)$. ■

From this theorem we get the following corollary.

Corollary 1.2.9 (Isomorphism theorem) *If N is a normal subgroup of G and if K is a subgroup of G, then:*

(1) *KN is a subgroup of G and $K \cap N$ is a normal subgroup of K.*

(2) *$KN/N \simeq K/K \cap N$.*

Proof (1) Since N is a normal subgroup of G, we have

$$KN = \bigcup_{x \in K} xN = \bigcup_{x \in K} Nx = NK.$$

Hence $(KN)(KN) = (KK)(NN) = KN$ and $(KN)^{-1} = N^{-1}K^{-1} = NK = KN$ and KN is a subgroup of G by theorem 1.2.2. It is easy to check that $K \cap N$ is a normal subgroup of K.

(2) Define a map f from K to KN/N by

$$f(a) = aN \quad \text{for } a \in K.$$

Then f is surjective and $\mathrm{Ker}\, f = K \cap N$ since $f(a) = N \Leftrightarrow a \in K \cap N$. The conclusion of the corollary now follows from theorem 1.2.8. ■

Let G be a group. Isomorphisms from G onto G are called automorphisms of G. The collection of all automorphisms of G is denoted by Aut G. If we define the product of two elements of Aut G to be their composition as maps, Aut G becomes a group: the automorphism group of G. For $a \in G$ we define a map σ_a from G into G by $\sigma_a(x) = x^{\sigma_a} = a^{-1}xa$. The map σ_a is an automorphism, called the inner automorphism corresponding with a. Let Inn (G) denote the collection of all inner automorphisms of G, i.e. Inn $(G) = \{\sigma \in \mathrm{Aut}\, G \mid \sigma = \sigma_a, \exists a \in G\}$. If $a, b \in G$, then $\sigma_{ab} = \sigma_a \sigma_b$ and $(\sigma_a)^{-1} = \sigma_{a^{-1}}$, hence Inn G is a subgroup of Aut G, called the subgroup of inner automorphisms of G. If $\tau \in \mathrm{Aut}\, G$ and $\sigma = \sigma_a \in \mathrm{Inn}\, G$, then $\tau^{-1}\sigma_a\tau = \sigma_{a^\tau}$, hence Aut $G \trianglerighteq$ Inn G. The quotient group Aut G/Inn G is called the group of outer automorphisms of G and denoted by

Out (G). Consider the surjective homomorphism from G onto Inn G defined by $a \to \sigma_a$. From $\sigma_a = 1 \Leftrightarrow \sigma_a(x) = x \; \forall x \in G \Leftrightarrow a^{-1}xa = x \; \forall x \in G$ we conclude that the kernel of this homomorphism is just the centre $Z(G)$ of G. Hence $G/Z(G) \simeq \text{Inn}\,(G)$. If H is a subgroup of G such that $H^\tau = H$ for all automorphisms $\tau \in \text{Aut}\, G$, then H is called a characteristic subgroup of G. Notation: $G \trianglerighteq H$. For example, $Z^{(i)}(G)$ and $D^{(i)}(G)$ $(i = 1, 2, \ldots)$ are all characteristic subgroups of G. From Aut $G \geq$ Inn G we conclude that a characteristic subgroup of G is necessarily normal. If $G \trianglerighteq H$ and if $H \trianglerighteq K$, then $G \trianglerighteq K$ because if a is an arbitrary element of G then σ_a is an automorphism of H.

As an example, let us determine the group of automorphisms of a group G of prime order p. As we saw before, G is a simple cyclic group. If α is a generating element of G, then G can be represented as: $G = \{1, \alpha, \ldots, \alpha^{p-1}\}$. Multiplication is given by $\alpha^i\alpha^j = \alpha^{i+j} = \alpha^k$, where k is the remainder of $i + j$ divided by p. Hence the group G is completely determined by p and therefore isomorphic to the additive group $\mathbb{Z}_p$. Now, let $f \in \text{Aut}\,(G)$. Then f is completely determined by its value on α, say $f(\alpha) = \alpha^i$. Here, i is not unambiguously determined, but since $\alpha^i = \alpha^j \Leftrightarrow \alpha^{i-j} = 1 \Leftrightarrow i \equiv j \pmod p$ and since moreover $i \not\equiv 0 \pmod p$, the element $a = (i)_p$ of $\mathbb{Z}_p^*$ is unambiguously determined by f. Conversely, an element $a \in \mathbb{Z}_p^*$ unambiguously determines an element $f_a \in \text{Aut}\, G$ such that $f_a(\alpha) = \alpha^i$ where $(a) = (i)_p$. For $a, b \in \mathbb{Z}_p^*$ we have $f_a f_b = f_{ab}$ and further $f_a = 1 \Leftrightarrow a = (1)_p$, hence the correspondence $f_a \leftrightarrow a$ sets up an isomorphism between Aut G and $\mathbb{Z}_p^*$. Since Inn $G = 1$, we have Aut $G =$ Out G. We sum up our results in the next theorem.

Theorem 1.2.10 *If G is a group of prime order p, then*

$$\text{Aut}\, G \simeq \mathbb{Z}_p^* \quad (= \textit{cyclic group of order } p - 1).$$

Since Inn $(G) = 1$, Out $G =$ Aut G.

For the rest of this section, let G be a finite group. Let π be a subset of the set of prime numbers, such that every prime factor of the order of G is contained in π (i.e. such that $|G|$ is a π-number), then G is called a π-group. Especially, if $\pi = \{p\}$, then G is called a p-group. The collection of all prime numbers not contained in π

is denoted by π' (and if $\pi = \{p\}$ by p'). A subgroup H of G such that H is a π-group and $|G:H|$ is a π'-number is called a Hall π-subgroup. Especially, if $\pi = \{p\}$, then H is called a Sylow p-subgroup. If H is a Hall π-subgroup of G and if K is a Hall π'-subgroup of G, then K is called a π-complementary subgroup of H or a π-complement of G. If H_1 and H_2 are normal π-subgroups of G then H_1H_2 is also a normal π-subgroup. ($H_1H_2/H_2 \simeq H_1/H_1 \cap H_2$ by corollary 1.2.9, hence H_1H_2/H_2 is a π-subgroup. Moreover, $|H_1H_2| = |H_1H_2 : H_2||H_2|$ by theorem 1.2.5, hence $|H_1H_2|$ is a π-number, i.e. H_1H_2 is a π-subgroup.) So we conclude that there exists a greatest normal π-subgroup in G, that we denote by $O_\pi(G)$ or simply by O_π. For $\sigma \in \operatorname{Aut} G$, $\sigma(O_\pi) = O_\pi^\sigma$ is also a π-subgroup. From $O_\pi^\sigma \trianglelefteq G^\sigma = G$, we conclude $O_\pi^\sigma = O_\pi$, i.e. $O_\pi \leqq G$. If π_1 and π_2 are subsets of the set of prime numbers, and if f is the natural homomorphism from G onto $\bar{G} = G/O_{\pi_1}$, the subgroup $f^{-1}(O_{\pi_2}(\bar{G}))$ is denoted by O_{π_1,π_2}. It is not difficult to check that $O_{\pi_1,\pi_2} \leqq G$. If $\pi_1, \pi_2, \ldots, \pi_r$ are subsets of the set of prime numbers, the inductively defined subgroups $O_{\pi_1,\ldots,\pi_r}$ are characteristic subgroups of G. Now, let H_1 and H_2 be normal subgroups of G, such that both G/H_1 and G/H_2 are π-groups, then $H_1 \cap H_2$ is also a normal subgroup of G and using theorem 1.2.5 and corollary 1.2.9 it is not difficult to prove that $G/H_1 \cap H_2$ is a π-group. We conclude that among all the subgroups N of G such that G/N is a π-group, there exists a smallest normal subgroup, that is denoted by $O^\pi(G)$ or simply by O^π. It follows immediately from the definition that $O^\pi \leqq G$. Just as for O_π, the subgroups $O^{\pi_1,\ldots,\pi_r}$ are defined inductively and easily seen to be characteristic.

1.2.3 Permutation groups and permutation representations

A one-to-one surjective map of a set Ω onto itself is called a permutation of Ω. The collection of all permutations of Ω is represented by S^Ω. For σ and $\tau \in S^\Omega$ the composition of $\sigma\tau$ of σ and τ is again an element of S^Ω. Hence taking compositions defines a multiplication on S^Ω and with this multiplication S^Ω becomes a group called the symmetric group on Ω. Subgroups of S^Ω are called permutation groups on Ω. If Ω is finite, say $|\Omega| = n$, S^Ω is often written as S_n and called the symmetric group of degree n, and permutation groups on Ω are also called permutation groups of degree n. The identity of S^Ω is the identity map of Ω and is represented by 1_Ω.

Let G be a group. If a homomorphism f from G into S^Ω is given, we say that G has a permutation representation on Ω, or that G operates on Ω. Notation: (G, Ω). Elements of Ω are called points and $|\Omega|$ is called the degree of the permutation representation. If σ is an element of G the image $a^{f(\sigma)}$ of a point a under the permutation $f(\sigma)$ is often simply denoted by a^σ. If Δ is a subset of Ω, the collection of all images of points of Δ under $f(\sigma)$ is denoted by Δ^σ, i.e. $\Delta^\sigma = \{a^\sigma | a \in \Delta\}$.

If G is a permutation group on Ω, a permutation representation of G on Ω is given by the inclusion homomorphism. In this case we identify G with (G, Ω), so we will use expressions such as: the permutation group (G, Ω). For example if G is an arbitrary group, and if N is a normal subgroup of G, every element σ of G determines an inner automorphism of G and this inner automorphism of G induces a permutation of N. Hence we have a permutation representation of G on N. The permutation of N determined by $\sigma \in G$ is called the conjugacy operation determined by σ. As another example, let H be a subgroup of G and let Ω denote the set of all left cosets of H. For $\sigma \in G$ we define a permutation of Ω by $(Hx)^\sigma = Hx\sigma$, giving a permutation representation of G on $(H \backslash G)$ called the left permutation representation of G corresponding with H and denoted by $(G, H \backslash G)$. If $\Omega = (G/H)$ and if $\sigma \in G$, we define a permutation of Ω by $(xH)^\sigma = \sigma^{-1}xH$, giving the right permutation representation $(G, G/H)$ of G corresponding with H. Especially, if $H = \{1\}$, we call $(G, H \backslash G)$ and $(G, G/H)$ respectively the left and right regular permutation representation of G.

Let G be a group operating on Ω. The relation $\sim$ defined on Ω by

$$a \sim b \Leftrightarrow a^\sigma = b \quad \text{for some } \sigma \in G$$

is an equivalence relation on Ω. The equivalence classes are called orbits (of Ω) under G, or G-orbits. The G-orbit containing the element a of Ω is $a^G = \{a^\sigma | \sigma \in G\}$. Let $\{\Omega_1, \ldots, \Omega_r\}$ be the collection of all different G-orbits of Ω, then Ω (as a set) can be represented as a direct sum: $\Omega = \sum_{i=1}^{r} \Omega_i$. This representation as a direct sum is called the orbit decomposition of Ω under G or the G-orbit decomposition. The G-orbit decomposition is unambiguously determined by G. If Ω itself is a G-orbit, (G, Ω) is called transitive. It follows immediately from the definition that this happens if and only if for every two elements a and $b \in \Omega$ there exists an element $\sigma \in G$ such that $a^\sigma = b$. If Ω_i is a G-orbit of Ω, then (G, Ω_i) is transitive, and (G, Ω_i) is called a transitive component of (G, Ω). If a is an element of Ω,

$$G_a = \{\sigma \in G \mid a^\sigma = a\}$$

is a subgroup of G, called the stabilizer of a. If Δ is a subset of Ω the subgroup $G_\Delta = \bigcap_{a \in \Delta} G_a$ is called the pointwise stabilizer of Δ. The set $\{\sigma \in G \mid \Delta^\sigma = \Delta\}$ is also a subgroup of G, that is denoted by $G_{\langle \Delta \rangle}$ and called the setwise stabilizer of Δ.

Theorem 1.2.11 *If G is a finite group operating on Ω and if Δ is a G-orbit containing the element a of Ω, then $|G| = |\Delta|\,|G_a|$.*

Proof Follows easily from

$$a^\sigma = a^\tau \Leftrightarrow \sigma\tau^{-1} \in G_a \Leftrightarrow G_a\sigma = G_a\tau. \qquad \blacksquare$$

A subset Δ of Ω that is invariant under all elements σ of G, i.e. $\Delta^\sigma = \Delta$ for all $\sigma \in G$, is called an invariant domain. Evidently, G acts on Δ and (G, Δ) is called a Δ-component of (G, Ω). Orbits are examples of invariant domains and conversely, every invariant domain is the direct sum (as a set) of a certain number of orbits.

Let (G, Ω) and (H, Γ) be permutation representations of the groups G and H respectively, let f be a homomorphism from G into H given by $f(\sigma) = \tilde{\sigma}$ and let φ be a mapping from Ω into Γ given by $\varphi(a) = \tilde{a}$. The pair (f, φ) is called a homomorphism from (G, Ω) into (H, Γ) if the following condition is satisfied:

$$\widetilde{a^\sigma} = \tilde{a}^{\tilde{\sigma}} \quad \forall a \in \Omega, \forall \sigma \in G.$$

Notation: $(G, \Omega) \precsim (H, \Gamma)$. If f is an isomorphism and if φ is a one-to-one surjective map, we say that (G, Ω) and (H, Γ) are isomorphic. Notation: $(G, \Omega) \simeq (H, \Gamma)$. We then have the following simple theorem:

Theorem 1.2.12 *For subgroups H_1 and H_2 of S^Ω we have:*

$$(H_1, \Omega) \simeq (H_2, \Omega) \Leftrightarrow H_1 \underset{S^\Omega}{\sim} H_2.$$

Proof $\Rightarrow$: Let us assume that $(H_1, \Omega) \simeq (H_2, \Omega)$ is given by an isomorphism f between H_1 and H_2 and a one-to-one surjective map φ from Ω onto Ω, i.e. $\varphi \in S^\Omega$. Now, if $h \in H_1$ we have $\varphi h(a) = (f(h))(\varphi(a))$ $(\forall a \in \Omega)$, hence $\varphi h \varphi^{-1}(a) = f(h)(a)$ $(\forall a \in \Omega)$. Therefore $h^\varphi = f(h)$ and $H_1^\varphi = H_2$. ■

$\Leftarrow$: Evident.

Let (G, Ω) be transitive. If the stabilizer G_a of $a \in \Omega$ is transitive on $\Omega - \{a\}$, then (G, Ω) is called doubly transitive. The property of being doubly transitive is independent of the choice of a.

1.2.4 Free groups, generators and fundamental relations

Now we want to consider a class of groups called free groups that is of special importance because free groups are, in a certain sense, the most general groups possible.

Suppose we are given n letters $a_1, \dots, a_n$. Using these letters we form the countable set of symbols: $\Lambda = \{a_i^{n_i} | n_i \in \mathbb{Z},\ i = 1, \dots, n\}$. For $r \geq 0$, we define the sets $\Lambda^{(r)}$ by

$$\Lambda^{(r)} = \{(x_1^{n_1}, \dots, x_r^{n_r}) \in \underbrace{\Lambda \times \dots \times \Lambda}_{r \text{ copies}} | x_i \neq x_{i+1},\ i = 1, \dots, r-1,$$

$$n_i \in \mathbb{Z}\} \quad (r > 0),$$

$$\Lambda^{(0)} = \{1\} \quad (r = 0).$$

The element $(x_1^{n_1}, \dots, x_r^{n_r})$ of $\Lambda^{(r)}$ will usually be denoted by $x_1^{n_1} \dots x_r^{n_r}$. Putting $F(n) = \bigcup_{r=0}^{\infty} \Lambda^{(r)}$, we want to define a multiplication on $F(n)$. Let α, β be elements of $F(n)$, then their product $\alpha\beta$ is defined as follows:

(i) If $\alpha = 1$, then $\alpha\beta = \beta$ and if $\beta = 1$, then $\alpha\beta = \alpha$.

(ii) If neither α nor β is equal to 1, we can write: $\alpha = x_1^{n_1} \dots x_r^{n_r}$, $\beta = y_1^{m_1} \dots y_t^{m_t}$. We distinguish the following cases:

$r = t$ and $x_r^{n_r} \dots x_1^{n_1} = y_1^{-m_1} \dots y_t^{-m_t}$, then $\alpha\beta = 1$;

$r < t$ and $x_r^{n_r} \dots x_1^{n_1} = y_1^{-m_1} \dots y_r^{-m_r}$, then $\alpha\beta = y_{r+1}^{m_{r+1}} \dots y_t^{m_t}$

$r > t$ and $x_r^{n_r} \dots x_{r-t+1}^{n_{r-t+1}} = y_1^{-m_1} \dots y_t^{-m_t}$, then $\alpha\beta = x_1^{n_1} \dots x_{r-t}^{n_{r-t}}$.

In all other cases there exists an $i \geq 0$ such that

$$x_r^{n_r} \dots x_{r-i+1}^{n_{r-i+1}} = y_1^{-m_1} \dots y_i^{-m_i}$$

and

$$x_{r-i}^{n_{r-i}} \neq y_{i+1}^{-m_{i+1}}.$$

Then

$$\alpha\beta = x_1^{n_1} \dots x_{r-i}^{n_{r-i}} y_{i+1}^{m_{i+1}} \dots y_t^{m_t} \quad \text{if } x_{r-i} \neq y_{i+1}$$

$$\alpha\beta = x_1^{n_1} \dots x_{r-i-1}^{n_{r-i-1}} x_{r-i}^{n_{r-i}+m_{i+1}} y_{i+2}^{m_{i+2}} \dots y_t^{m_t} \quad \text{if } x_{r-1} = y_{i+1}$$

(and hence $n_{r-i} \neq -m_{i+1}$). It is easy to check that $F(n)$ provided with this multiplication becomes a group called the free group generated by $\{a_1, \dots, a_n\}$. The identity of this group is 1 and the inverse of the element $x_1^{n_1} \dots x_t^{n_t}$ is $x_t^{-n_t} \dots x_1^{-n_1}$.

Let G be a group generated by the subset $\{b_1, \dots, b_n\} \subset G$. By assigning b_i to a_i $(i = 1, \dots, n)$ we get a map from $F(n)$ onto G, which is easily seen to be a homomorphism. Letting N denote the kernel of this homomorphism, we have $F(n)/N \simeq G$ by the homomorphism theorem. If $\{f_i(a_1, \dots, a_n) | i \in I\}$ is an arbitrary system of generators of N (as a normal subgroup of $F(n)$, i.e. $N = \langle\langle\{f_i(a_1, \dots, a_n) | i \in I\}\rangle\rangle$) we get the relations $\{f_i(b_1, \dots, b_n) = 1 | i \in I\}$, called fundamental relations for $\{b_1, \dots, b_n\}$. The group G is unambiguously determined by a set of generators and a set of fundamental relations, but of course G can have many different sets of generators and sets of fundamental relations. If $M = \{g_i(a_1, \dots, a_n) | i \in J\}$ is some subset of $F(n)$ and if we put $N = \langle\langle M \rangle\rangle$, then $G = F(n)/N$ is the group generated by n elements $b_1, \dots, b_n$ satisfying the fundamental relations $g_i(b_1, \dots, b_n) = 1$ $(i \in J)$. If H is also a group generated by n elements $c_1, \dots, c_n$ satisfying fundamental relations $g_i(c_1, \dots, c_n) = 1$ $(i \in J)$, then the kernel of the homomorphism from $F(n)$ onto H defined by $a_i \to c_i$ $(i = 1, \dots, n)$

contains N. Hence this homomorphism induces a natural homomorphism from G onto H.

1.3 Algebraic structures

Let R be a set. If certain laws of composition on R and operations of certain sets on R are given, R is said to carry an algebraic structure or to be an algebraic system. Group structure is one example of an algebraic structure. In this section we study some other algebraic structures that we will need in the rest of this book.

1.3.1 Algebraic structures and vector spaces

Suppose that we are given two operations, called addition and multiplication on the set R. For $a, b \in R$, the sum of a and b is denoted by $a + b$ and their product by ab. If the following conditions are satisfied:

(i) R is an Abelian group under addition;

(ii) $(a + b)c = ac + bc$ and $c(a + b) = ca + cb \quad \forall a, b, c \in R$;

then R is called a distributive ring. If in addition the following condition is satisfied:

(iii) $a(ab) = (aa)b$ and $(ba)a = b(aa) \quad \forall a, b \in R$;

then R is called an alternative ring. If moreover the following condition is satisfied:

(iv) $a(bc) = (ab)c, \quad \forall a, b, c$ (associativity);

then R is called an associative ring. If R is one of these rings and in addition the following condition is satisfied:

(v) there is a unit for the multiplication and every element of R, except 0, has an inverse with regard to this unit for multiplication, then R is called a distributive, alternative or associative field respectively. If the following condition is satisfied:

(vi) $ab = ba \quad \forall a, b \in R$;

then R is said to be commutative.

From now on, we shall refer to associative rings as 'rings', associative fields as 'skew fields' and commutative associative fields as 'fields'. If R has a finite number of elements, the resulting structures will be called finite: finite rings and so on. An element e of R satisfying $ea = ae = a$ for all $a \in R$ is called an identity of R. For associative rings, just as for groups, it is proved that if an identity exists it is uniquely determined. The identity of R (if it exists) is denoted by 1. If R has an identity and if $a \in R$, then an element n of R such that

$$ab = ba = 1$$

is called an inverse of a. In associative rings (just as in groups) the inverse of an element, if it exists, is unique. The inverse of a (if it exists) is denoted by a^{-1}. Finally, if in addition to (i), (iv) and (v) the following condition is satisfied:

(vii) $(a+b)c = ac + bc \quad \forall a, b, c \in R$;

then R is called a near field. In associative rings (hence also in near fields) multiplication satisfies the general associative law (see theorem 1.2.1).

Just as we defined subgroups in the case of groups, we can also define subrings, subfields and so on. For example, if S is a subset of the skew field R such that:

(1) for all $a, b \in S$, we have $a + b \in S$ and $ab \in S$, i.e. S is closed under addition and multiplication;

(2) S with the operations inherited from R is a skew field;

then S is called a skew subfield of R, and R is called an extension skew field of S.

$\mathbb{C}$, provided with the usual addition and multiplication, is a field. $\mathbb{R}$ and $\mathbb{Q}$ are closed under the addition and multiplication on $\mathbb{C}$ and easily seen to be subfields of $\mathbb{C}$. $\mathbb{Z}$ is closed under addition and multiplication on $\mathbb{C}$, but it is not a subfield (no inverse exists, except for 1 and -1). However, if we regard $\mathbb{C}$ as a ring, then $\mathbb{Z}$ is a subring of $\mathbb{C}$. If R is a ring and if S is a subset of R, then the subset of R consisting of all elements that commute with each element of S is denoted by $V_R(S)$, i.e.;

$$V_R(S) = \{a \in R \mid ax = xa, \forall x \in S\}.$$

It is easy to check that in fact $V_R(S)$ is a subring of R, called the commutator ring of S in R. The subset $V_R(R)$ is called the centre of R, is denoted by $Z(R)$ and is a commutative subring of R. If R

is a skew field, and if S is an arbitrary subset of R, then $V_R(S)$ is also a skew field.

For H and K Abelian groups, let $\mathrm{Hom}(H, K)$ denote the collection of all homomorphisms from H into K. For $f, g \in \mathrm{Hom}(H, K)$ define a mapping $f+g$ from H into K by

$$(f+g)(a) = f(a)g(a) \quad \forall a \in H.$$

Since K is Abelian we have for $a, b \in H$:

$$\begin{aligned}(f+g)(ab) &= f(ab)g(ab) = f(a)f(b)g(a)g(b)\\ &= f(a)g(a)f(b)g(b) = (f+g)(a)\cdot(f+g)(b),\end{aligned}$$

i.e. $f+g$ is an element of $\mathrm{Hom}(H, K)$. With this operation $\mathrm{Hom}(H, K)$ becomes an additive group. (In fact, the associativity follows from the associativity of K. The identity, or, since we are considering additive groups, the zero element, is the homomorphism f_0 defined by $f_0(a) = 1$ for all $a \in H$. The inverse of f is the homomorphism g defined by $g(a) = (f(a))^{-1}$ for all $a \in H$. Finally, the commutativity is a consequence of the commutativity of K.) Now, take $K = H$ and consider $\mathrm{Hom}(H, H)$. For $f, g \in \mathrm{Hom}(H, H)$ define a map fg from H into H by

$$(fg)(a) = g(f(a)) \quad \forall a \in H.$$

Since

$$\begin{aligned}(fg)(ab) &= g(f(ab)) = g(f(a)f(b)) = g(f(a))g(f(b))\\ &= (fg)(a)(fg)(b),\end{aligned}$$

$fg \in \mathrm{Hom}(H, H)$. It is easy to verify that $\mathrm{Hom}(H, H)$ with this addition and this multiplication becomes a ring. This ring has an identity, namely the homomorphism 1_H defined by $1_H(a) = a$ for all $a \in H$, and is called the ring of endomorphisms of the Abelian group H. For example, if H is a group of prime order p, then it is not difficult to see that every element of $\mathrm{Hom}(H, H)$, except the zero element, is an automorphism. Hence $\mathrm{Hom}(H, H) - \{0\} = \mathrm{Aut}\, H$. Since $\mathrm{Aut}\, H$ is an Abelian group of order $p-1$ (theorem 1.2.10), $\mathrm{Hom}(H, H)$ is a field with p elements.

Let V be an additive group, let R be a commutative ring operating on V and let this operation be denoted by λv for $\lambda \in R$ and $v \in V$.

Such an operation is called a scalar product. If the following conditions are satisfied:

$$\left.\begin{array}{l}\text{(i) } \lambda(v+v')=\lambda v+\lambda v' \\ \text{(ii) } (\lambda+\mu)v=\lambda v+\mu v \\ \text{(iii) } (\lambda\mu)v=\lambda(\mu v)\end{array}\right\} \quad \forall \lambda, \mu \in R, \forall v, v' \in V;$$

(iv) if R has a unit 1, then $1v=v \quad \forall v \in V$;

then V is called an R-module or R-vectorspace. If $v_1, \ldots, v_r$ are elements of V and if $\lambda_1, \ldots, \lambda_r$ are elements of R, then the element $\lambda_1 v_1 + \ldots + \lambda_r v_r$ is called a linear combination of $v_1, \ldots, v_r$ or an R-linear combination of $v_1, \ldots, v_r$. If V and W are R-modules, we define

$$\mathrm{Hom}_R(V, W) = \{ f \in \mathrm{Hom}(V, W) \mid f(\lambda v) = \lambda f(v), \forall \lambda \in R, \forall v \in V \}.$$

It is easy to check that $\mathrm{Hom}_R(V, W)$ is a subgroup of $\mathrm{Hom}(V, W)$ (regarded as an Abelian group). Defining the scalar product of $\lambda \in R$ and $f \in \mathrm{Hom}_R(V, W)$ by

$$(\lambda f)(a) = \lambda(f(a)) \quad \forall a \in V,$$

$\mathrm{Hom}_R(V, W)$ becomes an R-module. Hereafter $\mathrm{Hom}_R(V, W)$ is usually considered as an R-module. Now, take $W = V$ and consider $\mathrm{Hom}_R(V, V)$. For $f, g \in \mathrm{Hom}_R(V, V)$ we have

$$(fg)(\lambda a) = g(f(\lambda a)) = g(\lambda(f(a))) = \lambda(g(f(a))) = \lambda((fg)(a)).$$

Therefore $fg \in \mathrm{Hom}_R(V, V)$, i.e. $\mathrm{Hom}_R(V, V)$ is a subring of $\mathrm{Hom}(V, V)$. Moreover, for $f, g \in \mathrm{Hom}_R(V, V)$ and $\lambda \in R$, we have

$$\begin{aligned}(\lambda(fg))a &= \lambda((fg)(a)) = (fg)(\lambda a) = g(f(\lambda a)) \\ &= g((\lambda f)(a)) = ((\lambda f)g)(a) \quad \forall a \in V,\end{aligned}$$

hence $\lambda(fg) = (\lambda f)g$ and similarly $\lambda(fg) = f(\lambda g)$. Generally, if T is a ring and if R is a commutative ring, such that T, as an Abelian group, is an R-module and if moreover the following condition is satisfied:

$$\lambda(fg) = (\lambda f)g = f(\lambda g) \quad \forall \lambda \in R, \forall f, g \in T,$$

then T is called an R-algebra. So we can sum up the above results by saying that $\mathrm{Hom}_R(V, V)$ is an R-algebra.

An arrangement of nm elements of a commutative ring R:

$$\begin{bmatrix} \lambda_{11}, \dots, \lambda_{1m} \\ \lambda_{21}, \dots, \lambda_{2m} \\ \vdots \qquad \vdots \\ \lambda_{n1}, \dots, \lambda_{nm} \end{bmatrix}$$

is called an (n, m)-matrix with elements from R. The above matrix is usually represented by (λ_{ij}). The element λ_{ij} is called the (i, j)th entry of the matrix (λ_{ij}). The collection of all (n, m)-matrices with elements from R is denoted by $(R)_{(n,m)}$. For $\alpha = (\lambda_{ij})$, $\beta = (\mu_{ij}) \in (R)_{(n,m)}$ and $\lambda \in R$, we define the sum $\alpha + \beta$ by

$$\alpha + \beta = (v_{ij}) \quad \text{where } v_{ij} = \lambda_{ij} + \mu_{ij},$$

and the scalar product $\lambda\alpha$ by

$$\lambda a = (v'_{ij}) \quad \text{where } v'_{ij} = \lambda\lambda_{ij},$$

and with this addition and scalar multiplication $(R)_{(n,m)}$ becomes an R-module. Elements of $(R)_{(n,1)}$ and $(R)_{(1,n)}$ are called column vectors and row vectors respectively. Instead of $(R)_{(n,1)}$ we also write R^n; instead of $(R)_{(1,n)}$ also ${}^nR$. If $n = m$, we write $(R)_n$ for $(R)_{(n,n)}$. Elements of $(R)_n$ are called (square) matrices of degree n. For two elements $\alpha = (\lambda_{ij})$ and $\beta = (\mu_{ij})$ of $(R)_n$, we define the product $\alpha\beta$ by

$$\alpha\beta = (v_{ij}) \quad \text{where } v_{ij} = \sum_{k=1}^{n} \lambda_{ik}\mu_{kj}.$$

$(R)_n$, with this product and the previously defined addition and scalar multiplication, becomes an R-algebra, called the complete matrix ring of degree n over R. If R has an identity 1, then the matrix (δ_{ij}) is the identity of $(R)_n$.

Important examples of R-algebras are provided by group rings, which we proceed to define. Let G be a finite group, and let R be a commutative ring with identity. We define $R(G)$ to be the set of all maps of G into R. The map that maps $\sigma \in G$ onto $\lambda_\sigma \in R$ is denoted by $(\lambda_\sigma | \sigma \in G)$. For $\alpha = (\lambda_\sigma | \sigma \in G)$, $\beta = (\mu_\sigma | \sigma \in G) \in R(G)$ and $\lambda \in R$ we define the sum $\alpha + \beta$ by

$$\alpha + \beta = (\lambda_\sigma + \mu_\sigma | \sigma \in G),$$

and the product $\alpha\beta$ by

$$\alpha\beta = (v_\sigma | \sigma \in G) \quad \text{where } v_\sigma = \sum_{\tau\rho=\sigma} \lambda_\tau \mu_\rho,$$

and the scalar product $\lambda\alpha$ by

$$\lambda\alpha = (\lambda\lambda_\sigma | \sigma \in G).$$

It is easy to check that with this addition, multiplication and scalar multiplication $R(G)$ becomes an R-algebra, called the group ring of G over R. For $\alpha = (\lambda_\sigma | \sigma \in G)$ we call λ_σ the coefficient of σ in α. If, for $\tau \in G$, we denote the element $(\lambda_\sigma | \sigma \in G)$, which is defined by $\lambda_\sigma = 1$ if $\sigma = \tau$ and $\lambda_\sigma = 0$ if $\sigma \neq \tau$, also by τ, then multiplication on $R(G)$ is consistent with multiplication on G, i.e. G can be regarded as a subset of $R(G)$, that is a group under the multiplication of $R(G)$. Moreover, each element $\alpha = (\lambda_\sigma | \sigma \in G)$ can be represented in exactly one way as: $\alpha = \sum_{\sigma \in G} \lambda_\sigma \sigma$. The centre $Z(R(G))$ of $R(G)$ is easily seen to be a commutative R-algebra. To study $Z(R(G))$ in some more detail, let r be the number of conjugacy classes of G, let $K_1, \ldots, K_r$ represent the conjugacy classes of G and put $\bar{K}_i = \sum_{\sigma \in K_i} \sigma \in R(G)$ $(i = 1, \ldots, r)$. Then $Z(R(G)) = \{\sum_{i=1}^r \lambda_i \bar{K}_i | \lambda_i \in R\}$. (Since evidently $\tau \bar{K}_i = \bar{K}_i \tau$ for all $\tau \in G$, hence $\bar{K}_i \in Z(R(G))$ $(i = 1, \ldots, r)$ and $\{\sum_{i=1}^r \lambda_i \bar{K}_i | \lambda_i \in R\} \subset Z(R(G))$. Conversely, let $\alpha = \sum \lambda_\sigma \sigma \in Z(R(G))$, then $\alpha = \tau^{-1} \alpha \tau$ for all $\tau \in G$, i.e. $\sum \lambda_\sigma \sigma = \sum \lambda_\sigma \tau^{-1} \sigma \tau$. Therefore, elements that belong to the same conjugacy class, have the same coefficients and we can write $\alpha = \sum_{i=1}^r \lambda_i \bar{K}_i$ where $\lambda_i = \lambda_\sigma$ for some $\sigma \in K_i$.) Let $\sigma \in K_k$, then the number of elements of the set $\{(x, y) \in K_i \times K_j | xy = \sigma\}$ is only dependent on i, j and k, because for any $\tau \in G$ we have $xy = \sigma \Leftrightarrow x^\tau y^\tau = \sigma^\tau$. Denoting this number by c_{ijk} it is easy to prove that $\bar{K}_i \bar{K}_j = \sum c_{ijk} \bar{K}_k$. This implies that the ring structure of $Z(R(G))$ is completely determined by the conjugacy classes of G.

1.3.2 Rings and ideals

If R_1 and R_2 are rings and if f is a mapping from R_1 into R_2 satisfying:

$$\left.\begin{aligned} f(a+b) &= f(a) + f(b) \\ f(ab) &= f(a)f(b) \end{aligned}\right\} \quad \forall a, b \in R_1;$$

then f is called a homomorphism from R_1 into R_2. Notation: $R_1 \overset{\sim}{\to} R_2$. If f is one-to-one and surjective, then f is called an isomorphism. In this case R_1 and R_2 are called isomorphic. Notation: $R_1 \simeq R_2$. An isomorphism from R onto itself is called an automorphism of R. The collection of all automorphisms of R, with the product of two automorphisms defined to be their composition, is a group. This group is called the automorphism group of R and it is denoted by Aut R.

Let R be a ring. A subset T of R, that is closed under the operations of R and becomes a ring with the operations inherited from R, is called a subring of R, and R is called an extension ring of T. If S is a subset of R, then the intersection of all subrings of R containing S is a subring. This subring is called the subring generated by S and denoted by $\langle S \rangle$. A subring T of R satisfying:

$$\text{if } a \in T, \text{ then } xa \in T \text{ and } ax \in T \text{ for all } x \in R,$$

is called an ideal of R. If S is a subset of R, then the intersection of all ideals of R containing S is an ideal. This ideal is called the ideal generated by S and it is denoted by (S). Especially, the ideal generated by one element of R is called a principal ideal. For example, if R is a commutative ring with an identity, then the ideal generated by the element a is $\{xa \mid x \in R\}$ and the ideal generated by the two elements a and b is $\{xa + yb \mid x, y \in R\}$. If $\mathfrak{P}$ is an ideal of R, we can form the quotient group $R/\mathfrak{P}$, where we consider R as an additive group and $\mathfrak{P}$ as a subgroup of R. For two elements $\bar{a} = a + \mathfrak{P}$ and $\bar{b} = b + \mathfrak{P}$ we define the product $\bar{a}\bar{b}$ by

$$\bar{a}\bar{b} = ab + \mathfrak{P}.$$

(This definition is independent of the choice of the representants a and b.) With this product $R/\mathfrak{P}$ becomes a ring called the quotient ring of R with respect to $\mathfrak{P}$.

If R_1 and R_2 are rings and if f is a homomorphism from R_1 into R_2, then $f(R_1)$ is a subring of R_2 and the kernel of $f(=f^{-1}(0))$ is an ideal of R_1. It is easy to verify that the isomorphism for the additive group $R_1/\text{Ker } f \simeq f(R_1)$ is also an isomorphism for the rings. Hence:

Theorem 1.3.1 (Homomorphism theorem) *A homomorphism f from the ring R_1 into the ring R_2 induces an isomorphism $R_1/\text{Ker}\, f \simeq f(R_1)$.*

The only ideals of a field R are $\{0\}$ and R itself. If conversely a commutative ring R with identity has no other ideals than $\{0\}$ and R itself, then the ideal generated by any element $a \in R^{\#} = R - \{0\}$ is R, hence $ab = ba = 1$ for some $b \in R$ and therefore R is a field. We have proved:

Theorem 1.3.2 *If R is a commutative ring with identity, then R is a field if and only if the only ideals of R are $\{0\}$ and R itself.*

An ideal $\mathfrak{P}$ of R is called a maximal ideal if $\mathfrak{P} \neq R$ and the only ideals $\mathfrak{A}$ of R satisfying $\mathfrak{P} \subseteq \mathfrak{A} \subseteq R$ are $\mathfrak{P}$ and R. If $\mathfrak{P}$ is an arbitrary ideal of R, and if $\tilde{\mathfrak{A}}$ is an ideal of the quotient ring $R/\mathfrak{P}$, then $\mathfrak{A} = \{a \in R \mid a + \mathfrak{P} \in \tilde{\mathfrak{A}}\}$ is an ideal of R containing $\mathfrak{P}$. From this we conclude:

Theorem 1.3.3 *If $\mathfrak{P}$ is an ideal of the commutative ring R with identity, then $\mathfrak{P}$ is maximal if and only if $R/\mathfrak{P}$ is a field.*

Every ideal of $\mathbb{Z}$ is principal. (In fact, if $\mathfrak{A} \neq \{0\}$ is an ideal of $\mathbb{Z}$, $\mathfrak{A}$ contains positive numbers. Let n be the smallest of all the positive numbers contained in $\mathfrak{A}$, then we will prove that $\mathfrak{A}$ is generated by n. If a is any element of $\mathfrak{A}$, by dividing a by n we get $a = nb + c$ for some b and c with $0 \leq c < n$. Hence $c = a - nb \in \mathfrak{A}$ and therefore $c = 0$, i.e. $a = nb$.) For any ideal (n) of $\mathbb{Z}$ we denote the quotient ring $\mathbb{Z}/(n)$ by $\mathbb{Z}_n$. The addition and multiplication on $\mathbb{Z}/(n)$ agree with those defined in §1.1. If we take n to be a prime number p, then $\mathbb{Z}_p^* = \mathbb{Z}_p - \{0\}$ is a commutative group, i.e. $\mathbb{Z}_p$ is a field with p elements. We have already seen that the ring of endomorphisms $\text{Hom}(H, H)$ of a group H of prime order p is a field with p elements, and we can see here that $\text{Hom}(H, H) \simeq \mathbb{Z}_p$. In fact, let a be a generating element of H. Then an element f of $\text{Hom}(H, H)$ is determined by its value on a, say $f(a) = a^i$. From $a^i = a^j \Leftrightarrow i + (p) = j + (p)$, we see that the map assigning $i + (p) \in \mathbb{Z}_p$ to $f \in \text{Hom}(H, H)$ is well defined. It is easy to verify that this map is an isomorphism between $\mathbb{Z}_p$ and $\mathrm{Hom}(H, H)$.

1.3.3 Fields and polynomials

Let K be a field. Let $\tilde{K}$ denote the collection of all infinite sequences $(\lambda_i) = (\lambda_0, \lambda_1, \lambda_2, \ldots)$ of elements of K indexed by 0 and the natural numbers, such that all but a finite number of the $\lambda_0, \lambda_1, \lambda_2, \ldots$ are 0, i.e.

$$\tilde{K} = \{(\lambda_i) \mid \exists n \text{ such that } \lambda_{n+1} = \lambda_{n+2} = \ldots = 0\}.$$

For (λ_i) and (μ_i) elements of $\tilde{K}$ and for $\lambda \in K$ we define the sum of (λ_i) and (μ_i) by

$$(\lambda_i) + (\mu_i) = (\lambda_i + \mu_i) = (\lambda_0 + \mu_0, \lambda_1 + \mu_1, \ldots, \lambda_n + \mu_n, \ldots),$$

the product of (λ_i) and (μ_i) by

$$(\lambda_i)(\mu_i) = (v_i), \quad \text{where } v_i = \sum_{k+l=i} \lambda_k \mu_l,$$

and the product of λ and (λ_i) by

$$\lambda(\lambda_i) = (\lambda\lambda_i) = (\lambda\lambda_0, \lambda\lambda_1, \ldots). \tag{1.3.1}$$

$\tilde{K}$ provided with these operations becomes a K-algebra. We represent the zero-element $(0, 0, 0, \ldots)$, of $\tilde{K}$ by 0 and its identity $(1, 0, 0, \ldots)$ by 1. Putting $x = (0, 1, 0, \ldots)$, we have $x^2 = (0, 0, 1, 0, \ldots)$ and in general $x^n = (\delta_{i,n})$ and we define $x^0 = 1$. An element $(\lambda_0, \lambda_1, \ldots, \lambda_n, \ldots)$ of $\tilde{K}$ can be written as: $(\lambda_0, \lambda_1, \ldots) = \lambda_0 + \lambda_1 x + \ldots + \lambda_n x^n + \ldots$ (finite sum). Therefore the elements of $\tilde{K}$ are given by the expressions

$$\lambda_0 + \lambda_1 x + \ldots + \lambda_n x^n \quad (\lambda_i \in K).$$

$\tilde{K}$ is called the ring of polynomials in the variable x over R and denoted by $K[x]$. Elements of $K[x]$ are called polynomials with coefficients in K and are written as: $f(x) = \sum_{i=0}^{n} \lambda_i x^i$. It follows immediately from the definition, that

$$\sum_{i=0}^{n} \lambda_i x^i = 0 \Leftrightarrow \lambda_0 = \ldots = \lambda_n = 0.$$

For f, g arbitrary elements of $K[x]$, say $f=\sum_{i=0}^{n}\lambda_i x^i$ and $g=\sum_{i=0}^{n}\mu_i x^i$ (we can always make the upper limits the same by adding terms with coefficient 0) and for λ and arbitrary element of K, the definitions of addition, multiplication and scalar multiplication in the new notation become:

$$f(x)+g(x)=\sum_{i=0}^{n}(\lambda_i+\mu_i)x^i;$$

$$f(x)g(x)=\sum_{k}\left(\sum_{i+j=k}\lambda_i\mu_j\right)x^k;$$

$$\lambda f(x)=\sum_{i=0}^{n}(\lambda\lambda_i)x^i.$$

If $f(x)=\lambda_0+\lambda_1 x+\ldots+\lambda_n x^n\in K[x]$ and if $\lambda_n\neq 0$, then λ_n is called the leading coefficient of f and n is called the degree of f. The degree of f is denoted by $\deg f(x)$. Polynomials with leading coefficient equal to 1 are called monic. A polynomial $f(x)$ is called reducible if there exist polynomials $g_1(x)$ and $g_2(x)$, such that

$$f(x)=g_1(x)g_2(x) \text{ and } \deg g_1(x)\geq 1 \text{ and } \deg g_2(x)\geq 1.$$

The polynomials $g_1(x)$ and $g_2(x)$ are called divisors of f and f is called a multiple of $g_i(x)$. Polynomials that are not reducible are called irreducible. If L is an extension field of K, if $f(x)=\sum\lambda_i x^i\in K[x]$ and if $\alpha\in L$, then $f(\alpha)=\sum\lambda_i\alpha^i\in L$ is called the result of substituting α for x in $f(x)$. If $f(\alpha)=0$, then α is called a root of $f(x)$ in L. If $f(\alpha)=0$, then $\lambda f(\alpha)=0$ for all $\lambda\in K$. Therefore, if there exists a non-zero polynomial of which α is a root, then among all the polynomials of minimal degree of which α is a root, there exists a monic polynomial. This monic polynomial is called a minimal polynomial of $\alpha\in L$ in $K[x]$. If it exists at all it is unique, as we shall prove in the following theorem.

Theorem 1.3.4 *Let K be a field, then:*

(1) *If $f(x), g(x)\in K[x]$ have leading coefficients λ and μ respectively, then the leading coefficient of $f(x)g(x)$ equals $\lambda\mu$. Hence, if $f(x)\neq 0$ and $g(x)\neq 0$, then* $\deg f(x)g(x)=\deg f(x)+\deg g(x)$.

(2) *If $f(x)$ and $g(x)\in K[x]$, such that $g(x)\neq 0$, then there exist*

$h(x)$ and $r(x) \in K[x]$ satisfying

$$f(x) = g(x)h(x) + r(x) \quad \text{and } r(x) = 0 \text{ or } \deg r(x) < \deg g(x)$$

and $h(x)$ and $r(x)$ are uniquely determined.

(3) *For $f(x) \in K[x]$ and $\alpha \in K$, we have:*

$$f(\alpha) = 0 \Leftrightarrow f(x) = (x - \alpha)g(x) \quad \textit{for some } g(x) \in K[x].$$

Hence, the number of roots of f in K does not exceed the degree of f.

(4) *All ideals of $K[x]$ are principal ideals.*

(5) *If L is an extension field of K, if $\alpha \in L$, and if there exists a minimal polynomial of α in $K[x]$, then it is uniquely determined. Denoting this minimal polynomial by $f_0(x)$, we note the following properties of f_0: f_0 is irreducible, and if g is any polynomial in $K[x]$ with $g(\alpha) = 0$, then $g(x) = f_0(x)h(x)$ for some $h(x) \in K[x]$.*

Proof (1) Evident.

(2) Induction on n. If $f(x) = 0$, it suffices to put $h(x) = r(x) = 0$. Assume the statement to be true for all $f(x)$ with $\deg f(x) < n$, we prove it for $\deg f(x) = n$. If $n < \deg g$, then we can put $h(x) = 0$ and $r(x) = f(x)$, hence we may assume $n \geq \deg g(x)$. Let

$$f(x) = \lambda_0 + \ldots + \lambda_n x^n (\lambda_n \neq 0), \quad g(x) = \mu_0 + \ldots + \mu_m x^m (\mu_m \neq 0),$$

then, since $n \geq m$, we can put $\tilde{f}(x) = f(x) - \lambda_n \mu_m^{-1} x^{n-m} g(x)$. Since $\deg \tilde{f}(x) < n$, we have by the induction hypothesis: there exist elements $\tilde{h}(x)$ and $\tilde{r}(x) \in K[x]$ with $\tilde{r}(x) = 0$ or $\deg \tilde{r}(x) < \deg g(x)$ such that

$$\tilde{f}(x) = g(x)\tilde{h}(x) + \tilde{r}(x).$$

Therefore:

$$f(x) = g(x)h(x) + r(x) \quad \text{where: } h(x) = \lambda_n \mu_m^{-1} x^{n-m} + \tilde{h}(x)$$

and $r(x) = \tilde{r}(x)$. If

$$f(x) = g(x)h_1(x) + r_1(x) = g(x)h_2(x) + r_2(x),$$

then

$$g(x)(h_1(x) - h_2(x)) = r_2(x) - r_1(x).$$

Hence $h_1(x)-h_2(x)=0$ by (1), i.e. $h_1(x)=h_2(x)$ and therefore $r_1(x)=r_2(x)$.

(3) Follows immediately from (2).

(4) Let $\mathfrak{P}$ be an ideal of $K[x]$. We may assume $\mathfrak{P}\neq 0$. Let $g(x)$ be a non-zero polynomial of $\mathfrak{P}$ with minimal degree. For an arbitrary $f(x)\in\mathfrak{P}$, by (2) there exist $h(x)$ and $r(x)\in K[x]$ with $r(x)=0$ or $\deg r(x)<\deg g(x)$, such that $f(x)=g(x)h(x)+r(x)$. From $r(x)=f(x)-g(x)h(x)\in\mathfrak{P}$, we conclude that $r(x)=0$. Therefore, $\mathfrak{P}$ is generated by $g(x)$.

(5) The collection $\{f(x)\in K[x] \mid f(\alpha)=0\}$ is an ideal. Now apply (4). ■

For elements $f(x), g(x)\in K[x]$ we have

$$(f(x))=(g(x))\Leftrightarrow f(x)=\lambda g(x)\quad \text{for some } \lambda\in K\ (\lambda\neq 0).$$

Hence the generating element of an ideal of $K[x]$ is uniquely determined up to constant factors. If we require our generating element to be monic, then it is uniquely determined. For $f_1(x),\dots,f_r(x)\in K[x]$, a generating element of the ideal generated by $f_1(x),\dots,f_r(x)$ is called a greatest common divisor of $f_1(x),\dots,f_r(x)$ and a generating element of the ideal $(f_1(x))\cap\dots\cap(f_r(x))$ is called a least common multiple of $f_1(x),\dots,f_r(x)$. There is a uniquely determined monic least common multiple and also a uniquely determined monic greatest common divisor of $f_1(x),\dots,f_r(x)$. If $f(x)$ and $h(x)$ are such that $f(x)=g(x)h(x)$ for some $g(x)$, then $f(x)$ is called a multiple of $h(x)$ and $h(x)$ is called a divisor of $f(x)$. Notation: $h(x)\mid f(x)$. Theorems 1.3.5 and 1.3.6 follow almost immediately from these definitions.

Theorem 1.3.5 *If we denote the greatest common divisor and the least common multiple of $f_1(x),\dots,f_r(x)\in K[x]$ by $h_1(x)$ and $h_2(x)$ respectively, then we have for $h(x)\in K[x]$:*

(1) $h(x)\mid f_i(x)\quad (i=1,\dots,r)\Leftrightarrow h(x)\mid h_1(x)$;

(2) $f_i(x)\mid h(x)\quad (i=1,\dots,r)\Leftrightarrow h_2(x)\mid h(x)$.

Theorem 1.3.6 *For $f(x)\in K[x]$ and $\mathfrak{P}=(f(x))$ we have:*

$$\mathfrak{P} \text{ is maximal ideal} \Leftrightarrow f(x) \text{ is irreducible.}$$

1.3.4 Finite fields

Let K be a finite skew field and let $Z(K)$ represent as usual the centre of K, i.e.

$$Z(K)=\{a\in K\,|\,ax=xa,\ \forall x\in K\}.$$

Since $Z(K)$ contains the identity of K we have $Z(K)\neq\{0\}$. It is easy to check that $Z(K)$ is a subfield of K. Since K is finite, the subset $\{n1\,|\,n\in\mathbb{Z}\}$ is also finite, i.e. there are two different integers n_1 and n_2 such that $n_1 1=n_2 1$. Therefore there exists an integer $n\neq 0$, such that $n1=0$. Let p denote the smallest number of the set $\{n\in\mathbb{Z}\,|\,n>0 \text{ and } n1=0\}$, then p is a prime number. (For suppose $p=q_1q_2$, where $q_i\in\mathbb{Z}$ and $q_i \gneq 1$, then by the distributive law we have $(q_1 1)(q_2 1)=(q_1q_2)1=p1=0$, hence $q_1 1=0$ or $q_2 1=0$, contradicting the minimality of p.) This prime number p is called the characteristic of K. Let n be an arbitrary integer. If $p\,|\,n$, then $n=pm$ for some integer m and $n1=(p1)(m1)=0$. On the other hand, if $p\nmid n$, then p and n are relatively prime, i.e. there are integers t and s such that $pt+ns=1$. Therefore: $1=(pt+ns)1=(ns)1=(n1)(s1)$, i.e. $n1\neq 0$ and its inverse can be written as $s1$ for some $s\in\mathbb{Z}$. Hence every element $n1$ can be uniquely represented as $r1$ with $0\le r<p$ and $\{n1\,|\,n\in\mathbb{Z}\}=\{0,1,2\cdot 1,\dots,(p-1)1\}$ is a subfield of K. This field is called the prime field of K. Obviously, the prime field of K is contained in the centre of K. The structure of the prime field of a field K is completely determined by the characteristic of K. (Because the map from $\mathbb{Z}$ into the prime field of K defined by $n\to n\cdot 1$ is easily seen to be a surjective homomorphism with kernel (p). Hence the prime field of K is isomorphic to $\mathbb{Z}/(p)=\mathbb{Z}_p$ by theorem 1.3.1.) If $a\in K$ $(a\neq 0)$ and $n\in\mathbb{Z}$, then

$$na=0\Leftrightarrow (n1)a=0\Leftrightarrow n1=0\Leftrightarrow p\,|\,n,$$

hence every element $na\in K$ can be uniquely represented as ra with $0\le r<p$.

Theorem 1.3.7 *If K is a finite skew field and if F is a subfield of K, then $|K|=|F|^m$ for some $m\in\mathbb{N}$. If p is the characteristic of K then $|K|=p^n$ for some $n\in\mathbb{N}$. If $|K|=p^n$ and if F is a subfield of K, then $|F|=p^d$ where $d\,|\,n$.*

Proof For any r elements $a_1, \dots, a_r \in K$, $\{\sum_{i=1}^r \lambda_i a_i \mid \lambda_i \in F\}$ is a subgroup of the additive group underlying K. Since K is finite, there exists an r such that $K = \{\sum_{i=1}^r \lambda_i a_i \mid \lambda_i \in F\}$ with some $a_1, \dots, a_r \in K$. Let m denote the minimum of all such r, i.e. $K = \{\sum_{i=1}^m \lambda_i a_i \mid \lambda_i \in F\}$ with some $a_1, \dots, a_m \in K$ and it is impossible to find a smaller number of elements of K, such that K can be represented in this way. Now it suffices to prove that every element of K can be represented in exactly one way as $\sum_{i=1}^m \lambda_i a_i$ where $\lambda_i \in F$. Suppose $\sum_{i=1}^m \lambda_i a_i = \sum_{i=1}^m \mu_i a_i$, where $\lambda_i \neq \mu_i$ for at least one i, say $i = m$, then $\sum_{i=1}^m (\lambda_i - \mu_i) a_i = 0$ and $\lambda_m - \mu_m \neq 0$. Let ν denote the inverse of $\lambda_m - \mu_m$ in F, then: $\sum_{i=1}^{m-1} \nu(\lambda_i - \mu_i) a_i + a_m = 0$, i.e. $a_m = -\sum_{i=1}^{m-1} \nu(\lambda_i - \mu_i) a_i$ and $K = \{\sum_{i=1}^{m-1} \nu_i a_i \mid \nu_i \in F\}$, contradicting the minimality of m. If F_0 is the prime field of K, then $F_0 \subseteq F \subseteq K$, hence $|F| = |F_0|^d = p^d$ for some $d \in \mathbb{N}$ and $|K| = p^{dm}$.

Theorem 1.3.8 (Wedderburn) *Finite skew fields are commutative, i.e. every finite skew field is a field.*

Proof Let k be a finite skew field, and let Z be its centre. We want to show that the assumption $Z \neq K$ leads to a contradiction. For α an element of the multiplicative group $K^\# = K - \{0\}$, let C_α denote the conjugacy class containing α in $K^\#$ and let $C_{K^\#}(\alpha)$ be the centralizer of α in $K^\#$. If $\alpha \in Z$, then $C_K(\alpha) = K^\#$ and $|C_\alpha| = 1$ and if $\alpha \notin Z$, then $K^\# \supsetneq C_{K^\#}(\alpha)$. Here we note that $C_{K^\#}(\alpha) = V_K(\alpha) - \{0\}$ and $V_K(\alpha)$ is a skew field containing Z. Letting p denote the characteristic of K, we have $|Z| = p^r = q$ for some r, $|K| = q^n$ for some n ($n \neq 1$ by assumption) and $|V_K(\alpha)| = q^d$ for some d such that $d \mid n$. From $|C_\alpha| = |K^\# : C_{K^\#}(\alpha)|$ we conclude that

$$|C_\alpha| = \frac{q^n - 1}{q^d - 1}.$$

Therefore

$$q^n - 1 = |K^\#| = \sum_{\substack{C \text{ conjugacy} \\ \text{class of } K^\#}} |C| = q - 1 + \sum_{\substack{1 \le d < n \\ d \mid n}}' \frac{q^n - 1}{q^d - 1} \quad (n \neq 1). \tag{1.3.2}$$

(Here, Σ' denotes summation over some subset of the set of ds satisfying the conditions, repetition of the same d allowed.) For $1 \leq d < n$ and $d|n$ the nth cyclotomic polynomial $\Phi_n(X)$ satisfies $\Phi_n(X)|(X^n - 1)/(X^d - 1)$, hence $\Phi_n(q)|(q^n - 1)/(q^d - 1)$ and therefore by (1.3.2), $\Phi_n(q)|q - 1$, so $|\Phi_n(q)| \leq q - 1$. On the other hand, $\Phi_n(q) = \prod(q - \omega_i)$, where all ω_i are roots of unity and $\omega_i \neq 1$. Hence $|q - \omega_i| > q - 1$ and therefore $|\Phi_n(q)| = \prod|q - \omega_i| > q - 1$. Contradiction. ■

Theorem 1.3.9 *If K is a field, then every finite subgroup of $K^{\#} = K - \{0\}$ is cyclic. Especially, if K is finite, then $K^{\#}$ is cyclic.*

Proof The proof is similar to the proof for the special cases $K = \mathbb{C}$ and $K = \mathbb{Z}_p$ (theorem 1.2.4). The main points are; (i) in a finite Abelian group there exists an element, of which the order equals the least common multiple of the orders of all the elements of the group, and (ii) application of theorem 1.3.4(3). ■

Theorem 1.3.10 *If p is a prime number and if n is an arbitrary natural number, then there exists a unique finite field K such that $|K| = p^n$.*

Proof We first prove the uniqueness. Let K be a field with p^n elements and let F be the prime field of K. We notice that the set of roots of $x^{p^n} - x = 0$ in K equals K itself. For any $\alpha \in K^{\#}$ let $\varphi_\alpha(x)$ be the irreducible factor of $x^{p^n - 1} - 1$ in $F[x]$ satisfying $\varphi_\alpha(\alpha) = 0$ and let τ_α be a mapping from $F[x]$ to K defined by

$$\tau_\alpha(f(x)) = f(\alpha) \quad f(x) \in F[x].$$

Then it is easy to see that τ_α is a homomorphism, kernel $\tau_\alpha = (\varphi_\alpha(x))$ and by theorem 1.3.1 $F[x]/(\varphi_\alpha(x))$ is isomorphic to a subfield of K. Hence if α is a generating element of the cyclic group $K^{\#}$, then $F[x]/(\varphi_\alpha(x)) \simeq K$ and degree $\varphi_\alpha(x) = n$. Now let $\varphi(x)$ be any irreducible factor of $\Phi_{p^n - 1}(x)$ in $F[x]$ of degree n and let $\beta \in K$ be a root of $\varphi(x) = 0$. Then, since $|F[x]/(\varphi(x))| = p^n$, τ_β is surjective and $K \simeq F[x]/(\varphi(x))$. Thus K, if it exists, is uniquely determined.

To prove the existence of K we first notice the following statement: Let L be a field of characteristic p and assume

$|\{\alpha\in L \mid \alpha^{p^n}=\alpha\}|<p^n$. Then there exists a finite field M such that $M \supsetneqq L$ and $\{\alpha\in M \mid \alpha^{p^n}=\alpha\} \supsetneqq \{\alpha\in L \mid \alpha^{p^n}=\alpha\}$. (Since, from the assumption, there exists an irreducible factor $\varphi(x)$ of $x^{p^n}-x$ in $L[x]$ whose degree ≥ 2. By theorems 1.3.3 and 1.3.6, $L[x]/(\varphi(x))$ is a field. We denote this field by M and $x+(\varphi(x))\in M$ by β. Since for $a, b\in L$

$$a \neq b \Leftrightarrow a+(\varphi(x)) \neq b+(\varphi(x))$$

we can naturally identify L with a subfield of M and $\varphi(x)$ can be considered to be an element of $M[x]$. Since $\varphi(\beta)=\varphi(x)+(\varphi(x))=0+(\varphi(x))$, β is a root of $\varphi(x)=0$ which proves the statement.) Starting with the case $L=$ prime field of characteristic p and using the above statement repeatedly, we can get a field M such that

$$|\{\alpha\in M \mid \alpha^{p^n}=\alpha\}|=p^n.$$

Put $K=\{\alpha\in M \mid \alpha^{p^n}=\alpha\}$. For $\alpha, \beta\in K$

$$(\alpha\beta)^{p^n}=\alpha^{p^n}\beta^{p^n}=\alpha\beta$$
$$(\alpha+\beta)^{p^n}=\sum\tbinom{p^n}{i}\alpha^i\beta^{p^n-i}=\alpha^{p^n}+\beta^{p^n}=\alpha+\beta,$$

i.e. $\alpha\beta, \alpha+\beta\in K$. This concludes that K is a field with p^n elements. ■

The field with p^n elements (where p is prime) will be denoted by F_{p^n}. Notice that F_p and $\mathbb{Z}_p$ are identical.

Theorem 1.3.11 *The automorphism group of F_{p^n} is a cyclic group of order n.*

Proof The map σ assigning a^p to every element $a\in F_{p^n}$ is an automorphism of F_{p^n}. (Since $(a+b)^p=\sum\binom{p}{i}a^ib^{p-i}=a^p+b^p$, $(ab)^p=a^pb^p$ and $a^p=0\Leftrightarrow a=0$.) Since $\sigma^n(a)=a^{p^n}=a$ for all $a\in F_{p^n}$, $\sigma^n=1$ and since for $1\leq l<n$ the polynomial $x^{p^l}-x$ has at most p^l roots, $\sigma^l\neq 1$, i.e. $\langle\sigma\rangle$ is a cyclic subgroup of order n of Aut F_{p^n}. Now let α be a generating element of $F^{\#}_{p^n}$, then, as we proved in the first part of the proof of theorem 1.3.10, there exists an nth degree irreducible polynomial $\varphi(x)\in F_p(x)$ (note that F_p is the prime field of F_{p^n}), of which α is a root. Let τ be an arbitrary automorphism

of F_{p^n}, then τ maps every element of F_p onto itself (since $1^\tau = 1$ and $(i)^\tau = (\underbrace{1 + \ldots + 1}_{i})^\tau = 1^\tau + \ldots + 1^\tau = i$ for $0 \leq i \leq p-1$). Hence α^τ is a root of $\varphi(x)$ for all τ and therefore: $|\text{Aut } F_{p^n}| \leq n$. Since Aut F_{p^n} contains the subgroup $\langle \sigma \rangle$ of order n, we conclude: Aut $F_{p^n} = \langle \sigma \rangle$. ■

Theorem 1.3.12 *Let K be a finite field. Every element $\alpha \in K$ can be represented as the sum of two squares: $\alpha = \beta^2 + \gamma^2$ for some $\beta, \gamma \in K$. If a, b, c are elements of $K^\#$, then there exist elements $\alpha, \beta, \gamma \in K$ (not all of them zero) such that $a\alpha^2 + b\beta^2 + c\gamma^2 = 0$.*

Proof Put $|K| = q = p^n$, where p is prime. If $p = 2$, then $K^\#$ is a cyclic group of order $2^n - 1$ (odd for all n) and it is easy to see that every element of K is the square of some element of K. So we may assume $p \neq 2$. Putting $K^{\#2} = \{\alpha^2 | \alpha \in K^\#\}$, we have that $K^{\#2}$ is a subgroup of $K^\#$ and further, since $q - 1$ is even, that $[K^\# : K^{\#2}] = 2$. If $\alpha \in K^{\#2}$, then $\alpha = \beta^2$ for some $\beta \in K$, hence we may assume $\alpha \notin K^{\#2}$. Suppose there exists $\alpha_0 \notin K^{\#2}$ such that $\alpha_0 = \beta_0^2 + \gamma_0^2$ with $\beta_0, \gamma_0 \in K^\#$, then $\alpha \in K^{\#2}\alpha_0$, i.e. $\alpha = \beta^2 + \gamma^2$ for some β and $\gamma \in K^\#$. So it suffices to show that the assumption that all sums of two elements of $K^{\#2}$ belong to $K^{\#2}$ leads to a contradiction. From the assumption $\{0\} \cup K^{\#2}$ becomes a subfield of K, but on the other hand we have

$$|\{0\} \cup K^{\#2}| = 1 + \frac{q-1}{2} = \frac{q+1}{2} \not\equiv 0 \pmod{p}.$$

Contradiction. The proof of the second statement is almost evident from the first. ■

Next we want to prove that the nth cyclotomic polynomial $\Phi_n(X)$ is irreducible over $\mathbb{Q}$. (We will need this property of $\Phi_n(X)$ in lemma 2.10.16.) Suppose that $\Phi_n(X)$ is reducible over $\mathbb{Q}$, then by lemma 1.1.5 there exist non-constant monic polynomials $f_1(X)$ and $f_2(X) \in \mathbb{Z}[X]$ such that $\Phi_n(X) = f_1(X)f_2(X)$ and $f_1(X)$ is irreducible over $\mathbb{Q}$. If ω is a root of $f_1(X)$, then ω is a primitive nth root of unity, hence

$$\Phi_n(X) = \prod_{\substack{1 \le l < n \\ (l,n)=1}} (X - \omega^l).$$

Therefore, there exists an l with $1 \le l < n$ and $(l, n) = 1$ such that ω^l is no root of $f_1(X)$. Let l_0 represent the smallest l with this property. Since ω is a root of $f_1(X)$, we have $l_0 > 1$. Let p be a prime factor of l_0. Since $\omega^{l_0/p}$ is a root of $f_1(X)$ and since $f_1(X)$ is irreducible over $\mathbb{Q}$, every polynomial of $\mathbb{Q}[X]$ that has $\omega^{l_0/p}$ as a root is divisible by $f_1(X)$. In particular, putting $g(X) = f_2(X^p)$, we have $g(\omega^{l_0/p}) = f_2(\omega^{l_0}) = 0$, hence $f_1(X)|g(X)$. Generally, if $f(X)$ is a polynomial of $\mathbb{Z}[X]$, let $\overline{f(X)}$ represent the polynomial of $F_p[X]$ obtained by regarding the coefficients of $f(X)$ as congruence classes mod p. For $a \in F_p$ we have $a^p = a$ (theorem 1.3.9), hence for

$$\overline{f(X)} = a_n X^n + a_{n-1} X^{n-1} + \dots + a_0 \in F_p[X],$$

we have

$$\begin{aligned}\overline{(f(X))^p} &= (a_n X^n + a_{n-1} X^{n-1} + \dots + a_0)^p \\ &= a_n^p X^{pn} + a_{n-1}^p X^{p(n-1)} + \dots + a_0^p \\ &= a_n X^{pn} + a_{n-1} X^{p(n-1)} + \dots + a_0 = \overline{f(X^p)}.\end{aligned}$$

From $f_1(X)|g(X)$ we conclude: $\overline{f_1(X)}|\overline{g(X)} = \overline{f_2(X^p)} = \overline{f_2(X)}^p$. As we saw in the proof of theorem 1.3.10, it is possible to construct an extension field K of F_p, such that $\overline{f_1(X)}$ has a root α in K. Hence α is a multiple root of $\overline{f_1(X)}\,\overline{f_2(X)} = X^n - 1$ considered as an element of $F_p[X]$. To prove that $X^n - 1$ has no multiple roots (and hence that $\Phi_n(X)$ is irreducible over $\mathbb{Q}$) we need the following definition and lemma: For $f(X) \in K[X]$, where K is a field, we define $\tilde{f}(X, Y) \in K[X, Y]$ by

$$\tilde{f}(X, Y) = \frac{f(X+Y) - f(X)}{Y}$$

and $f(X)' \in K[X]$ by

$$f(X)' = \tilde{f}(X, 0).$$

$f(X)'$ is called the derivative of f.

Lemma 1.3.13 *For $f(X), g(X) \in K[X]$, we have*

(i) $(f(X)+g(X))' = f(X)' + g(X)'$,

$(f(X)g(X))' = f(X)'g(X) + f(X)g(X)'$,

(ii) $(a_m X^m + a_{m-1} X^{m-1} + \dots + a_0)'$

$= m a_m X^{m-1} + (m-1) a_{m-1} X^{m-2} + \dots + a_1$,

(iii) *If α is a multiple root of $f(X)$, then α is a root of $f(X)'$.*

Proof (i) and (ii) follow easily from the definition; we prove (iii). Since α is a multiple root of $f(X)$, we have

$$f(X) = (X-\alpha)^2 g(X) \text{ for some } g(X) \in K[X],$$

hence $f(X)' = (X-\alpha)\{2g(X) + (X-\alpha)g(X)'\}$ and α is a root of $f(X)'$. ■

It is now easy to prove that $X^n - 1$, regarded as a polynomial of $F_p[X]$, has no multiple roots. Suppose that α is a multiple root of $X^n - 1$, then α is also a root of $(X^n-1)' = nX^{n-1}$, i.e. $n\alpha^{n-1} = 0$ (in F_p). However, from $(n, l_0) = 1$, we have $(n, p) = 1$ and hence $n\alpha^{n-1} \neq 0$ in F_p. We have proved:

Theorem 1.3.14 *The nth cyclotomic polynomial $\Phi_n(X)$ is irreducible over $\mathbb{Q}$.*

1.3.5 Alternative rings

Let R be an alternative ring. For $x, y, z \in R$, we denote the element $(xy)z - x(yz)$ by (x, y, z). The following properties follow easily from the definition.

Lemma 1.3.15 (i) $(x, x, y) = 0, (x, y, y) = 0$.

(ii) $(x, y, z) = -(y, x, z) = (y, z, x) = -(z, y, x)$.

(iii) *If two out of three elements x, y, z are the same, then* $(x, y, z) = 0$.

(iv) $x(y, z, u) - (xy, z, u) + (x, yz, u) - (x, y, zu) + (x, y, z)u = 0$.

(v) $(ab, b, c) = b(a, b, c)$.

Proof (i) Evident.

(ii) $0 = (x+y, x+y, z) = (x, x, z) + (x, y, z) + (y, x, z) + (y, y, z) = (x, y, z) + (y, x, z)$. The rest of the proof is similar.

(iii) Follows from (i) and (ii).

(iv) Straightforward computation.

(v) Substitute $x = a$, $y = z = b$ and $u = c$ in (iv). Next substitute $x = y = b$, $z = c$ and $u = a$ in (iv) and apply (i), (ii) and (iii) to simplify and compare the two resulting identities. ∎

Let $x_1, \dots, x_m$ (not necessarily different) be a finite sequence of certain elements in R, we define monomials in $x_1, \dots, x_m$ inductively:

(i) the only monomial in x_1 is x_1;

(ii) the monomials in $x_1, \dots, x_m$ are the products in R of a monomial in $x_1, \dots, x_{m_1}$ and a monomial in $x_{m_1+1}, \dots, x_m (1 \leq m_1 < m)$. To denote a monomial in $x_1, \dots, x_m$ we use a symbol $M(x_1, \dots, x_m)$, and m is called the length of $M(x_1, \dots, x_m)$.

For example, the only monomial in x_1, x_2 is x_1x_2, the only monomials in x_1, x_2, x_3 are $(x_1x_2)x_3$ and $x_1(x_2x_3)$ and so on. If S is a subset of R such that $x_1, \dots, x_n \in S$, then any monomial in $x_1, \dots, x_n$ is called a monomial of S and we define $\langle S \rangle$ by

$$\langle S \rangle = \left\{ \sum_{i=1}^{n} M_i \,\middle|\, n \in \mathbb{N} \text{ and } M_i \text{ is a monomial of } S \right\} \cup \{0\}.$$

$\langle S \rangle$ provided with the addition and multiplication of R is an alternative ring.

Lemma 1.3.16 *If S is a subset of R consisting of two elements, then $\langle S \rangle$ is a ring.*

Proof Since $\langle S \rangle$ is an alternative ring, the only thing that has to be shown is that multiplication on $\langle S \rangle$ is associative. Since the elements of $\langle S \rangle$ are sums of monomials of S, it suffices to prove the associativity for monomials of S, i.e. we have to prove that a monomial $M(x_1, \dots, x_m)$ is uniquely determined by the elements $x_1, \dots, x_m$ (in this order). We use induction on m. For $m = 1$ this is trivial, so let us assume that the statement is true for all monomials of length less than m. By definition $M(x_1, \dots, x_m)$ can be written as the product of two monomials $M(x_1, \dots, x_l)$ and

$M(x_{l+1}, \ldots, x_m)$ where $1 \leq l < m$:

$$M(x_1, \ldots, x_m) = M(x_1, \ldots, x_l)M(x_{l+1}, \ldots, x_m).$$

It suffices to prove that assuming $l < m - 1$ there exists an l' such that $l < l' \leq m - 1$ and

$$M(x_1, \ldots, x_l)M(x_{l+1}, \ldots, x_m) = M(x_1, \ldots, x_{l'})M(x_{l'+1}, \ldots, x_m)$$

for certain monomials $M(x_1, \ldots, x_{l'})$ and $M(x_{l'+1}, \ldots, x_m)$. (Because by repeating this procedure a finite number of times, we arrive at a monomial $M(x_1, \ldots, x_{m-1})$ such that $M(x_1, \ldots, x_m) = M(x_1, \ldots, x_{m-1})x_m$ while $M(x_1, \ldots, x_{m-1})$ is uniquely determined by $x_1, \ldots, x_{m-1}$ by the induction hypothesis.) There are three cases to consider:

(i) $x_l = x_{l+1}$;
(ii) $x_l \neq x_{l+1}$ and $x_{l+1} = x_m$;
(iii) $x_l \neq x_{l+1}$ and $x_{l+1} \neq x_m$.

Putting $S = \{a, b\}$ $(a, b \in R)$, the three possibilities can be rephrased as:

(i) $x_l = x_{l+1} = a$;
(ii) $x_l = b$ and $x_{l+1} = x_m = a$;
(iii) $x_l = x_m = b$ and $x_{l+1} = a$.

We consider these three cases separately.

(i) Putting (the uniquely determined monomial)

$$M(x_1, \ldots, x_{l-1}) = x$$

and (the uniquely determined monomial) $M(x_{l+2}, \ldots, x_m) = y$, we have

$$M(x_1, \ldots, x_l) = xa \quad \text{and} \quad M(x_{l+1}, \ldots, x_m) = ay.$$

Since $(xa)y$ and $x(ay)$ are both monomials in $x_1, \ldots, x_{l-1}, x_{l+1}, \ldots, x_m$, by the induction hypothesis we have $(xa)y = x(ay)$, i.e. $(x, a, y) = 0$. By lemma 1.3.15(v) $(xa, a, y) = a(x, a, y)$, hence $(xa, a, y) = 0$, i.e. $(xa)(ay) = ((xa)a)y$ and

$$M(x_1, \ldots, x_l)M(x_{l+1}, \ldots, x_m) = M(x_1, \ldots, x_{l+1})M(x_{l+2}, \ldots, x_m).$$

(ii) Putting $M(x_1, \ldots, x_l) = y$ and $M(x_{l+2}, \ldots, x_{m-1}) = x$, we have $M(x_{l+1}, \dots, x_m) = axa$. Since $(xa)y$ and $x(ay)$ are both monomials in $x_{l+2}, \dots, x_m, x_1, \dots, x_l$, we have $(x, a, y) = 0$, hence $(xa, a, y) = 0$ and also $(y, a, xa) = 0$ by lemma 1.3.15 (ii). Hence

$$y(axa) = (ya)(xa)$$

and

$$M(x_1, \dots, x_l)M(x_{l+1}, \dots, x_m) = M(x_1, \dots, x_{l+1})M(x_{l+2}, \dots, x_m).$$

(iii) Putting $x = M(x_1, \dots, x_{l-1})$ and $y = M(x_{l+1}, \dots, x_{m-1})$ we have $M(x_1, \dots, x_l) = xb$ and $M(x_{l+1}, \dots, x_m) = yb$. From $(x, b, y) = 0$ by the same argument as above, we conclude: $(xb, y, b) = 0$. Hence $(xb)(yb) = (xby)b$, i.e.

$$M(x_1, \dots, x_l)M(x_{l+1}, \dots, x_m) = M(x_1, \dots, x_{m-1})x_m. \qquad \blacksquare$$

Corollary 1.3.17 *For $a \in R$, the monomial $M(a, \dots, a)$ of length m is uniquely determined by a and m. This uniquely determined monomial is denoted by a^m.*

Theorem 1.3.18 (Artin–Zorn) *A finite alternative field is a field.*

Proof Let R be a finite alternative field. Since R is finite, there exists a finite subset S of R such that $R = \langle S \rangle$. We proceed by induction on the number of elements of S. If there exists a subset S of R with $|S| = 2$ and $\langle S \rangle = R$, then R is associative by lemma 1.3.16 and hence a finite skew field. Therefore R is a finite field by theorem 1.3.8. So we assume $|S| = m > 2$. Let $S = \{a_1, \dots, a_m\}$ for certain $a_1, \dots, a_m \in R$. Let us put $S_1 = \{a_1, a_2\}$ and $R_1 = \langle S_1 \rangle$. For $a \in R_1$ $(a \neq 0)$, we have $a, a^2, a^3, \dots \in R_1$, hence $a^k = a^{k+l}$ for certain positive integers k and l. By corollary 1.3.17 we have $a^k(1 - a^l) = 0$, i.e. $a^l = 1$. Therefore, $1 \in R_1$ and the inverses of non-zero elements of R_1 are elements of R_1, i.e. R_1 is an alternative field, and it follows that R_1 is a finite field as $|S_1| = 2$. Let b be a generating element of the cyclic group $R_1^{\#}$, and let $\tilde{S} = \{b, a_3, \dots, a_m\}$, then $R = \langle \tilde{S} \rangle$ and $|\tilde{S}| = m - 1$. Hence R is a finite field by the induction hypothesis. $\blacksquare$

1.4 Vector spaces

In this section we consider the basic properties of vector spaces over a field K (i.e. of K-modules).

1.4.1 Dimension and basis

Let V be a K-module. If U is a subgroup of the additive group underlying V satisfying the condition:

$$\lambda U \subseteq U \quad \forall \lambda \in K;$$

then U is a K-module, called a K-submodule of the K-module V (or a vector subspace of the vector space V). If we define a scalar multiplication on the additive group V/U by

$$\lambda(v+U) = \lambda v + U \quad \lambda \in K, v \in V,$$

then V/U becomes a K-module. (The definition of the scalar product is independent from the choice of the representant v of $\bar{v} = v + U$.) This K-module is called the quotient K-module of V with respect to U or the quotient vector space of V with respect to U. For a subset T of V the collection of all linear combinations of elements of T (with coefficients in K) is a K-submodule of V, that is denoted by $\langle T \rangle$ and called the K-submodule generated by T.

For two K-modules V_1 and V_2 $\mathrm{Hom}_K(V_1, V_2)$ is a K-module as we already saw in §1.3. Elements of $\mathrm{Hom}_K(V_1, V_2)$ are called linear maps from V_1 into V_2. Especially, when $V_1 = V_2 = V$ elements of $\mathrm{Hom}_K(V, V)$ are called linear transformations of V.

Elements $v_1, \ldots, v_m$ of the K-module V are called linearly independent if the following condition is satisfied:

$$\text{if} \qquad \sum_{i=1}^{m} \lambda_i v_i = 0 \quad \text{for } \lambda_i \in K, \quad \text{then } \lambda_1 = \ldots = \lambda_m = 0.$$

If $v_1, \ldots, v_m$ are not linearly independent, they are called linearly dependent. The greatest natural number of the set: $\{m \in \mathbb{N} \mid \text{there exist } m \text{ linearly independent elements in } V\}$ is called the dimension of V. The dimension of V is denoted by $\dim_K V$. If there is no maximum, we put $\dim_K V = \infty$. The elements $v_1, \dots, v_n$ are called a (K-) basis for V, if $v_1, \dots, v_n$ are linearly independent and if every element of V can be represented as a linear combination of $v_1, \dots, v_n$ with coefficients in K.

Lemma 1.4.1 *Let $v_1, \dots, v_n$ be linearly independent elements of V. If $U = \langle v_1, \dots, v_r \rangle$ for some $r < n$, then the elements $\bar{v}_{r+1}, \dots, \bar{v}_n$ of V/U are linearly independent.*

Proof Let $\lambda_{r+1}, \dots, \lambda_n \in K$ be such that

$$\lambda_{r+1}\bar{v}_{r+1} + \dots + \lambda_n \bar{v}_n = 0,$$

i.e.

$$\lambda_{r+1} v_{r+1} + \dots + \lambda_n v_n \in U.$$

Then there exist $\lambda_1, \dots, \lambda_r \in K$ such that

$$\lambda_{r+1} v_{r+1} + \dots + \lambda_n v_n = \lambda_1 v_1 + \dots + \lambda_r v_r.$$

Since $v_1, \dots, v_n$ are linearly independent, we conclude that $\lambda_1 = \dots = \lambda_n = 0$. ■

Theorem 1.4.2 *If $v_1, \dots, v_n$ is a K-basis of V, then* $\dim_K V = n$.

Proof The proof is by induction on n. For $n = 1$, the statement is evident. Let us assume that the statement is true for $1, 2, \dots, n-1$. It is clear from the definition that $\dim_K V \geq n$, so it suffices to show that the assumption $\dim_K V > n$ leads to a contradiction. If $\dim_K V > n$, then there are $n+1$ linearly independent elements $u_1, \dots, u_{n+1}$ in V. Since $v_1, \dots, v_n$ form a basis of V, u_1 can be represented as

$$u_1 = \sum_{i=1}^{n} \lambda_i v_i,$$

where at least one of the coefficients, say λ_1, is not equal to zero. Then $u_1, v_2, \dots, v_n$ form a basis of V. Putting $U = \langle u_1 \rangle = \{\lambda u_1 \mid \lambda \in K\}$ we can consider the quotient K-module $\bar{V} = V/U$. According to lemma 1.4.1 $\bar{v}_2, \dots, \bar{v}_n$ are a K-basis of $\bar{V}$, while $\bar{u}_2, \dots, \bar{u}_{n+1}$ are linearly independent. Hence $n \leq n-1$ by the induction hypothesis. Contradiction. ■

The following lemma is almost trivial. We leave its proof to the reader.

Lemma 1.4.3 *Let V be a K-module.*

(1) *If $\dim_K V = n$ and if $v_1, \dots, v_n$ are linearly independent elements of V, then $v_1, \dots, v_n$ form a basis of V, i.e. every element of V can be represented as a linear combination of $v_1, \dots, v_n$ in a unique way.*

(2) *If $v_1, \dots, v_n$ are elements of V and if every element of V can be represented in a unique way as a linear combination of $v_1, \dots, v_n$, then $v_1, \dots, v_n$ form a basis of V.*

Choosing a basis $v_1, \dots, v_n$ of V, every element $v \in V$ determines a unique row vector $(\lambda_1, \dots, \lambda_n)$ such that $v = \lambda_1 v_1 + \dots + \lambda_n v_n$. It is easy to see that this correspondence sets up an isomorphism between V and ${}^nK$, both regarded as K-modules. The vector $(\lambda_1, \dots, \lambda_n)$ is called the coordinate vector of v, the scalars $\lambda_1, \dots, \lambda_n$ are called the coordinates of v (with respect to the basis $v_1, \dots, v_n$).

Let V_1 and V_2 be an m-dimensional and an n-dimensional K-module respectively and let $\{v_1, \dots, v_m\}$ and $\{u_1, \dots, u_n\}$ be bases of V_1 and V_2 respectively. For $f \in \operatorname{Hom}_K(V_1, V_2)$ there are uniquely determined elements $\lambda_{ij} \in K$ such that

$$v_i^f = f(v_i) = \sum_{j=1}^{n} \lambda_{ij} u_j \quad (i = 1, \dots, m),$$

i.e. f determines a unique (m, n)-matrix (λ_{ij}) with elements from K. Conversely, given an (m, n)-matrix (μ_{ij}) with elements from K, the map $g: V_1 \to V_2$ defined by

$$g\left(\sum_{i=1}^{m} \lambda_i v_i\right) = \sum_{j=1}^{n}\left(\sum_{i=1}^{m} \lambda_i \mu_{ij}\right) u_j$$

is an element of $\operatorname{Hom}_K(V_1, V_2)$. It is easy to see that these two correspondences are each other's inverses, i.e. there is a one-to-one correspondence between K-modules $\operatorname{Hom}_K(V_1, V_2)$ and $(K)_{(m,n)}$. Moreover, this correspondence is an isomorphism as K-modules, as is also easy to check. The element of $(K)_{(m,n)}$ corresponding with the element f of $\operatorname{Hom}_K(V_1, V_2)$ is called the matrix representation of f (with respect to the bases $v_1, \dots, v_m$ and $u_1, \dots, u_n$ of V_1 respectively V_2). Now, if V_3 is an l-dimensional K-module and if

$f \in \mathrm{Hom}_K(V_1, V_2)$ and $g \in \mathrm{Hom}_K(V_2, V_3)$, then their composition fg is an element of $\mathrm{Hom}_K(V_1, V_3)$. Choosing bases for V_1, V_2 and V_3, let f have the matrix representation A and let g have the matrix representation B, then fg has the matrix representation AB with respect to the chosen bases in V_1 and V_3. Especially, in the case that V_1, V_2 and V_3 are the same K-module which is denoted by V, the isomorphism, defined by a basis of V, between K-modules $\mathrm{Hom}_K(V, V)$ and $(K)_n$ is also an isomorphism between rings $\mathrm{Hom}_K(V, V)$ and $(K)_n$. Generally, a mapping from a K-algebra, R_1, into another K-algebra, R_2, is called an algebra homomorphism if it is a homomorphism for the K-modules and a homomorphism for the rings underlying R_1 and R_2. So, the above isomorphism between $\mathrm{Hom}_K(V, V)$ and $(K)_n$ is in fact an algebra isomorphism. Returning to $\mathrm{Hom}_K(V_1, V_2)$ and $(K)_{(m,n)}$ we remark that the elements $e_{ij} \in (K)_{(m,n)}$ $(1 \leq i \leq m, 1 \leq j \leq n)$, where e_{ij} is defined to be the matrix with 1 in the (i, j) entry and 0 elsewhere, form a basis for K-module $(K)_{(m,n)}$, hence $\dim_K \mathrm{Hom}_K(V_1, V_2) = \dim_K(K)_{(m,n)} = mn$. The elements of the matrix corresponding with some $f \in \mathrm{Hom}_K(V_1, V_2)$ are just the coordinates of f with respect to the basis of $\mathrm{Hom}_K(V_1, V_2)$ determined by the basis $\{e_{ij}\}$ of $(K)_{(m,n)}$.

Consider the group ring $K(G)$, where G is some finite group. Then G, considered as a subset of $K(G)$, is a basis for K-algebra $K(G)$, i.e. $\dim_K K(G) = |G|$. The dimension of the centre $Z(K(G))$ of $K(G)$ is equal to the number of conjugate classes in G. Letting $K_1, \ldots, K_r$ represent the conjugate classes of G, it is easy to prove that $\bar{K}_1, \ldots, \bar{K}_r$ form a basis of $Z(K(G))$.

If R is a K-algebra and if $v_1, \ldots, v_n$ form a basis of R regarded as a K-module, then there exist n^3 elements $c_{ijk} \in K$ $(1 \leq i, j, k \leq n)$ such that: $v_i v_j = \sum_{k=1}^n c_{ijk} v_k$. From $(v_i v_j) v_k = v_i(v_j v_k)$ we get relations for these n^3 elements c_{ijk}:

$$\sum_{l=1}^{n} c_{ijl} c_{lkh} = \sum_{l=1}^{n} c_{ilh} c_{jkl} \quad (1 \leq i, j, k, h \leq n). \tag{1.4.1}$$

Conversely if V is a K-module, if $v_1, \ldots, v_n$ form a basis for V and if n^3 elements $c_{ijk} \in K$ $(1 \leq i, j, k \leq n)$ are given, such that (1.4.1) is satisfied, then V can be made into a K-algebra by defining the product of two elements $\sum \lambda_i v_i$ and $\sum \mu_i v_i$ of V by

$$(\sum \lambda_i v_i)(\sum \mu_i v_i) = \sum_k \left(\sum_{i,j} \lambda_i \mu_j c_{ijk} \right) v_k .$$

So, the multiplicative structure of a K-algebra R is completely determined by a basis of R as K-module and n^3 elements c_{ijk} of K satisfying (1.4.1). The elements c_{ijk} of K are called the structural constants of the K-algebra R.

Let U_1 and U_2 be K-submodules of the K-module V. We say that V is the direct sum of U_1 and U_2 if every element of V can be written in exactly one way as the sum of an element of U_1 and an element of U_2. Notation: $V = U_1 \oplus U_2$. If $v_1, \dots, v_m$ are linearly independent elements of V such that $\langle v_1, \dots, v_m \rangle \neq V$, then for every element $u \in V - \langle v_1, \dots, v_m \rangle$ the elements $v_1, \dots, v_m$, u are linearly independent. Therefore, if V is finite dimensional, it is possible to find elements $v_{m+1}, \dots, v_n$ such that $v_1, \dots, v_m, v_{m+1}, \dots, v_n$ form a basis for V. We say that we have extended $v_1, \dots, v_m$ to a basis of V. Since $\langle v_1, \dots, v_m \rangle \cap \langle v_{m+1}, \dots, v_n \rangle = \{0\}$, we have $V = \langle v_1, \dots, v_m \rangle \oplus \langle v_{m+1}, \dots, v_n \rangle$. The following theorem is almost evident from the above argument:

Theorem 1.4.4 *Let V be a K-module.*

(1) *If W is a K-submodule of V, then* $\dim_K W \leq \dim_K V$.

(2) *If W is a K-submodule of V such that* $\dim_K W \lneq \dim_K V$, *then there exists a K-submodule W' of V such that $V = W \oplus W'$ where W' is called a complementary K-submodule of W. Furthermore, $V/W \simeq W'$ (as K-modules) and* $\dim_K V = \dim_K W + \dim_K W'$.

(3) *If W_1 and W_2 are K-submodules of V, then*

$$\dim_K(W_1 + W_2) + \dim_K(W_1 \cap W_2) = \dim_K W_1 + \dim_K W_2,$$

where $W_1 + W_2$ is the K-submodule of V generated by W_1 and W_2.

1.4.2 $\Gamma L(V)$, $GL(V)$, $SL(V)$

Let V be an n-dimensional K-module. Let θ be an automorphism of K and let $f: V \to V$ be a mapping satisfying

$$f(v+v') = f(v) + f(v') \quad \forall v, v' \in V$$
$$f(\lambda v) = \lambda^{\theta} f(v) \quad \forall \lambda \in K, \forall v \in V$$

then f is called a (θ-) semi-linear transformation and θ is called the companion automorphism of f. If moreover f is bijective, then f is called a non-singular semi-linear transformation. For $\theta = 1$ we get linear transformations.

If Γ denotes the collection of all semi-linear transformations, then $\Gamma \supseteq \mathrm{Hom}_K(V, V)$, but Γ is not necessarily a ring (since the sum of two elements of Γ is not necessarily an element of Γ). However, if f is a θ_1-semi-linear transformation and if g is a θ_2-semi-linear transformation, then their composition fg satisfies

$$(fg)(\lambda v) = \lambda^{\theta_1 \theta_2}(fg)(v),$$

that is, fg is a $\theta_1\theta_2$-semi-linear transformation. Therefore, the collection of all non-singular semi-linear transformations of V becomes a group under composition. This group is denoted by $\Gamma L(V)$ and called the general semi-linear group. The intersection $\Gamma L(V) \cap \mathrm{Hom}_K(V, V)$, i.e. the collection of all non-singular linear transformations of V, is a subgroup of $\Gamma L(V)$, denoted by $GL(V)$, and called the general linear group. By assigning to every $f \in \Gamma L(V)$ its companion automorphism of K we get a homomorphism from $\Gamma L(V)$ into $\mathrm{Aut}\, K$. The kernel of this homomorphism is $GL(V)$, hence $GL(V) \trianglelefteq \Gamma L(V)$. To see that this homomorphism is surjective, choose a K-basis $v_1, \ldots, v_n$ of V and for $\theta \in \mathrm{Aut}\, K$ define $f: V \to V$ by

$$f(\textstyle\sum \lambda_i v_i) = \sum \lambda_i^{\theta} v_i \quad \lambda_i \in K.$$

Then $f \in \Gamma L(V)$ and the companion automorphism of f is just θ. Hence:

$$\Gamma L(V)/GL(V) \simeq \mathrm{Aut}\, K.$$

We find that we can set up an isomorphism between $\mathrm{Hom}_K(V, V)$ and $(K)_n$ by choosing a basis $v_1, \ldots, v_n$ and assigning its matrix representation with respect to this basis to every element of $\mathrm{Hom}_K(V, V)$. The image of $GL(V)$ in $(K)_n$ under this isomorphism is denoted by $GL(n, K)$. Hence $GL(n, K)$ with matrix multiplication is a group isomorphic with $GL(V)$. From the definition it is easy to see that an element A of $(K)_n$ belongs to $GL(n, K)$ if and only if the row vectors of A are linearly independent, and also if and only if there exists an element B of $(K)_n$ such that $BA = AB = E$, where E is the identity element of $GL(n, K)$, i.e. $E = (\delta_{ij})$.

Now we want to derive a few properties of $GL(n, K)$. For r and $s \in \mathbb{N}$ such that $1 \leq r \neq s \leq n$ and for $\lambda \in K$ the element $A_{r,s:\lambda} = (\mu_{ij})$ is defined by:

$$\mu_{ij} = \begin{cases} 1 & \text{if} \quad \text{if } i = j \\ \lambda & \text{if } i = r \text{ and } j = s \\ 0 & \text{elsewhere.} \end{cases}$$

From $A_{r,s:\lambda} A_{r,s:-\lambda} = A_{r,s:-\lambda} A_{r,s:\lambda} = E$ we conclude that $A_{r,s:\lambda} \in GL(n, K)$. The subgroup of $GL(n, K)$ generated by all such matrices $A_{r,s:\lambda}$ is denoted by $SL(n, K)$; i.e.

$$SL(n, K) = \langle \{ A_{r,s:\lambda} \mid 1 \leq r \neq s \leq n \text{ and } \lambda \in K \} \rangle.$$

Then we have the following properties:

(*a*) Let for $\mu \in K$ $(\mu \neq 0)$ the element $P_\mu = (\mu_{ij})$ of $GL(n, K)$ be defined by

$$\mu_{ij} = \begin{cases} 1 & \text{if } i = j < n \\ \mu & \text{if } i = j = n \\ 0 & \text{elsewhere.} \end{cases}$$

If A is an arbitrary element of $GL(n, K)$ then there exist B_1 and $B_2 \in SL(n, K)$ and $\mu \in K$ $(\mu \neq 0)$ such that

$$B_1 A B_2 = P_\mu. \tag{1.4.2}$$

Proof We first show that if $n \ngeq 1$ there exist C_1 and C_2 in $SL(n, K)$ such that $C_1 A C_2 = (v_{ij})$ where $v_{11} = 1$ and $v_{i1} = v_{1j} = 0$ $(\forall i, j \ngeq 1)$. Put $A = (\lambda_{ij})$. If $\lambda_{11} = 1$, $C_1 = \prod_{i=2}^n A_{i,1:-\lambda_{i1}}$ and $C_2 = \prod_{i=2}^n A_{1,i:-\lambda_{1i}}$ satisfy the requirement. If $\lambda_{11} \neq 1$, there are two cases to consider:

(i) $\lambda_{1i} \neq 0$ for some $i \geq 2$. Then the $(1, 1)$ entry of $AA_{i,1:(1-\lambda_{11})/\lambda_{1i}}$ equals 1, and we are reduced to the case $\lambda_{11} = 1$.

(ii) $\lambda_{1i} = 0$ for all $i \geq 2$. Since $A \in GL(n, K)$, we must have $\lambda_{11} \neq 0$. Hence $AA_{1,2:1}$ has a (1, 1) entry that is not equal to 1 and a (1, 2) entry that is not equal to 0 and we are reduced to the previous case.

By repeating this procedure we get form (1.4.2) in (*a*) and the only thing that remains to be shown is that $\mu \neq 0$. Let σ be the element in $GL(V)$ corresponding with P_μ. Since $V^\sigma = V$, there exist $\lambda_1, \ldots, \lambda_n$ such that

$$\sigma(\sum \lambda_i v_i) = v_n.$$

On the other hand, it follows from the definition of σ that

$$\sigma(\sum \lambda_i v_i) = \sum_{i=1}^{n-1} \lambda_i v_i + \lambda_n \mu v_n.$$

Hence $\lambda_1 = \ldots = \lambda_{n-1} = 0$ and $\lambda_n \mu = 1$, so $\mu \neq 0$. ■

For an arbitrary element A of $(K)_n$ not contained in $GL(n, K)$ we can prove the following statement by a similar (rather rougher) method to that in the proof of (*a*). We leave it to the reader.

(*a'*) Let $A \in (K)_n$ but $\notin GL(n, K)$. Then there exists $B \in SL(n, K)$ such that $BA = (\mu_{ij})$ where $\mu_{nj} = 0$ for $1 \leq j \leq n$.

(*b*) An arbitrary element A of $GL(n, K)$ can be written as $A = BP_\mu$ where $B \in SL(n, K)$ and $\mu \neq 0$. (We call this a standard factorization of A temporarily. The uniqueness of this factorization will be proved in the proof of theorem 1.4.5.)

Proof By (*a*) there are B_1 and $B_2 \in SL(n, K)$ and $\mu \in K$ $(\mu \neq 0)$ such that $A = B_1 P_\mu B_2$. For the generating elements $A_{i,j:\lambda}$ of $SL(n, K)$ we have

$$P_\mu A_{i,j:\lambda} = \begin{cases} A_{i,n:\mu^{-1}\lambda} P_\mu & \text{if } j = n \\ A_{n,j:\mu\lambda} P_\mu & \text{if } i = n \\ A_{i,j:\lambda} P_\mu & \text{if } i \neq n \text{ and } j \neq n. \end{cases} \tag{1.4.3}$$

Therefore, there exists a $B_3 \in SL(n, K)$ such that: $P_\mu B_2 = B_3 P_\mu$ and hence:

$$A = B_1 P_\mu B_2 = B_1 B_3 P_\mu = BP_\mu. \qquad \blacksquare$$

(*c*) $GL(n, K) \trianglerighteq SL(n, K)$.

Proof Let S denote the set of generating elements of $SL(n, K)$, then $P_\mu^{-1} S P_\mu = S$ by (1.4.3). Use (*a*) to finish the proof. $\blacksquare$

After choosing a basis for V, we can assign a matrix to every element of $GL(V)$, but this assignment is dependent on the choice of the basis. Let $v_1, \ldots, v_n$ and $u_1, \ldots, u_n$ be two bases of V and let $f \in GL(V)$ be represented by $A = (\lambda_{ij})$ with respect to the first basis and by $B = (\mu_{ij})$ with respect to the second basis, then

$$f(v_i) = \sum \lambda_{ij} v_j \text{ and } f(u_i) = \sum \mu_{ij} u_j.$$

Since both $v_1, \ldots, v_n$ and $u_1, \ldots, u_n$ are bases for V, there exist matrices $P = (\nu_{ij})$ and $Q = (\varepsilon_{ij})$ such that

$$u_i = \sum \nu_{ij} v_j \text{ and } v_i = \sum \varepsilon_{ij} u_j.$$

The following relations are direct consequences of the definitions:

(i) $PQ = QP = E$, hence $P, Q \in GL(n, K)$ and $Q^{-1} = P$;

(ii) $B = PAQ = Q^{-1} AQ$.

In other words, a change of basis corresponds with changing to a conjugate element for the matrix representation of a certain mapping. Since $GL(n, K) \trianglerighteq SL(n, K)$ by (*c*), the subgroup of $GL(V)$ corresponding with $SL(n, K)$ is unambiguously determined. This subgroup is denoted by $SL(V)$ and called the special linear group. By looking at the generating elements of $SL(V)$ (or $SL(n, K)$) it is easy to prove that the centre of $GL(V)$ equals $\{\lambda 1_V \mid \lambda \in K, \lambda \neq 0\}$ (1_V denotes the identity linear transformation of V). The centre of $GL(V)$ is denoted by $Z(V)$. Since the centre of a group is always a characteristic subgroup and since $\Gamma L(V) \trianglerighteq GL(V)$, we have $\Gamma L(V) \trianglerighteq Z(V)$. The reader may prove that the centre of $SL(V)$ is equal to $Z(V) \cap SL(V)$.

The determinant of a matrix consisting of elements of K is defined in exactly the same way as the determinant of a matrix with real or complex entries. However, since we do not want to assume any knowledge beyond the real or complex case, we will define the determinant of a matrix with elements in an arbitrary field K. Our treatment follows Dieudonné's, which was intended to define the determinant of matrices with elements in a skew field.

A mapping $f_n:(K)_n \to K$ satisfying the following properties:

(1) $f_n(E)=1$;

(2) if X is the matrix resulting by multiplying all elements of the kth row of A by μ, then: $f_n(X)=\mu f_n(A)$, ($\forall A \in (K)_n$ and $\forall \mu \in K$);

(3) if X is the matrix resulting by adding the jth row of A to the kth row of A ($k \neq j$), then $f_n(X)=f_n(A)$;

is called a determinant mapping.

We start with two simple consequences from this definition:

Statement 1.4.5 *If X is the matrix obtained by interchanging the kth and the jth row of $A(k \neq j)$, then $f_n(X)=-f_n(A)$.*

Proof Let B be the matrix obtained by adding the jth row to the kth row of A. Let C be the matrix obtained from B by multiplying the kth row by -1. Let D be the matrix obtained by adding the kth row to the jth row of C. Let F be the matrix obtained by multiplying the kth row of D by -1. Let G be the matrix obtained by adding the jth row to the kth row of F and finally let H be the matrix obtained by multiplying the jth row of G by -1. Then $H=X$ and $f_n(X)=-f_n(G)=-f_n(F)=f_n(D)=f_n(C)=-f_n(B)=-f_n(A)$). ■

The proof of the following statement is similar:

Statement 1.4.6 *If X is the matrix obtained from A by adding λ times the jth row to the kth row ($k \neq j$), then $f_n(X)=f_n(A)$ ($\lambda \in K$).*

Theorem 1.4.7 *For $n \geq 1$ there exists exactly one determinant mapping.*

Proof We prove by induction on n. For $n=1$ we put $f_1((a))=a$ for all $(a)\in(K)_1$. In this case f_1 is obviously a determinant mapping and uniquely determined. Let us assume that f_{n-1} exists and is uniquely determined. Let A be an arbitrary element of $(K)_n$. For the case $A \notin GL(n, K)$, we put $f_n(A) = 0$. Let us assume $A \in GL(n, K)$ and let us denote the ith row vector of A by a_i. Since $A \in GL(n, K)$, there are uniquely determined $\lambda_1, \ldots, \lambda_n \in K$ such that

$$\sum \lambda_i a_i = (1, 0, \ldots, 0).$$

At least one of the $\lambda_1, \ldots, \lambda_n$, say λ_i, is not equal to 0. Let A_i denote the matrix of rank $n-1$ obtained from A by deleting the first column and the ith row and define $f_n(A)$ by

$$f_n(A) = (-1)^{i+1} \lambda_i^{-1} f_{n-1}(A_i).$$

We have to show that this definition of $f_n(A)$ is independent from the choice of i. Suppose that also $\lambda_j \neq 0$. Let B be the matrix obtained from A by multiplying all the elements of the jth row by λ_j, let C be the matrix obtained from B by adding $\sum_{\nu \neq i,j} \lambda_\nu a_\nu$ to the jth row and let D be the matrix obtained from A by interchanging the jth row and the ith row. Using existence and uniqueness of f_{n-1} (by the induction hypothesis), we have

$$\lambda_j f_{n-1}(A_i) = f_{n-1}(B_i)$$

which, by statement 1.4.6

$$= f_{n-1}(C_i) = -\lambda_i f_{n-1}(D_i)$$

which, by statement 1.4.5

$$= (-\lambda_i)(-1)^{i-j-1} f_{n-1}(A_j).$$

Hence:

$$(-1)^{i+1} \lambda_i^{-1} f_{n-1}(A_i) = (-1)^{j+1} \lambda_j^{-1} f_{n-1}(A_j).$$

Next, we have to verify that conditions (1), (2) and (3) are satisfied. Condition (1) is evident, and conditions (2) and (3) are evidently satisfied for matrices $A \notin GL(n, K)$. So suppose $A \in GL(n, K)$. If $\mu = 0$, $X \notin GL(n, K)$, i.e. $f_n(X) = 0 = \mu f_n(A)$. For $\mu \neq 0$, the role of $\lambda_1, \ldots, \lambda_n$ for A is played by $\lambda_1, \ldots, \mu^{-1}\lambda_k, \ldots, \lambda_n$ for X. Again, let $\lambda_i \neq 0$. If $k \neq i$, then

$$f_n(X) = (-1)^{i+1} \lambda_i^{-1} f_{n-1}(X_i) = (-1)^{i+1} \lambda_i^{-1} \mu f_{n-1}(A_i) = \mu f_n(A).$$

And if $k = i$, then

$$f_n(X) = (-1)^{i+1} (\mu^{-1} \lambda_i)^{-1} f_{n-1}(A_i) = \mu f_n(A),$$

which verifies condition (2). Condition (3) is verified in a similar way. So we have proved the existence of f_n. Finally we prove the uniqueness of f_n. For $A \in GL(n, K)$ let $A = BP_\mu$ be a standard factorization of A. Then we have $f_n(A) = \mu$. (For, by property (3) $f_n(A_{r,s:\lambda}X) = f_n(X)$ for an arbitrary $X \in (K)_n$. Hence we have $f_n(A) = f_n(P_\mu)$, and this is equal to $\mu f_n(E) = \mu$ by properties (1) and (2).) For $A \in (K)_n$ but $\notin GL(n, K)$ we have $f_n(A) = 0$. (For by statement (a') A is written as $A = BP$ for some $B \in SL(n, K)$ and for some $P = (\mu_{ij})$ where $\mu_{nj} = 0$, $1 \le j \le n$. In the same way we have $f_n(A) = f_n(P)$. By property (3) we have $f_n(P) = 0$.) This has shown the uniqueness of f_n, and at the same time the uniqueness of μ appearing in a standard factorization of $A = BP_\mu$. ■

From the proof of the last theorem we derive:
(a) if $A, B \in (K)_n$, then $f_n(AB) = f_n(A) f_n(B)$;
(b) $GL(n, K) = \{X \in (K)_n \mid f_n(X) \neq 0\}$;
(c) $SL(n, K) = \{X \in (K)_n \mid f_n(X) = 1\}$;
(d) $GL(n, K)/SL(n, K) (\simeq GL(V)/(SL(V)) \simeq K^\# = K - \{0\}$.
We often use notations $GL(n, q)$ and $SL(n, q)$ instead of $GL(n, K)$ and $SL(n, K)$ if $|K| = q$. Instead of $f_n(X)$ we will write det X. We call det X the determinant of X.

As we saw before, the effect of a change of basis of V for the matrix representation of some element of $\mathrm{Hom}_K(V, V)$ is to change it into a conjugate element of $(K)_n$. Therefore, the determinant of the matrix of some element σ of $\mathrm{Hom}_K(V, V)$ is independent from the choice of the basis of V, i.e. we can define det σ as the determinant of an arbitrary matrix representation of σ. Hence:

$$GL(V) = \{\sigma \in \mathrm{Hom}_K(V, V) \mid \det \sigma \neq 0\};$$
$$SL(V) = \{\sigma \in \mathrm{Hom}_K(V, V) \mid \det \sigma = 1\}.$$

1.5 Geometric structures

1.5.1 Block design

A pair $(\Omega, \mathfrak{B})$ consisting of a finite set Ω and a collection $\mathfrak{B}$ of subsets of Ω (i.e. $\mathfrak{B} \subseteq 2^\Omega$) is called a (finite) geometric structure or an incidence structure. The elements of Ω are called points and the elements of $\mathfrak{B}$ are called blocks of the geometric structure. For elements $p_1, \ldots, p_r$ of Ω we put

$$\mathfrak{B}_{p_i} = \{B \in \mathfrak{B} \mid p_i \in B\}$$

and

$$\mathfrak{B}_{p_1,\ldots,p_r} = \bigcap_{i=1}^{r} \mathfrak{B}_{p_i}.$$

If a geometric structure $(\Omega, \mathfrak{B})$ satisfies the following conditions:

(1) $\varnothing \notin \mathfrak{B}$ and $\Omega \notin \mathfrak{B}$;

(2) all sets of $\mathfrak{B}$ have the same number of elements;

(3) for every pair of elements p_1 and p_2 of $\Omega (p_1 \neq p_2)$ the number of elements of $\mathfrak{B}_{p_1,p_2}$ is the same and not equal to zero;

then it is called a block design (balanced incomplete block design) or simply a design. The number of elements of each set of $\mathfrak{B}$ determined by (2) is denoted by k, the number of elements of each set $\mathfrak{B}_{p_1,p_2} (p_1 \neq p_2)$ determined by (3) is denoted by λ and the numbers of elements of Ω and $\mathfrak{B}$ are denoted by v and b respectively. We notice that $v > k \geq 2$ from the conditions (1) and (3).

For any $p \in \Omega$ the number of elements of $\mathfrak{B}_p$ is $(v-1)\lambda/(k-1)$ and hence is independent from p. (Proof: Let us denote the number of elements of $\{(B, q) \mid B \in \mathfrak{B}_p, q \neq p \text{ and } q \in B\}$ by t_p. Since there are $|\mathfrak{B}_p|$ choices for B and $k-1$ choices for q, after having chosen B we have: $t_p = |\mathfrak{B}_p| \cdot (k-1)$. On the other hand, there are $v-1$ choices for q and λ choices for B after having chosen q, hence $t_p = (v-1)\lambda$ and the desired result follows.) This number is represented by r. The five constants (v, b, k, r, λ) are called the parameters of the block design $(\Omega, \mathfrak{B})$. In this section we will use the letters v, b, k, r, and λ for the parameters unless stated otherwise. The following relations between the parameters are fundamental:

Theorem 1.5.1 (1) $vr = bk$

(2) $(v-1)\lambda = (k-1)r$

(3) $v \gneq k$ and $r \gneq \lambda$.

Proof (1) Let us denote the number of elements of the set

$\{(B, p) | B \in \mathfrak{B} \text{ and } p \in B\}$ by t. Since there are b choices for B and k choices for p once B is chosen we have that $t = bk$. On the other hand, there are v choices for p and r choices for B once p is chosen, hence $t = vr$, i.e. $vr = bk$.

(2) Already proved.

(3) The condition (1) in the definition of block design implies $v \gneqq k$. From this and (2) it follows that $r \gneqq \lambda$. ■

It follows from this theorem, that if v, k and λ are known b and r are determined by:

$$r = \frac{(v-1)\lambda}{k-1}, \quad b = \frac{v(v-1)\lambda}{k(k-1)}. \tag{1.5.1}$$

A block design $D = (\Omega, \mathfrak{B})$ with parameters v, k and λ (and hence r and b given by (1.5.1)) is called a (v, k, λ)-design and $r - \lambda (> 0)$ is called the order of this design.

By conditions (1) and (3) in the definition of block design we have $v - 1 \geq k \geq 2$. If $k = 2$, then $\lambda = 1$ and $b = \frac{1}{2}v(v-1)$ by (1.5.1), hence in this case every subset of two elements of Ω is a block. If $v - 1 = k$, then we have $\lambda = v - 2$ and $b = v$ by (1.5.1), hence every subset of Ω consisting of $v - 1$ elements is a block. Generally, if t is such that $2 \leq t \leq v - 1$ and if we define $\mathfrak{B}$ to be the collection of all subsets of Ω consisting of t elements, then $(\Omega, \mathfrak{B})$ is a block design with parameters $(v, {}_vC_t, t, {}_{v-1}C_{t-1}, {}_{v-2}C_{t-2})$. These designs have many special properties that fall outside the scope of a general study of designs. They are called trivial designs. So, if a design is non-trivial, then the inequalities $2 < k < v - 1$ surely hold.

If $(\Omega, \mathfrak{B})$ is an incidence structure, if $\Omega = \{p_1, \ldots, p_r\}$ and if $\mathfrak{B} = \{B_1, \ldots, B_b\}$, then the (v, b)-matrix $A = (t_{ij})$ defined by

$$t_{ij} = \begin{cases} 1 & \text{if } p_i \in B_j \\ 0 & \text{if } p_i \notin B_j, \end{cases}$$

is called the incidence matrix of $(\Omega, \mathfrak{B})$. If we choose another way of enumerating the elements of Ω and $\mathfrak{B}$, the resulting incidence matrix can be obtained from A by changing the order of the rows and columns of A. So, the incidence matrix of $(\Omega, \mathfrak{B})$ is uniquely determined apart from the order of the rows and columns. Conversely, every (v, b)-matrix with entries 0 or 1 determines an incidence structure $(\Omega, \mathfrak{B})$ with this matrix as an incidence matrix. If A represents the incidence matrix of a block design with parameters (v, b, k, r, λ), then A is a (v, b)-matrix and has the following properties:

(i) every column of A has k entries that are equal to 1 and $v - k$ entries that are equal to 0;

(ii)

$$A^t A = \begin{bmatrix} r & \lambda & \cdots & \lambda \\ \lambda & r & \ddots & \vdots \\ \vdots & \ddots & \ddots & \lambda \\ \lambda & \cdots & \lambda & r \end{bmatrix}, \text{ where } {}^tA \text{ is the transposed matrix of } A.$$

Conversely, it is easy to check that a (v, b)-matrix A satisfying (i) and (ii) with $v \gneq k$ and $r > \lambda > 0$ defines a (v, k, λ)-block design.

As an example of how we can derive properties of block designs from properties of incidence matrices we prove the following fundamental inequality:

Theorem 1.5.2 (Fischer's inequality) *If* $(\Omega, \mathfrak{B})$ *is a block design with parameters* (v, b, k, r, λ), *then*

$$v \leq b \text{ and } k \leq r.$$

Proof Let A represent the incidence matrix of $(\Omega, \mathfrak{B})$, then

$$\det A^t A = (r + (v-1)\lambda) \begin{vmatrix} 1 & 1 & \cdots & \cdots & 1 \\ \lambda & r & \lambda & \cdots & \lambda \\ \vdots & \ddots & \ddots & \ddots & \vdots \\ \vdots & & \ddots & \ddots & \lambda \\ \lambda & \cdots & \cdots & \lambda & r \end{vmatrix}$$

$$= (r + (v-1)\lambda) \begin{vmatrix} 1 & 0 & \cdots & 0 \\ \lambda & r - \lambda & \ddots & \vdots \\ \vdots & \ddots & \ddots & 0 \\ \lambda & \cdots & \lambda & r - \lambda \end{vmatrix}$$

$$= (r + (v-1)\lambda)(r - \lambda)^{v-1} \neq 0.$$

Hence v = number of rows of $A \leq$ number of columns of $A = b$. Combining $v \leq b$ with theorem 1.5.1 (1) we get $k \leq r$. ■

Theorem 1.5.3 *For a block design with parameters* (v, b, k, r, λ) *the following conditions are equivalent:*

(1) $v = b$;
(2) $k = r$;
(3) *for any pair of blocks* B_1 *and* $B_2 (B_1 \neq B_2)$ *the number of elements of* $B_1 \cap B_2$ *is the same.*

Proof The equivalence of (1) and (2) is clear from theorem 1.5.1. We will prove the equivalence of (1) and (3). First, suppose $v = b$, then the incidence matrix A is a square matrix such that $\det A \neq 0$ and $A^t A = (r - \lambda)I + \lambda J$ (where I is the identity matrix and J the matrix the entries of which are all 1). From $AJ = kJ$ and $JA = kJ$ we get

$$
\begin{aligned}
{}^t AA &= A^{-1} A^t AA = A^{-1}((r-\lambda)I + \lambda J)A = (r-\lambda)I + \lambda A^{-1} JA \\
&= (r-\lambda)I + k\lambda A^{-1} J = (r-\lambda)I + \lambda J \\
&= \begin{bmatrix} r & \lambda \cdots\cdots & \lambda \\ \lambda & r \ddots & \vdots \\ & \ddots & \lambda \\ \lambda \cdots & \lambda & r \end{bmatrix}.
\end{aligned}
$$

From this, and the definition of the matrix A, $|B_1 \cap B_2| = \lambda$ for all pairs $B_1, B_2 (B_1 \neq B_2)$. Conversely, suppose that (3) is satisfied. For $p_1, p_2 \in \Omega (p_1 \neq p_2)$ we have $\mathfrak{B}_{p_1} \neq \mathfrak{B}_{p_2}$ since $r \neq \lambda$. Therefore $(\tilde{\Omega}, \tilde{\mathfrak{B}})$ with $\tilde{\Omega} = \mathfrak{B}$ and $\tilde{\mathfrak{B}} = \{\mathfrak{B}_p | p \in \Omega\}$ is a block design and the parameters of this block design are (b, v, r, k, λ). Applying Fischer's inequality to $(\Omega, \mathfrak{B})$ and $(\tilde{\Omega}, \tilde{\mathfrak{B}})$ we get $v \leq b$ and $b \leq v$ respectively, i.e. $b = v$. ■

A design satisfying conditions in theorem 1.5.3 is called symmetric. If $D = (\Omega, \mathfrak{B})$ is a symmetric block design, then, as we saw in the above proof, we can define a new block design $\tilde{D} = (\tilde{\Omega}, \tilde{\mathfrak{B}})$ by putting $\tilde{\Omega} = \mathfrak{B}$ and $\tilde{\mathfrak{B}} = \Omega$ – if we identify the point p with the subset $\mathfrak{B}_p = \{B \in \mathfrak{B} | p \in B\}$. The block design $\tilde{D}$ is called the dual block design of

D. The design $\tilde{D}$ is also symmetric and it has the same parameters as D. If $D = (\Omega, \mathfrak{B})$ is a non-trivial symmetric (v, k, λ)-design, and if $\mathfrak{B}^c = \{\Omega - B \mid B \in \mathfrak{B}\}$, then $D^c = (\Omega, \mathfrak{B}^c)$ is a symmetric $(v, v-k, v-2k+\lambda)$-design. (Because: For $B_1, B_2 \in \mathfrak{B}(B_1 \neq B_2)$ we have $|B_1 \cup B_2| = 2k - \lambda$ by theorem 1.5.3 and therefore:

$$\begin{aligned}|(\Omega - B_1) \cap (\Omega - B_2)| &= |\Omega - (B_1 \cup B_2)| \\ &= v - 2k + \lambda.\end{aligned}$$

Since D is non-trivial, we have $2 < k < v - 1$, hence $\lambda < k - 1$ because $\lambda(v-1) = k(k-1)$ by theorem 1.5.1. Rewriting $\lambda(v-1) = k(k-1)$ we have $\lambda(v-k) = (k-\lambda)(k-1)$, and therefore $v - k > k - \lambda$, i.e. $|(\Omega - B_1) \cap (\Omega - B_2)| > 0$. So, the number of elements of the intersection of two different elements of $\mathfrak{B}^c$ is constant and not equal to zero. Put $\tilde{\Omega} = \mathfrak{B}^c$ and put $\tilde{\mathfrak{B}}^c = \{C_p \mid p \in \Omega\}$ where $C_p = \{\Omega - B \mid B \not\ni p\}$. Then the above argument shows that $(\tilde{\Omega}, \tilde{\mathfrak{B}}^c)$ is a design. Since $|\tilde{\Omega}| = |\tilde{\mathfrak{B}}^c|$, this is symmetric and the parameters are $(v, v-k, v-2k+\lambda)$. Since $(\Omega, \mathfrak{B}^c)$ is the dual block design of $(\tilde{\Omega}, \tilde{\mathfrak{B}}^c)$, $(\Omega, \mathfrak{B}^c)$ is a symmetric $(v, v-k, v-2k+\lambda)$-design.) D^c is called the complementary design of D. Since $(v-k) - (v-2k+\lambda) = k - \lambda$, the order of D equals the order of D^c. It is evident that D^c is non-trivial.

Theorem 1.5.4 *If $D = (\Omega, \mathfrak{B})$ is a non-trivial, symmetric (v, k, λ)-design and if $n = k - \lambda$ is the order of D, then*

$$n^2 + n + 1 \geq v \geq 4n - 1.$$

Proof By considering D^c instead of D if necessary, we may assume $v \geq 2k$. Putting $\lambda' = v - 2k + \lambda$, we have

$$\lambda\lambda' = \lambda(v - 2k + \lambda) = k^2 - k + \lambda - 2k\lambda + \lambda^2 = n^2 - n$$

and

$$\lambda + \lambda' = v - 2k + 2\lambda = v - 2n.$$

From $(\lambda + \lambda')^2 \geq 4\lambda\lambda'$ we get $(v - 2n)^2 \geq 4n(n-1)$. Since D is non-trivial, we have $n \geq 2$, hence $4n(n-1)$ is not a square, and so we get $(v-2n)^2 \geq 4n(n-1) + 1 = (2n-1)^2$. From this last inequality we conclude: $v \geq 4n - 1$. From $2k \leq v$ we have

$$0 \leq v - 2k + \lambda - 1 = v - 2n - (\lambda + 1).$$

Multiplying both sides by $\lambda - 1$ we get

$$0 \le (\lambda - 1)v - 2(\lambda - 1)n - (\lambda^2 - 1) = \lambda(v - 2n - \lambda) - (v - 2n - 1)$$
$$= n(n-1) - (v - 2n - 1),$$

hence $v \le n^2 + n + 1$. ■

Let D be a non-trivial symmetric (v, k, λ)-design and we shall examine what happens if v is actually equal to either the lowest or the highest bound given by theorem 1.5.4. First consider the case $v = 4n - 1$.

From $k(k-1) = \lambda(4n-2) = 2\lambda(2n-1) = 2(k-n)(2n-1)$, we get $0 = k^2 - k(4n-1) + 2n(2n-1) = (k-2n)(k-2n+1)$, hence $k = 2n$ or $k = 2n - 1$. So, D is either a $(4n-1, 2n, n)$- or a $(4n-1, 2n-1, n-1)$-design. Notice, that if D is a $(4n-1, 2n, n)$-design, then D^c is a $(4n-1, 2n-1, n-1)$-design and conversely. A symmetric $(4n-1, 2n-1, n-1)$-design is called a Hadamard design of order n. It is conjectured that there exist Hadamard designs of order n for all $n \ge 2$ but this conjecture is still unproved. Next we want to consider the case $v = n^2 + n + 1$. From $v - 2n - 1 = n(n-1) = \lambda(v - 2n - \lambda)$, we get $\lambda^2 - (v-2n)\lambda + v - 2n - 1 = 0$, hence $(\lambda - 1)\{\lambda - (v - 2n - 1)\} = 0$, i.e. $\lambda = 1$ or $\lambda = v - 2n - 1$. Notice that if D is such that $\lambda = 1$ then D^c satisfies $\lambda = v - 2n - 1$ and conversely.

Let D be an arbitrary $(v, k, 1)$-design. By theorems 1.5.1 and 1.5.2 we have $v \ge k^2 - k + 1$ and hence by theorem 1.5.3 we have the following theorem:

Theorem 1.5.5 *For a $(v, k, 1)$-design the following conditions are equivalent:*
(1) $v = b$;
(1′) $r = k$;
(2) $v = k^2 - k + 1 (= n^2 + n + 1$ *where* $n = k - 1)$;
(3) *the intersection of any two different blocks contains exactly one point.*

Non-trivial $(v, k, 1)$-block designs satisfying the conditions of the above theorem (i.e. symmetric designs with $\lambda = 1$ and $k > 2$)

are called projective planes. We shall return to the study of projective planes in chapter 5.

Theorem 1.5.6 *Let $D = (\Omega, \mathfrak{B})$ be a block design with parameters $(v, b, k, r, 1)$. Then we have:*

(1) $b \gneq v \Rightarrow v \geq k^2$.

(2) *The following conditions are equivalent:*

(*a*) $v = k^2$

(*b*) $r = k + 1$

(*c*) $b = k^2 + k$.

(*d*) *For any block B and point p not contained in B, there exists exactly one block that contains p and has no points in common with B.*

Proof (1) From $b \gneq v$ we get $r > k$, hence $r \geq k + 1$. Substitution in theorem 1.5.1 (2) gives the desired result.

(2) The equivalence of (*a*), (*b*) and (c) is clear from theorem 1.5.1. We will prove the equivalence of (*b*) and (*d*).

(b)$\Rightarrow$(*d*): For any point $q \in B$ there exists exactly one block containing p and q, and with different points of B there correspond different blocks. So, the number of blocks containing p and having a non-empty intersection with B is equal to $|B| = k$. Since $|\mathfrak{B}_p| = k + 1$, there is exactly one block containing p and having an empty intersection with B.

(*d*)$\Rightarrow$(*b*): Similar. ■

$(v, k, 1)$-designs satisfying (one of those) conditions (*a*), (*b*), (*c*) and (*d*), i.e. designs such that $\lambda = 1$ and $v = k^2$, are called affine planes.

Let $(\Omega, \mathfrak{B})$ be an arbitrary design. Let $\tilde{\Omega}$ be a subset of Ω, then the block design $(\tilde{\Omega}, \tilde{\mathfrak{B}})$ is called a subdesign of $(\Omega, \mathfrak{B})$ if

$$\tilde{\mathfrak{B}} = \{B \cap \tilde{\Omega} \mid B \in \mathfrak{B} \text{ and } |B \cap \tilde{\Omega}| \geq 2\}.$$

Let $(\Omega, \mathfrak{B})$ be a projective plane with parameters $(v, k, 1)$ and let B_0 be an element of $\mathfrak{B}$. Putting $\tilde{\Omega} = \Omega - B_0$ and $\tilde{\mathfrak{B}} = \{B \cap \tilde{\Omega} \mid B \in \mathfrak{B}$ and $B \neq B_0\}$, $(\tilde{\Omega}, \tilde{\mathfrak{B}})$ is an affine plane with parameters $(v - k, v - 1, k - 1, k, 1)$. Conversely, let $(\tilde{\Omega}, \tilde{\mathfrak{B}})$ be an affine plane with parameters $(\tilde{v}, \tilde{b}, \tilde{k}, \tilde{r}, 1)$. We define an equivalence relation $\sim$ on

$\tilde{\mathfrak{B}}$ by

$$\tilde{B}_1 \sim \tilde{B}_2 \Leftrightarrow \tilde{B}_1 \cap \tilde{B}_2 = \varnothing \text{ or } \tilde{B}_1 = \tilde{B}_2 \quad (\tilde{B}_1, \tilde{B}_2 \in \tilde{\mathfrak{B}})$$

From the definition of an affine plane, it is clear that the number of equivalence classes equals $\tilde{r} = \tilde{k} + 1$. Let $C_1, C_2, \dots, C_{\tilde{r}}$ represent the equivalence classes and put

$$\Omega = \tilde{\Omega} \cup \{C_1, \dots . C_{\tilde{r}}\},$$
$$\mathfrak{B} = \{B \subset \Omega \,|\, B = \{C_1, \dots, C_{\tilde{r}}\} \text{ or } B = \tilde{B} \cup \{C_i\} \text{ for some } \tilde{B} \in C_i\}.$$

Then $(\Omega, \mathfrak{B})$ is a projective plane with parameters

$$\tilde{v} + \tilde{r} = v = b = \tilde{b} + 1, \quad \tilde{k} + 1 = k = r = \tilde{r} \text{ and } \lambda = 1.$$

It is easy to verify that the above procedures to construct an affine plane from a projective plane and vice versa are each other's inverse. Hence:

Theorem 1.5.7 *If $(\Omega, \mathfrak{B})$ is a projective plane and if B_0 is an element of $\mathfrak{B}$, then $(\tilde{\Omega}, \tilde{\mathfrak{B}})$ with $\tilde{\Omega} = \Omega - B_0$ and $\tilde{\mathfrak{B}} = \{B \cap \tilde{\Omega} \,|\, B \in \mathfrak{B}$ and $B \neq B_0\}$ is an affine plane and every affine plane can be obtained in this way.*

For designs with $\lambda = 1$, there is exactly one block containing any pair of different points p_1 and p_2. In this case, blocks are called lines and the line containing p_1 and p_2 is denoted by $\overline{p_1 p_2}$.

1.5.2 Block designs and groups of automorphisms

Let $(\Omega, \mathfrak{B})$ be an incidence structure. For σ a permutation of Ω (i.e. $\sigma \in S^{\Omega}$) and Δ a subset of Ω (i.e. $\Delta \in 2^{\Omega}$), we put as usual $\Delta^{\sigma} = \{p^{\sigma} \,|\, p \in \Delta\}$. Clearly, σ induces a permutation of 2^{Ω}. If $\mathfrak{B}^{\sigma} = \mathfrak{B}$, σ is called an automorphism of $(\Omega, \mathfrak{B})$. The collection of all automorphisms of $(\Omega, \mathfrak{B})$ forms a group, called the group of automorphisms of $(\Omega, \mathfrak{B})$ and denoted by Aut $(\Omega, \mathfrak{B})$.

As we shall see in chapters 5 and 6, the automorphism group of a design is closely related with the structure of that design. Here we limit ourselves to a few basic facts.

Theorem 1.5.8 (Generalization of Fischer's inequality) *Let* $\mathbf{D} = (\Omega, \mathfrak{B})$ *be a block design. A subgroup* G *of* Aut $\mathbf{D}$ *can be regarded as a permutation group on* Ω *and as a permutation group on* $\mathfrak{B}$. *We have:*

(1) *the number of orbits of* $(G, \Omega) \leq$ *the number of orbits of* $(G, \mathfrak{B})$;

(2) *if* $\mathbf{D}$ *is symmetric, then the above inequality becomes an equality.*

(For $G = 1$, we get theorem 1.5.2 and theorem 1.5.3.)

Proof Let $\Omega = \Omega^{(1)} + \ldots + \Omega^{(t)}$ and $\mathfrak{B} = \mathfrak{B}^{(1)} + \ldots + \mathfrak{B}^{(s)}$ be the orbit decompositions of Ω and $\mathfrak{B}$ under G respectively. For $B \in \mathfrak{B}$ we have $(B \cap \Omega^{(i)})^\sigma = B^\sigma \cap \Omega^{(i)} (\forall \sigma \in G)$, hence for $B \in \mathfrak{B}^{(j)}$, $|B \cap \Omega^{(i)}|$ is independent from the choice of B, so we can put $c_{ij} = |B \cap \Omega^{(i)}|$. In the same way $|\mathfrak{B}_p \cap \mathfrak{B}^{(j)}|$ is the same for all $p \in \Omega^{(i)}$ and we denote this number by d_{ji}. For $1 \leq i, j \leq t$ let us consider the following number N:

$$N = |\{(p, B, q) \mid p \in \Omega^{(i)}, q \in \Omega^{(j)}, B \in \mathfrak{B} \text{ and } p, q \in B\}|.$$

On one hand we can compute

$$\begin{aligned} N &= \sum_k \sum_{q \in \Omega^{(j)}} \sum_{B \in \mathfrak{B}_q \cap \mathfrak{B}^{(k)}} |\{(p, B, q) \mid p \in \Omega^{(i)} \cap B\}| \\ &= \sum_k |\Omega^{(j)}| d_{kj} c_{ik} \end{aligned}$$

and on the other hand

$$N = \begin{cases} |\Omega^{(i)}||\Omega^{(j)}|\lambda & (\text{for } i \neq j) \\ |\Omega^{(i)}|(|\Omega^{(i)}| - 1)\lambda + |\Omega^{(i)}| r & (\text{for } i = j). \end{cases}$$

So we get

$$\sum_k d_{kj} c_{ik} = \omega_i \lambda + n \delta_{ij}$$

where $\omega_i = |\Omega^{(i)}|$ and $n = r - \lambda$. Putting $C = (c_{ij})$ and $D = (d_{ij})$ C is a $(t \times s)$ matrix and D is an $(s \times t)$ matrix such that

$$CD = \begin{bmatrix} \omega_1\lambda + n & \omega_1\lambda \cdots\cdots & \omega_1\lambda \\ \omega_2\lambda & \omega_2\lambda + n \cdots & \omega_2\lambda \\ \vdots & \ddots & \vdots \\ \omega_t\lambda & \cdots\cdots\cdots & \omega_t\lambda + n \end{bmatrix}.$$

Hence

$$\det CD = \begin{vmatrix} v\lambda+n & v\lambda+n & \cdots\cdots & v\lambda+n \\ \omega_2\lambda & \omega_2\lambda+n & & \vdots \\ \vdots & & \ddots & \vdots \\ \omega_t\lambda & \cdots\cdots & \cdots\cdots & \omega_t\lambda+n \end{vmatrix}$$

$$= (v\lambda+n)\begin{vmatrix} 1 & 0 & \cdots & 0 \\ \omega_2\lambda & n & \ddots & \vdots \\ \vdots & 0 & \ddots & \vdots \\ \vdots & \vdots & \ddots & 0 \\ \omega_t\lambda & 0\ldots0 & & n \end{vmatrix}$$

$$= (v\lambda+n)n^{t-1} \neq 0.$$

Therefore $t \le s$, proving (1). To prove (2), consider the dual design $(\tilde{\Omega}, \tilde{\mathfrak{B}})$ of $(\Omega, \mathfrak{B})$ as in the proof of theorem 1.5.3, then G can be regarded as a subgroup of Aut $(\tilde{\Omega}, \tilde{\mathfrak{B}})$. (Since if $\sigma \in G$ is such that $B^\sigma = B$ for all $B \in \mathfrak{B}$, then for any point $p \in \Omega$ we have $p^\sigma = (\bigcap_{B \in \mathfrak{B}_p} B) = p$, i.e. $\sigma = 1$, from which the statement follows.) Therefore, $s \le t$ by (1), and so $s = t$. ■

Theorem 1.5.9 *If* $\mathbf{D} = (\Omega, \mathfrak{B})$ *is a symmetric design and if* $G \le$ Aut $\mathbf{D}$, *then*

(G, Ω) *is doubly transitive* $\Leftrightarrow$ $(G, \mathfrak{B})$ *is doubly transitive.*

Proof $\Rightarrow$: $(G, \mathfrak{B})$ is transitive by theorem 1.5.8. Therefore, for $B \in \mathfrak{B}$ and $a \in \Omega$ we have $|G : G_{\langle B \rangle}| = |\mathfrak{B}| = |\Omega| = |G : G_a|$. Since the number of orbits of (G_a, Ω) equals 2, the number of orbits of $(G_a, \mathfrak{B})$ also equals 2 by theorem 1.5.8. Hence, $\mathfrak{B}_a = \{X \in \mathfrak{B} \mid a \in X\}$ and $\mathfrak{B} - \mathfrak{B}_a$ are the orbits of $(G_a, \mathfrak{B})$. From $|G_a : G_{a,\langle B \rangle}| = |G_{\langle B \rangle} : G_{a,\langle B \rangle}|$, we have

$$|G_{\langle B \rangle} : G_{a,\langle B \rangle}| = \begin{cases} |\mathfrak{B}_a| = |B| & \text{if } a \in B \\ |\mathfrak{B} - \mathfrak{B}_a| = |\Omega - B| & \text{if } a \notin B, \end{cases}$$

hence $(G_{\langle B \rangle}, B)$ and $(G_{\langle B \rangle}, \Omega - B)$ are transitive. Therefore the number of orbits of $(G_{\langle B \rangle}, \Omega)$ equals 2 and so the number of orbits of $(G_{\langle B \rangle}, \mathfrak{B})$ also equals 2 by theorem 1.5.8, i.e. $(G, \mathfrak{B})$ is doubly transitive.

$\Leftarrow$: Let $\tilde{\mathbf{D}} = (\tilde{\Omega}, \tilde{\mathfrak{B}})$ be the dual design of $\mathbf{D}$. From the definition of $\tilde{\mathbf{D}}$ it is clear that $\operatorname{Aut} \mathbf{D} = \operatorname{Aut} \tilde{\mathbf{D}}$, so the result follows by applying the above proof to $\tilde{\mathbf{D}}$. ■

Theorem 1.5.10 *Let* $\mathbf{D} = (\Omega, \mathfrak{B})$ *be a symmetric design. For* $\sigma \in \operatorname{Aut} \mathbf{D}$ *we have*

$$|\{p \in \Omega \mid p^\sigma = p\}| = |\{B \in \mathfrak{B} \mid B^\sigma = B\}|.$$

Proof Let $A = (a_{ij})$ be the incidence matrix of $\mathbf{D}$, i.e. putting $\Omega = \{p_1, \ldots, p_v\}$ and $\mathfrak{B} = \{B_1, \ldots, B_v\}$

$$a_{ij} = \begin{cases} 1 & \text{if } p_i \in B_j \\ 0 & \text{if } p_i \notin B_j \end{cases}.$$

We define matrices $C = (c_{ij})$ and $D = (d_{ij})$ of degree v by

$$c_{ij} = \begin{cases} 1 & \text{if } p_i^\sigma = p_j \\ 0 & \text{if } p_i^\sigma \neq p_j \end{cases},$$

$$d_{ij} = \begin{cases} 1 & \text{if } B_i^\sigma = B_j \\ 0 & \text{if } B_i^\sigma \neq B_j \end{cases}.$$

Then it is easy to check that

$$C^t C = D^t D = E \text{ and } CAD = A,$$

hence

$$A^{-1} C A = D^{-1} = {}^t D.$$

Therefore $\operatorname{trace} A^{-1} C A = \operatorname{trace} {}^t D = \operatorname{trace} D$, hence $\operatorname{trace} C = \operatorname{trace} D$. Since

$$\operatorname{trace} C = |\{p \in \Omega \mid p^\sigma = p\}|, \quad \operatorname{trace} D = |\{B \in \mathfrak{B} \mid B^\sigma = B\}|,$$

the theorem follows. ■

Theorem 1.5.11 *Let* $\mathbf{D} = (\Omega, \mathfrak{B})$ *be a design and let* $G \leq \operatorname{Aut} \mathbf{D}$

be such that (G, Ω) and $(G, \mathfrak{B})$ are both transitive. Then, for $p_0 \in \Omega$ and $B_0 \in \mathfrak{B}$ we have:

(1) *the number of orbits of $(G_{p_0}, \Omega) \leq$ the number of orbits of $(G_{\langle B_0 \rangle}, \mathfrak{B})$;*

(2) *if* **D** *is symmetric, the equality sign holds in* (1).

Proof For $\sigma \in G$ and $(p, B) \in \Omega \times \mathfrak{B}$ we define $(p, B)^\sigma = (p^\sigma, B^\sigma)$. By doing so G can be regarded as a permutation group on $\Omega \times \mathfrak{B}$. Since (G, Ω) is transitive, the number of orbits of $(G, \Omega \times \mathfrak{B})$ equals the number of orbits of $(G_{p_0}, \mathfrak{B})$ and since $(G, \mathfrak{B})$ is transitive, it also equals the number of orbits of $(G_{\langle B_0 \rangle}, \Omega)$. Therefore, by theorem 1.5.8:

$$\begin{aligned} &\text{the number of orbits of } (G_{p_0}, \Omega) \\ &\leq \text{the number of orbits of } (G_{p_0}, \mathfrak{B}) \\ &= \text{the number of orbits of } (G_{\langle B_0 \rangle}, \Omega) \\ &\leq \text{the number of orbits of } (G_{\langle B_0 \rangle}, \mathfrak{B}). \end{aligned}$$

(2) is a consequence of (1) and theorem 1.5.8 (2). ■

Let σ be an automorphism of $\mathbf{D} = (\Omega, \mathfrak{B})$. If there is a point a of **D** such that σ leaves all blocks containing a fixed then σ is called central and a is called a centre of σ. If **D** is a design with $\lambda = 1$, then every central automorphism that is not the identity has exactly one centre. (Proof: Suppose that a and b are different centres of σ. For $x \notin \overline{ab}$, we have: $x = \overline{ax} \cap \overline{bx}$, hence $x^\sigma = \overline{ax^\sigma} \cap \overline{bx^\sigma} = \overline{ax} \cap \overline{bx} = x$, i.e. σ leaves all points not on $\overline{ab}$ fixed. If $k = 2$, then σ is the identity. Assume $k \geq 3$. For $x \in \overline{ab}$, pick a line l different from $\overline{ab}$ containing x. Since l contains two points not on $\overline{ab}$ we get $l^\sigma = l$. Hence: $x^\sigma = l^\sigma \cap \overline{ab}^\sigma = l \cap \overline{ab} = x$, i.e. σ is the identity.)

Now, let **D** be a design. Suppose that $(\operatorname{Aut} \mathbf{D}, \Omega)$ is doubly transitive. Then putting $\overline{xy} = \bigcap_{\substack{x, y \in B \\ B \in \mathfrak{B}}} B$ for $x, y \in \Omega$, $\overline{xy}$ is completely determined by any pair of its points by 2-transitivity of $(\operatorname{Aut} \mathbf{D}, \Omega)$. Putting $\tilde{\mathfrak{B}} = \{l \in \Omega \mid l = \overline{xy} \text{ for some } x, y \in \Omega\}$, the design $\tilde{\mathbf{D}} = (\Omega, \tilde{\mathfrak{B}})$ is such that $\lambda = 1$ and automorphisms of **D** can be regarded as automorphisms of $\tilde{\mathbf{D}}$. If $\sigma \neq 1$ is a central automorphism of **D**, then σ is also a central automorphism of $\tilde{\mathbf{D}}$ by the definition of central automorphisms, hence σ has exactly one centre. Thus we have proved:

Theorem 1.5.12 *If there is a subgroup G of* Aut **D**, *such that* (G, Ω) *is doubly transitive, then the centre of any non-identity central automorphism of* **D** *is uniquely determined.*

1.5.3 Projective and affine spaces

Let K be a finite field and let $V = V(n+1, K)$ $(n \geq 2)$ be a $(n+1)$-dimensional K-module. Let Ω represent the collection of all 1-dimensional K-submodules of V. Elements of Ω can be represented as:

$$\langle u \rangle = \{\lambda u \mid \lambda \in K\} \text{ where } u \neq 0 \text{ is an element of } V;$$

and

$$\langle u_1 \rangle = \langle u_2 \rangle \Leftrightarrow u_1 = \lambda u_2 \quad \text{for some } \lambda \in K.$$

For a K-submodule U of V, the collection of all elements of Ω contained in U is denoted by $[U]$. If U is zero dimensional, then we put $[U] = \emptyset$. Since

$$[U_1] = [U_2] \Leftrightarrow U_1 = U_2,$$

there is a one-to-one correspondence between the collection of submodules of V and the collection of those subsets of Ω that can be represented as $[U]$ for some submodule U of V. Defining

$$\mathfrak{B} = \{[U] \subset \Omega \mid U \text{ is a submodule of } V\},$$

we arrive at an incidence structure $(\Omega, \mathfrak{B})$, called the n-dimensional projective space over K and denoted by $\mathbf{P}(n, K)$, $\mathbf{P}(K)$ or simply $\mathbf{P}$. Putting

$$\mathfrak{B}^{(i)} = \{[U] \mid U \text{ is an } (i+1)\text{-dimensional submodule of } V\},$$

$\mathfrak{B}$ (as a set) is the direct sum of $\mathfrak{B}^{(-1)}, \mathfrak{B}^{(0)}, \ldots, \mathfrak{B}^{(n-1)}$ and $\mathfrak{B}^{(n)}$. The elements of $\mathfrak{B}^{(i)}$ are called i-dimensional subspaces of the projective space $\mathbf{P}$. Especially, elements of $\mathfrak{B}^{(0)} (= \Omega)$, $\mathfrak{B}^{(1)}$, $\mathfrak{B}^{(2)}$ and $\mathfrak{B}^{(n-1)}$ are called points, lines, planes and hyperplanes of $\mathbf{P}$ respectively. The incidence structure $(\Omega, \mathfrak{B}^{(i)})$ $(i = -1, 0, \ldots, n)$ is denoted by $\mathbf{P}_i(n, K)$, $\mathbf{P}_i(K)$ or simply $\mathbf{P}_i$.

Theorem 1.5.13 *For* $1 \leq i \leq n-1$, $\mathbf{P}_i(n, K)$ *is a block design with* *parameters:*

$$v = \frac{q^{n+1} - 1}{q - 1}, \quad b = N_{i+1}(n+1, q), \quad k = \frac{q^{i+1} - 1}{q - 1}$$
$$r = N_i(n, q) \text{ and } \lambda = N_{i-1}(n-1, q),$$

where $q = |K|$ *and* $N_e(d, q) = \prod_{j=1}^{e} (q^{d-j+1} - 1)/(q^j - 1)$ *for* $e \geq 1$ *and* $N_0(d, q) = 1$.

Proof Putting $V^{\#} = V - \{0\}$, we have $|V^{\#}| = q^{n+1} - 1$, and $v = |\Omega| = (q^{n+1} - 1)/(q - 1)$ since $\langle u_1 \rangle = \langle u_2 \rangle \Leftrightarrow u_1 = \lambda u_2$ for some $\lambda \in K$. Since $[U] \in \mathfrak{B}^{(i)}$ is the collection of all 1-dimensional subspaces of the $(i+1)$-dimensional K-module U, we find in the same way as above that $|[U]| = (q^{i+1} - 1)/(q - 1)$ and this is independent from U, i.e. $k = (q^{i+1} - 1)/(q - 1)$. Next, let $\langle u_1 \rangle$ and $\langle u_2 \rangle$ be two different points of Ω. We choose a complement V_0 of $\langle u_1, u_2 \rangle$ in V, i.e. V_0 is a K-submodule of V such that $V = \langle u_1, u_2 \rangle \oplus V_0$. Since $\langle u_1, u_2 \rangle$ is two-dimensional, V_0 is $(n-1)$-dimensional. For an $(i+1)$-dimensional subspace U of V that contains $\langle u_1, u_2 \rangle$ we have $U = \langle u_1, u_2 \rangle \oplus (U \cap V_0)$ and $U \cap V_0$ is an $(i-1)$-dimensional subspace of V_0. Conversely, $\langle u_1, u_2 \rangle \oplus U_0$ is an $(i+1)$-dimensional subspace of V for all $(i-1)$-dimensional subspaces U_0 of V_0. Therefore, the number of $(i+1)$-dimensional subspaces of V containing $\langle u_1, u_2 \rangle$ equals the number of $(i+1)$-dimensional subspaces of an $(n-1)$-dimensional K-module, and is therefore independent of the choice of u_1 and u_2. This proves that $\mathbf{P}_i(n, K)$ is a block design. To finish the proof it suffices to compute λ, since b and r can then be computed by theorem 1.5.1. Let $N_e(d, q)$ denote the number of e-dimensional K-submodules of a d-dimensional K-module W. We shall prove that

$$N_e(d, q) = \prod_{i=1}^{e} (q^{d-i+1} - 1)/(q^i - 1)$$

by induction on e. For $e = 1$ the formula is true as we showed in the first line of the proof. Assuming the formula for e we shall prove it for $e + 1$. Let us evaluate the number of elements of the set

$$\{(U_1, U_2) \mid U_1 \text{ and } U_2 \text{ are } (e+1)\text{- and } e\text{-dimensional subspaces of } W \text{ such that } U_1 \supset U_2\}$$

in two different ways. Since there are $N_{e+1}(d, q)$ choices for U_1 and since there are $N_e(e+1, q)$ choices for U_2 after U_1 is chosen, the number of elements of the above set equals $N_{e+1}(d, q) \cdot N_e(e+1, q)$. On the other hand, there are $N_e(d, q)$ choices for U_2 and for each U_2 there are as many choices for U_1 as there are 1-dimensional subspaces of W/U_2, i.e. $N_1(d-e, q)$ choices. Hence the number of elements of the above set also equals $N_e(d, q) \cdot N_1(d-e, q)$. Therefore:

$$N_{e+1}(d, q) = N_e(d, q) \cdot N_1(d-e, q)/N_e(e+1, q).$$

Using this equality and the induction hypothesis, we get the desired expression for $N_{e+1}(d, q)$. It is now clear from what has been already proved that $\lambda = N_{i-1}(n-1, q)$. ■

These block designs, $\mathbf{P}_i(n, K)$, are called projective designs; the points of $\mathbf{P}_i(n, K)$ are the points of $\mathbf{P}(n, K)$ and the blocks of $\mathbf{P}_i(n, K)$ are the i-dimensional subspaces of $\mathbf{P}(n, K)$. $\mathbf{P}_{n-1}(n, K)$ is a symmetric design with parameters

$$v = b = \frac{q^{n+1} - 1}{q - 1}, \quad k = r = \frac{q^n - 1}{q - 1} \quad \text{and} \quad \lambda = \frac{q^{n-1} - 1}{q - 1}.$$

$\mathbf{P}_1(n, K)$ is a $((q^{n+1} - 1)/(q-1), q+1, 1)$-design with the following properties:

(i) For two different points p_1 and p_2 of Ω, there is exactly one block containing p_1 and p_2. (Proof: Putting $p_i = \langle v_i \rangle$ for $i = 1, 2$ $\langle v_1, v_2 \rangle$ is two-dimensional and $[\langle v_1, v_2 \rangle]$ is the only block containing p_1 and p_2). This uniquely determined block is called the line through p_1 and p_2 and denoted by $p_1 p_2$.

(ii) Each line contains at least three points. (Proof: the number of points on a line equals $q + 1 \geq 3$).

(iii) If p_1, p_2, p_3, p_4 and p_5 are all different points of Ω such that $p_3 \notin \overline{p_1 p_2}, p_4 \in \overline{p_1 p_3}$ and $p_5 \in \overline{p_2 p_3}$, then $\overline{p_1 p_2} \cap \overline{p_4 p_5} \neq \varnothing$. (Proof: put $p_i = \langle v_i \rangle$ for $i = 1, \ldots, 5$, $U = \langle v_1, v_2, v_3 \rangle$, $U_1 = \langle v_1, v_2 \rangle$ and $U_2 = \langle v_4, v_5 \rangle$. Then $\dim_K U = 3$, $\dim_K U_1 = \dim_K U_2 = 2$ and $U = U_1 + U_2$, hence $\dim_K (U_1 \cap U_2) = 1$ by theorem 1.4.4 and so $\overline{p_1 p_2} \cap \overline{p_4 p_5} \neq \varnothing$.)

Let $[H]$ be a hyperplane of $\mathbf{P}(n, K)$. The incidence structure

$(\tilde{\Omega}, \tilde{\mathfrak{B}})$ with

$$\tilde{\Omega} = \Omega - [H]$$
$$\tilde{\mathfrak{B}} = \{\tilde{B} \mid \tilde{B} = [B] - [B \cap H], B \nsubseteq H \text{ and } [B] \in \mathfrak{B}\}$$

is called the n-dimensional affine space over K determined by the hyperplane $[H]$ of $\mathbf{P}(n, K)$ and denoted by $\mathbf{A} = \mathbf{A}(n, K)$. It is easy to check that $\mathbf{A}(n, K)$ is essentially independent from the choice of the hyperplane $[H]$. Putting

$$\tilde{\mathfrak{B}}^{(i)} = \{\tilde{B} \mid \tilde{B} = [B] - [B \cap H], B \nsubseteq H \text{ and } [B] \in \mathfrak{B}^{(i)}\}$$

we get that $\tilde{\mathfrak{B}}$ (as a set) is the direct sum of $\tilde{\Omega} = \tilde{\mathfrak{B}}^{(0)}, \tilde{\mathfrak{B}}^{(1)}, \ldots, \tilde{\mathfrak{B}}^{(n)}$. Elements of $\tilde{\mathfrak{B}}^{(i)}$ are called i-dimensional subspaces of $\mathbf{A}$. Especially, elements of $\tilde{\Omega} = \tilde{\mathfrak{B}}^{(0)}, \tilde{\mathfrak{B}}^{(1)}, \tilde{\mathfrak{B}}^{(2)}$ and $\tilde{\mathfrak{B}}^{(n-1)}$ are called points, lines, planes and hyperplanes of $\mathbf{A}(n, K)$ respectively. For $i = 0, \ldots, n$ the incidence structure $(\tilde{\Omega}, \tilde{\mathfrak{B}}^{(i)})$ is denoted by $\mathbf{A}_i(n, K)$.

Theorem 1.5.14 *$\mathbf{A}_i(n, K)$ is a block design for $1 \leq i \leq n-1$ with parameters:*

$$v = q^n, \quad b = q^{n-i} N_i(n, q), \quad k = q^i, \quad r = N_i(n, q), \quad \lambda = N_{i-1}(n-1, q),$$

where q and $N_e(d, q)$ have the same meaning as in theorem 1.5.13.

Proof. First we have

$$v = |\tilde{\Omega}| = |\Omega - [H]| = \frac{q^{n+1} - 1}{q - 1} - \frac{q^n - 1}{q - 1} = q^n.$$

If $[U] \in \mathfrak{B}^{(i)}$ such that $U \nsubseteq H$, then $U \cap H$ is an i-dimensional K-submodule of H by theorem 1.4.4. Conversely, if U_0 is an i-dimensional K-submodule of H, then every $(i+1)$-dimensional K-submodule U of V such that $U_0 \subsetneq U \nsubseteq H$ can be written as $U = \langle U_0, u \rangle$ for some $u \notin H$. Hence the number of those K-submodules equals $(q^{n+1} - q^n)/(q^{i+1} - q^i) = q^{n-i}$ and so $b = |\tilde{\mathfrak{B}}^{(i)}| = q^{n-i} N_i(n, q)$. Since every $B \in \tilde{\mathfrak{B}}^{(i)}$ can be written as: $B = [U] - [U \cap H]$ for some $U \in \mathfrak{B}^{(i)}$ with $U \nsubseteq H$, we have

$$k = |B| = \frac{q^{i+1} - 1}{q - 1} - \frac{q^i - 1}{q - 1} = q^i.$$

The rest of the proof is almost obvious and left to the reader. ■

The $\mathbf{A}_i(n, K)$ are called affine designs: the points of $\mathbf{A}_i(n, K)$ are the points of $\mathbf{A}(n, K)$ and the blocks of $\mathbf{A}_i(n, K)$ are the i-dimensional subspaces of $\mathbf{A}(n, K)$. The planes $\mathbf{P}_1(2, K)$ and $\mathbf{A}_1(2, K)$ are examples of projective and affine planes respectively, as defined in §1.5.1. They are called the projective plane defined over K and the affine plane defined over K respectively.

Finally, we want to describe an alternative way of looking at $\mathbf{A}(n, K)$. Let H be a hyperplane of V and let e be an element of V that is not contained in H. Since $V = \langle e \rangle \oplus H$, for every element $\langle v \rangle$ of $\mathbf{A}$ such that $v \notin H$ there is a uniquely determined element $\tilde{v}$ of H satisfying: $\langle v \rangle = \langle e + \tilde{v} \rangle$. This sets up a one-to-one correspondence between points of $\mathbf{A}$ and points of H. In this correspondence the r-dimensional subspace $[U] - [U \cap H]$ of $\mathbf{A}$, where U is an $(r+1)$-dimensional K-submodule of V, corresponds with

$$\tilde{U} = \{\tilde{v} \in H \mid e + \tilde{v} \in U\} \subseteq H.$$

If we pick one element $\tilde{v}_0$ of $\tilde{U}$ then we have for an element $\tilde{v}$ of H:

$$\tilde{v} \in \tilde{U} \Leftrightarrow e + \tilde{v} \in U \Leftrightarrow \tilde{v} - \tilde{v}_0 \in U \cap H \Leftrightarrow \tilde{v} \in \tilde{v}_0 + (U \cap H).$$

Therefore: $\tilde{U} = \tilde{v}_0 + (U \cap H)$. In other words, to every r-dimensional subspace $[U] - [U \cap H]$ of $\mathbf{A}$ there corresponds a unique residue class with respect to the r-dimensional K-submodule $U \cap H$ of H. Conversely, if H_0 is an arbitrary r-dimensional K-submodule of H and if $\tilde{u}_0 + H_0$ is a residue class with respect to H_0, then $U = \langle e + \tilde{u}_0, H_0 \rangle$ is an $(r+1)$-dimensional K-submodule of V determining an r-dimensional subspace $[U] - [U \cap H]$ of $\mathbf{A}$. Since those two correspondences clearly are inverse to each other, we have arrived at a one-to-one correspondence between the r-dimensional subspaces of $\mathbf{A}$ and the residue classes with respect to r-dimensional K-submodules of H. Therefore, the n-dimensional affine space $A(n, K)$ over K can be identified with the incidence structure $(\tilde{\Omega}, \tilde{\mathfrak{B}})$ where $\tilde{\Omega}$ is the collection of elements of the n-dimensional K-module H and

$$\tilde{\mathfrak{B}} = \bigcup_r \Big(\bigcup_U H/U\Big),$$

where U runs over all r-dimensional K-submodules of H.

1.5.4 The fundamental theorem of projective geometry

We want to determine the groups of automorphisms of $\mathbf{P}_i(n, K)$ and $\mathbf{A}_i(n, K)$ for $1 \leq i \leq n-1$. We fix a hyperplane $[H]$ of $\mathbf{P}(n, K)$ and use this hyperplane to define $\mathbf{A}(n, K)$. If σ is an automorphism of $\mathbf{P}_i(n, K) = (\Omega, \mathfrak{B}^{(i)})$ satisfying $[H]^\sigma = [H]$, then $\tilde{\Omega}^\sigma = \tilde{\Omega}$ and $\tilde{\mathfrak{B}}^{(i)\sigma} = \tilde{\mathfrak{B}}^{(i)}$, i.e. σ induces an automorphism $\tilde{\sigma}$ of $\mathbf{A}_i(n, K) = (\tilde{\Omega}, \tilde{\mathfrak{B}}^{(i)})$.

Now we will show that every automorphism of $\mathbf{A}_i(n, K)$ can be obtained in this way. First note that, if W is an $(r+1)$-dimensional K-submodule of V not contained in H, then $|[W] - [W \cap H]| = q^r$, where $q = |K|$, and the K-submodule of V generated by $W - (W \cap H)$ is W. Let $\tilde{\tau}$ be an automorphism of $\mathbf{A}_i(n, K)$. Let $\mathfrak{S}_r$ denote the collection of all $(r+1)$-dimensional K-submodules of V not contained in H. An element U of $\mathfrak{S}_i$ determines the element $[U] - [U \cap H]$ of $\tilde{\mathfrak{B}}^{(i)}$. By defining $U' \in \mathfrak{S}_i$ by

$$([U] - [U \cap H])^{\tilde{\tau}} = [U'] - [U' \cap H],$$

we set up a correspondence $U \to U'$, that is easily seen to be a permutation of $\mathfrak{S}_i$. We denote this permutation by τ', i.e. $U^{\tau'} = U'$. Next, if P is an element of $\mathfrak{S}_j$ for some $j \leq i$, then P can be represented as the intersection of a certain number of elements of $\mathfrak{S}_i$, say: $P = U_1 \cap \ldots \cap U_t$ $(U_k \in \mathfrak{S}_i)$. Defining $P^{\tau'}$ by

$$P^{\tau'} = U_1^{\tau'} \cap \ldots \cap U_t^{\tau'},$$

$P^{\tau'}$ is a K-submodule of V and:

$$\begin{aligned}[P^{\tau'}] - [P^{\tau'} \cap H] &= \bigcap_{k=1}^{t} ([U_k^{\tau'}] - [U_k^{\tau'} \cap H]) \\ &= \bigcap_{k=1}^{t} ([U_k] - [U_k \cap H])^{\tilde{\tau}} = ([P] - [P \cap H])^{\tilde{\tau}}.\end{aligned}$$

Hence $P^{\tau'} \in \mathfrak{S}_j$ and $P^{\tau'}$ is independent from the choice of the $U_1, \ldots, U_t$. Especially, for $P \in \mathfrak{S}_1$, $([P] - [P \cap H])^{\tilde{\tau}}$ is an element of $\tilde{\mathfrak{B}}^{(1)}$ and $\tilde{\tau}$ induces an automorphism of $\mathbf{A}_1(n, K)$, i.e. $\tilde{\tau}$ maps lines of $\mathbf{A}(n, K)$ onto lines. Therefore, if α is an arbitrary subspace of $\mathbf{A}(n, K)$, $\alpha^{\tilde{\tau}} = \{\langle v \rangle^{\tilde{\tau}} | \langle v \rangle \in \alpha\}$ is also a subspace of $\mathbf{A}(n, K)$ and the dimension of α equals the dimension of $\alpha^{\tilde{\tau}}$ by the remark at the beginning of the proof. In other words, if P is an element of

$\mathfrak{S}_j$ for $1 \le j \le n$, the element $P^{\tau'}$ of $\mathfrak{S}_j$ is uniquely determined by

$$([P]-[P\cap H])^{\tilde{\tau}} = ([P^{\tau'}]-[P^{\tau'}\cap H]).$$

We will prove that $\tilde{\tau}$ defines an automorphism of $\mathbf{P}_j(n, K)$ for $1 \le j \le n$. For $\langle v \rangle \in H$ we choose an element P of $\mathfrak{S}_1$ containing v, and we define $\langle v \rangle^{\tau} \in H$ by

$$\langle v \rangle^{\tau} = P^{\tau'} \cap H.$$

This definition depends only on $\langle v \rangle$, not on the choice of P. (For, suppose that P_1 and P_2 are different elements of $\mathfrak{S}_1$ containing v, then $P_1^{\tau'} \neq P_2^{\tau'}$, $P_1^{\tau'}, P_2^{\tau'} \subseteq (P_1 + P_2)^{\tau'}$ and $\dim_K (P_1 + P_2)^{\tau'} = \dim_K (P_1 + P_2) = 3$. Hence $P_1^{\tau'} \cap P_2^{\tau'} \neq 0$. From $P_1 \cap P_2 \subseteq H$ we get $P_1^{\tau'} \cap P_2^{\tau'} \subseteq H$. Therefore $H \cap P_1^{\tau'} = H \cap P_2^{\tau'}$.) Next we define a permutation τ of $\tilde{\Omega} = \Omega - [H]$ by putting

$$\langle v \rangle^{\tau} = \langle v \rangle^{\tilde{\tau}} \quad \forall \langle v \rangle \in \tilde{\Omega}.$$

From the above it is clear that τ is an automorphism of $\mathbf{P}_j(n, K)$, that the automorphism of $\mathbf{A}_i(n, K)$ induced by τ is just $\tilde{\tau}$ and finally that the mapping from Aut $\mathbf{A}_i(n, K)$ into Aut $\mathbf{P}_i(n, K)$ defined by $\tilde{\tau} \to \tau$ is a monomorphism. We sum up our results in the following theorem:

Theorem 1.5.15 *Let $[H]$ be a hyperplane of $\mathbf{P}(n, K)$ and let $\mathbf{A}(n, K)$ be the affine space determined by H. If τ is an automorphism of the design $\mathbf{P}_i(n, K)$ satisfying $[H]^{\tau} = [H]$, then τ induces an automorphism of $\mathbf{A}_i(n, K)$ and every automorphism of $\mathbf{A}_i(n, K)$ can be obtained in this way. The correspondence $\tau \to \tilde{\tau}$ defines an isomorphism between* Aut $\mathbf{A}_i(n, K)$ *and* $G = \{\tau \in \text{Aut } \mathbf{P}_i(n, K) \mid [H]^{\tau} = [H]\}$, *a subgroup of* Aut $\mathbf{P}_i(n, K)$.

The next theorem describes the structure of Aut $\mathbf{P}_i(n, K)$.

Theorem 1.5.16 (Fundamental theorem of projective geometry) *For $n \ge 2$, we have:*

$$\text{Aut}\,\mathbf{P}_1(n, K) \simeq \text{Aut}\,\mathbf{P}_2(n, K) \simeq \dots$$
$$\simeq \text{Aut}\,\mathbf{P}_{n-1}(n, K) (\simeq \text{Aut}\,\mathbf{P}(n, K)) \simeq \Gamma L(V)/Z(V).$$

Proof (*a*) We first show that Aut $\mathbf{P}_1(n, K) \simeq$ Aut $\mathbf{P}_i(n, K)$ $(2 \leq i \leq n-1)$. We already know that Aut $\mathbf{P}_1(n, K)$ and Aut $\mathbf{P}_i(n, K)$ are both subgroups of the symmetric group S^Ω on Ω. So, it suffices to show that every element of S^Ω that induces a permutation of $\mathfrak{B}^{(1)}$ induces a permutation of $\mathfrak{B}^{(i)}$ and conversely. Take $\sigma \in \mathrm{Aut}\,\mathbf{P}_1(n, K)$. For $v \in V$ $(v \neq 0)$ put $\langle v \rangle^\sigma = \langle \tilde{v} \rangle$. If $v_1, \ldots, v_r$ are linearly independent, so are $\tilde{v}_1, \ldots, \tilde{v}_r$ and $[\langle v_1, \ldots, v_r \rangle]^\sigma = [\langle \tilde{v}_1, \ldots, \tilde{v}_r \rangle]$. (We prove this by induction on r. For $r = 2$ this is true because $\sigma \in$ Aut $\mathbf{P}_1$. Suppose it is true for $r-1$. From

$$\tilde{v}_r \in \langle \tilde{v}_1, \ldots, \tilde{v}_{r-1} \rangle \Leftrightarrow \langle \tilde{v}_r \rangle \in [\langle \tilde{v}_1, \ldots, \tilde{v}_{r-1} \rangle] \Leftrightarrow \langle v_r \rangle \in [\langle v_1, \ldots, v_{r-1} \rangle],$$

we conclude that if $v_1, \ldots, v_r$ are linearly independent so are $\tilde{v}_1, \ldots, \tilde{v}_r$. An arbitrary element $u \in \langle v_1, \ldots, v_r \rangle$ can be written as $u = u_1 + u_2$ with $u_1 \in \langle v_1, \ldots, v_{r-1} \rangle$ and $u_2 \in \langle v_r \rangle$. Then $\tilde{u} \in \langle \tilde{u}_1, \tilde{u}_2 \rangle$ by the result for $r = 2$ and $\tilde{u}_1 \in \langle \tilde{v}_1, \ldots, \tilde{v}_{r-1} \rangle$ by the induction hypothesis. Therefore $\tilde{u} \in \langle \tilde{v}_1, \ldots, \tilde{v}_r \rangle$ and $[\langle v_1, \ldots, v_r \rangle]^\sigma = [\langle \tilde{v}_1, \ldots, \tilde{v}_r \rangle]$.) So we have proved that $\sigma \in \mathrm{Aut}\,\mathbf{P}_1(n, K)$ induces a permutation of $\mathfrak{B}^{(i)}$. Conversely, let σ be an element of Aut $\mathbf{P}_i(n, K)$ and put $[U]^\sigma = [\tilde{U}] \in \mathfrak{B}^{(i)}$ for $[U] \in \mathfrak{B}^{(i)}$. If we represent $[\langle v_1, v_2 \rangle] \in \mathfrak{B}^{(1)}$ by $\langle v_1, v_2 \rangle = \bigcap_j U_j$ for certain $U_j \in \mathfrak{B}^{(i)}$, then

$$[\langle v_1, v_2 \rangle]^\sigma = [\bigcap_j U_j]^\sigma = \bigcap_j [U_j]^\sigma = \bigcap_j [\tilde{U}_j] = [\bigcap_j \tilde{U}_j] \in \mathfrak{B}^{(1)}.$$

This proves that σ induces a permutation of $\mathfrak{B}^{(1)}$.

(*b*) Next we prove that Aut $\mathbf{P}_1(n, K) \simeq \Gamma L(V)/Z(V)$. For $f \in \Gamma L(V)$ the element $\tilde{f}$ of Aut $\mathbf{P}_1(n, K)$ is defined as follows. By definition of f we have

$$(\lambda v)^f = \lambda^\sigma v^f \quad \text{for } \lambda \in K \text{ and } v \in V,$$

where σ is an automorphism of K, hence for $\langle v \rangle \in \Omega$ the element $\langle v^f \rangle$ is unambiguously determined. Putting $\langle v \rangle^{\tilde{f}} = \langle v^f \rangle$, $\tilde{f}$ is a permutation of Ω. It is easy to check that $\tilde{f}$ induces a permutation of $\mathfrak{B}^{(1)}$ and that the map from $\Gamma L(V)$ into Aut $\mathbf{P}_1(n, K)$ given by $f \to \tilde{f}$ is a homomorphism. If $\tilde{f} \in$ Aut $\mathbf{P}_1(n, K)$ is the identity element of Aut $\mathbf{P}_1(n, K)$, then

$$v^f = \lambda_v v \quad \text{for some } \lambda_v \in K.$$

Choosing a basis $v_1, \ldots, v_{n+1}$ for V, we have:

$$\lambda_{v_1} = \ldots = \lambda_{v_{n+1}} = \lambda_{v_1 + \ldots + v_{n+1}}$$

and denoting this element of K by λ we conclude that

$$v^f = \lambda v \quad \forall v \in V, \text{i.e.} \quad f \in Z(V).$$

Thus to finish the proof it suffices to show that all elements of Aut $\mathbf{P}_1$ can be obtained from elements of $\Gamma L(V)$ in the way described above. Choose $\varphi \in \text{Aut}\,\mathbf{P}_1$. We divide the construction of an element $f \in \Gamma L(V)$ such that $\tilde{f} = \varphi$ in several steps.

(i) Let $v_1, \ldots, v_r \in V$ be linearly independent. If we put $\langle v_i \rangle^\varphi = \langle \tilde{v}_i \rangle$, then $\tilde{v}_1, \ldots, \tilde{v}_r$ are linearly independent and $[\langle v_1, \ldots, v_r \rangle]^\varphi = [\langle \tilde{v}_1, \ldots, \tilde{v}_r \rangle]$ (The proof is the same as under (*a*)).

(ii) Let $v_1, \ldots, v_{n+1}$ be a basis for V, then we can choose $u_1, \ldots, u_{n+1} \in V$ such that

$$\langle v_i \rangle^\varphi = \langle u_i \rangle \ (1 \leq i \leq n+1), \text{ and}$$
$$\langle v_1 + v_i \rangle^\varphi = \langle u_1 + u_i \rangle \ (2 \leq i \leq n+1).$$

(Proof: First choose arbitrary $v_1', \ldots, v_{n+1}'$ such that $\langle v_i \rangle^\varphi = \langle v_i' \rangle$ and put $u_1 = v_1'$. Since $[\langle v_1, v_i \rangle]^\varphi = [\langle u_1, v_i' \rangle]$ there exists a $\mu \in K$ such that $\langle v_1 + v_i \rangle^\varphi = \langle u_1 + \mu v_i' \rangle$ $(2 \leq i \leq n+1)$, hence it suffices to put $u_i = \mu v_i'$.)

(iii) If $\{i_2, \ldots, i_r\} \subseteq \{2, \ldots, n+1\}$, then

$$\langle v_1 + v_{i_2} + \ldots + v_{i_r} \rangle^\varphi = \langle u_1 + u_{i_2} + \ldots + u_{i_r} \rangle.$$

(Proof: Induction on r. For $r = 2$ this is proved under (ii). Assume the statement to hold for $2, \ldots, r-1$, then

$$\langle v_1 + v_{i_2} + \ldots + v_{i_{r-1}} \rangle^\varphi = \langle u_1 + u_{i_2} + \ldots + u_{i_{r-1}} \rangle,$$
$$\langle v_1 + v_{i_3} + \ldots v_{i_r} \rangle^\varphi = \langle u_1 + u_{i_3} + \ldots + u_{i_r} \rangle.$$

From

$$\langle v_1 + v_{i_2} + \ldots + v_{i_r} \rangle = [\langle v_1 + v_{i_2} + \ldots + v_{i_{r-1}}, v_{i_r} \rangle] \cap [\langle v_1 + v_{i_3} + \ldots + v_{i_r} \rangle, v_{i_2}],$$

we conclude

$$\begin{aligned}\langle v_1 + v_{i_2} + \ldots + v_{i_r}\rangle^\varphi &= [\langle u_1 + u_{i_2} + \ldots + u_{i_{r-1}}, u_{i_r}\rangle] \\ &\cap [\langle u_1 + u_{i_3} + \ldots + u_{i_r}, u_{i_2}\rangle] \\ &= \langle u_1 + u_{i_2} + \ldots + u_{i_r}\rangle.)\end{aligned}$$

(iv) If $\{i_1, \ldots, i_r\} \subseteq \{1, \ldots, n+1\}$, then

$$\langle v_{i_1} + \ldots + v_{i_r}\rangle^\varphi = \langle u_{i_1} + \ldots + u_{i_r}\rangle.$$

(Proof: If $1 \in \{i_1, \ldots, i_r\}$ this follows from (iii), so let us assume $1 \notin \{i_1, \ldots, i_r\}$. We prove the statement for $r = 2$, since the proof for $r > 2$ is similar to the proof of (iii). From $\mathfrak{B}^{(1)\varphi} = \mathfrak{B}^{(1)}$ we have $[\langle v_{i_1}, v_{i_2}\rangle]^\varphi = [\langle u_{i_1}, u_{i_2}\rangle]$ and $\langle v_{i_1} + v_{i_2}\rangle^\varphi = \langle u_{i_1} + \lambda u_{i_2}\rangle$ for some $\lambda \in K$. From

$$\langle v_1 + v_{i_1} + v_{i_2}\rangle = [\langle v_1, v_{i_1} + v_{i_2}\rangle] \cap [\langle v_1 + v_{i_1}, v_{i_2}\rangle],$$

we conclude

$$\begin{aligned}\langle v_1 + v_{i_1} + v_{i_2}\rangle^\varphi &= [\langle v_1, v_{i_1} + v_{i_2}\rangle]^\varphi \cap [\langle v_1 + v_{i_1}, v_{i_2}\rangle]^\varphi \\ &= [\langle u_1, u_{i_1} + \lambda u_{i_2}\rangle] \cap [\langle u_1 + u_{i_2}, u_{i_2}\rangle] \\ &= \langle u_1 + u_{i_1} + \lambda u_{i_2}\rangle.\end{aligned}$$

On the other hand we have by (iii):

$$\langle v_1 + v_{i_1} + v_{i_2}\rangle^\varphi = \langle u_1 + u_{i_1} + u_{i_2}\rangle,$$

hence

$$\langle v_{i_1} + v_{i_2}\rangle^\varphi = \langle u_{i_1} + u_{i_2}\rangle.)$$

(v) For $\lambda \in K$ we have $\langle v_1 + \lambda v_2\rangle \in \langle v_1, v_2\rangle$, hence $\langle v_1 + \lambda v_2\rangle^\varphi = \langle u_1 + \tilde{\lambda} u_2\rangle$ for some uniquely determined $\tilde{\lambda} \in K$. Defining $\sigma : K \to K$ by $\lambda^\sigma = \tilde{\lambda}$ we have

$$\langle \lambda_1 v_1 + \ldots + \lambda_{n+1} v_{n+1}\rangle^\varphi = \langle \lambda_1^\sigma u_1 + \ldots + \lambda_{n+1}^\sigma u_{n+1}\rangle.$$

(Proof: For $i > 1$ we put $\langle v_1 + \lambda v_i\rangle^\varphi = \langle u_1 + \lambda' u_i\rangle$. From $[\langle v_1 + \lambda v_2, v_i\rangle] \cap [\langle v_1 + \lambda v_i, v_2\rangle] = \langle v_1 + \lambda(v_2 + v_i)\rangle$, we conclude: $\langle u_1 + \lambda^\sigma u_2 + \lambda' u_i\rangle = \langle u_1 + \lambda''(u_2 + u_i)\rangle$ for some $\lambda'' \in K$, and this shows that $\lambda' = \lambda'' = \lambda^\sigma$. So we have $\langle v_1 + v_i\rangle^\varphi = \langle u_1 + \lambda^\sigma u_i\rangle$. For $i, j > 1, i \neq j$, we put $\langle v_i + \lambda v_j\rangle^\varphi = \langle u_i + \lambda' u_j\rangle$.

From

$$\langle v_1 + v_i + \lambda v_j \rangle = [\langle v_i, v_1 + \lambda v_j \rangle] \cap [\langle v_1 + v_i, v_j \rangle] \cap [\langle v_1, v_i + \lambda v_j \rangle]$$

we conclude:

$$\langle v_1 + v_i + \lambda v_j \rangle^\varphi = \langle u_1 + u_i + \lambda^\sigma u_j \rangle = \langle u_1 + u_i + \lambda' u_j \rangle,$$

hence $\lambda^\sigma = \lambda'$. Therefore, we have $\langle v_i + \lambda v_j \rangle^\varphi = \langle u_i + \lambda^\sigma u_j \rangle$ for all i, j such that $1 \leq i, j \leq n+1$ and $i \neq j$. For $i \neq j$ let us put $\langle \lambda v_i + \mu v_j \rangle^\varphi = \langle \lambda' u_i + \mu' u_j \rangle$. For $k \neq i$ and $k \neq j$, we have

$$\langle v_k + \lambda v_i + \mu v_j \rangle = [\langle v_k + \lambda v_i, v_j \rangle] \cap [\langle v_k + \mu v_j, v_i \rangle] \cap [\langle v_k, \lambda v_i + \mu v_j \rangle],$$

hence:

$$\langle v_k + \lambda v_i + \mu v_j \rangle^\varphi = [\langle u_k + \lambda^\sigma u_i, u_j \rangle] \cap [u_k + \mu^\sigma u_j, u_i \rangle] \cap [\langle u_k, \lambda' u_i + \mu' u_j \rangle].$$

Therefore:

$$\langle \lambda' u_i + \mu' \hat{u}_j \rangle = \langle \lambda^\sigma u_i + \mu^\sigma u_j \rangle,$$

i.e.

$$\langle \lambda v_i + \mu v_j \rangle^\varphi = \langle \lambda^\sigma u_i + \mu^\sigma u_j \rangle.$$

Let us prove $\langle \lambda_1 v_1 + \ldots + \lambda_r v_r \rangle^\varphi = \langle \lambda_1^\sigma u_1 + \ldots + \lambda_r^\sigma u_r \rangle$ by induction on r. We just proved that this is true for $r = 2$. Assuming it to be true for $2, \ldots, r-1$, and using

$$\begin{aligned} &\langle \lambda_1 v_1 + \ldots + \lambda_r v_r \rangle \\ &\quad = [\langle \lambda_1 v_1 + \ldots + \lambda_{r-1} v_{r-1}, v_r \rangle] \cap [\langle v_1, \lambda_2 v_2 + \ldots + \lambda_r v_r \rangle], \end{aligned}$$

we have

$$\begin{aligned} &\langle \lambda_1 v_1 + \ldots + \lambda_r v_r \rangle^\varphi \\ &\quad = [\langle \lambda_1^\sigma u_1 + \ldots + \lambda_{r-1}^\sigma u_{r-1}, u_r \rangle] \cap [\langle u_1, \lambda_2^\sigma u_2 + \ldots + \lambda_r^\sigma u_r \rangle] \\ &\quad = \langle \lambda_1^\sigma u_1 + \ldots + \lambda_r^\sigma u_r \rangle.) \end{aligned}$$

(vi) The map $\sigma : K \to K$ is an automorphism. (Proof: From the definition it is clear that $\lambda \neq 0 \Leftrightarrow \lambda^\sigma \neq 0$. For $\lambda \neq 0$ we have $\langle \lambda v_1 + v_2 \rangle = \langle v_1 + \lambda^{-1} v_2 \rangle$, hence $\langle \lambda^\sigma u_1 + u_2 \rangle = \langle u_1 + (\lambda^{-1})^\sigma u_2 \rangle$.

Therefore $(\lambda^\sigma)^{-1} = (\lambda^{-1})^\sigma$. From $\langle v_1 + \lambda\mu v_2 \rangle = \langle \lambda^{-1} v_1 + \mu v_2 \rangle$ we conclude in the same way: $(\lambda\mu)^\sigma = ((\lambda^{-1})^\sigma)^{-1} \mu^\sigma = \lambda^\sigma \mu^\sigma$. Finally, from

$$\langle v_1 + (\lambda + \mu) v_2 + v_3 \rangle = [\langle v_1 + \lambda v_2, v_3 + \mu v_2 \rangle] \cap [\langle v_1 + v_3, v_2 \rangle]$$

we have

$$\langle u_1 + (\lambda + \mu)^\sigma u_2 + u_3 \rangle = \langle u_1 + (\lambda^\sigma + \mu^\sigma) u_2 + u_3 \rangle$$

and so $(\lambda + \mu)^\sigma = \lambda^\sigma + \mu^\sigma$.) Now we define $f : V \to V$ by

$$(\Sigma \lambda_i v_i) f = \Sigma \lambda_i^\sigma u_i.$$

It is easy to check that $f \in \Gamma L(V)$, and $\tilde{f} = \varphi$. ∎

2

Fundamental properties of finite groups

2.1 The Sylow theorems

We saw in theorem 1.2.5 that the order of a subgroup of a finite group is a divisor of the order of the group. However, if a divisor d of the order of the group is given, then it is not always true that there exists a subgroup the order of which equals d. The following theorem guarantees the existence of subgroups of certain special orders of a given finite group.

Theorem 2.1.1 (Sylow) *Let G be a finite group. If p is a prime number and if m is the natural number such that p^m is the highest power of p dividing the order of G, then:*

(1) *G contains a subgroup of order p^m. Subgroups of order p^m are called Sylow p-subgroups or simply Sylow subgroups of G.*

(2) *Every p-subgroup of G is contained in some Sylow p-subgroup of G.*

(3) *The Sylow p-subgroups of G are conjugate.*

(4) *The number r of Sylow p-subgroups of G satisfies $r \equiv 1 \pmod{p}$.*

Proof (1) Let Ω be the collection of all subsets of G consisting of p^m elements. An element σ of G induces a permutation of Ω given by $S \to S\sigma$ $(S \in \Omega)$, i.e. G operates on Ω.

Let g represent the order of G, then

$$|\Omega| = \frac{g!}{(g-p^m)!\,p^m!}.$$

Since $p^m \mathrm{T}\, g$, g can be written $p^m g'$ with g' an integer not divisible by p. Then we have

$$\frac{g!}{(g-p^m)!p^m!} = \frac{g'p^m(g'p^m-1)\dots(g'p^m-p^m+1)}{p^m!}$$
$$= \prod_{i=0}^{p^m-1} \frac{g'p^m-i}{p^m-i}.$$

It is easy to show that $(g'p^m - i)_p = (p^m - i)_p, \forall i (0 \le i \le p^m - 1)$ and we can conclude that $|\Omega| \not\equiv 0 \pmod p$. Hence there exists an orbit Γ of (G, Ω) such that $|\Gamma| \not\equiv 0 \pmod p$. For $S \in \Gamma$ we have $|G| = |\Gamma||G_S| \equiv 0 \pmod{p^m}$, hence $|G_S| \equiv 0 \pmod{p^m}$. From $S = SG_S \supseteq \alpha G_S \, (\alpha \in S)$ we conclude $p^m = |S| \ge |\alpha G_S| = |G_S|$, hence G_S is a subgroup of G of order p^m.

(2) We use the same notation as in (1). Let P be an arbitrary p-subgroup of G. According to theorem 1.2.11 the length of each orbit of (P, Γ) is a divisor of $|P|$. Therefore there exists an orbit of length 1 among those orbits, i.e. there exists an element of Γ that is left fixed by P. Representing this element by S, we conclude that P is a subgroup of the Sylow p-subgroup G_S of G.

(3) Applying the argument of (2) to a Sylow p-group P we see that we can write $P = G_S$ for some $S \in \Gamma$. Since (G, Γ) is transitive, there exists for every two elements S_1 and S_2 of Γ an element σ of G such that $S_1\sigma = S_2$. Therefore $G_{S_2} = G_{S_1\sigma} = G_{S_1}^{\sigma}$, i.e. all Sylow p-subgroups are conjugate.

(4) Let Δ be the collection of all Sylow p-subgroups of G. An element $\sigma \in G$ determines a mapping from Δ into Δ defined by $P \to \sigma^{-1}P\sigma \, (P \in \Delta)$, i.e. G operates on Δ. Let H be one of the Sylow p-subgroups of G. Since the lengths of the orbits of (H, Δ) are divisors of $|H|$, they are powers of p. Clearly, $H (\in \Delta)$ is a fixed point of (H, Δ). If $P (\in \Delta)$ is a fixed point of (H, Δ), then $H \le \mathcal{N}_G(P)$ since $\sigma^{-1}P\sigma = P$ for all $\sigma \in H$. Since H and P are Sylow p-subgroups of $\mathcal{N}_G(P)$, H and P are $\mathcal{N}_G(P)$-conjugate by (3), hence $H = P$. Therefore (H, Δ) has exactly one fixed point and the lengths of all other orbits are divisible by p, i.e. $|\Delta| = r \equiv 1 \pmod p$. ∎

The following theorem follows easily from the Sylow theorems.

Theorem 2.1.2 *Let P be a Sylow p-subgroup of G and let Q be a p-subgroup of G.*

(1) *$Q \leq P \Leftrightarrow Q \leq \mathcal{N}_G(P)$.*

(2) *If r is the number of Sylow p-subgroups containing Q, then $r \equiv 1 \pmod p$.*

(3) *If N is a normal subgroup of G, then $P \cap N$ and PN/N are Sylow p-subgroups of N and G/N respectively.*

(4) *If H is a subgroup of G containing $\mathcal{N}_G(P)$ then $\mathcal{N}_G(H) = H$.*

Proof (1) $\Rightarrow$: Trivial. $\Leftarrow$: Since $\mathcal{N}_G(P) \rhd P$, we have $\mathcal{N}_G(P) \geq QP$ and $QP \rhd P$, and so $QP/P \simeq Q/Q \cap P$ by the isomorphism theorem, hence QP is a p-subgroup of G and $QP = P$.

(2) This is a generalization of theorem 2.1.1(4), and the proof is similar. Let Δ be the collection of all Sylow p-subgroups of G, then Q operates on Δ by means of the conjugacy operation. The lengths of all orbits of (Q, Δ) are powers of p. By theorem 2.1.1(4) $|\Delta| \equiv 1 \pmod p$, hence the number of orbits of length 1 of (Q, Δ) is congruent to $1 \pmod p$. On the other hand:
the number of orbits of length 1
= the number of Sylow p-subgroups P such that $\mathcal{N}_G(P) \geq Q$
which, by part (1) of this theorem
= the number of Sylow p-subgroups P such that $P \geq Q$.
Combining these two results we get the desired result.

(3) Since $G \trianglerighteq N$, we have $G \geq NP \trianglerighteq N$ and $|NP : N| = |P : P \cap N|$ by the isomorphism theorem. Then

$$|NP : P| = \frac{|NP : P \cap N|}{|P : P \cap N|} = \frac{|NP : N|\,|N : P \cap N|}{|P : P \cap N|} = |N : P \cap N|.$$

Hence $|N : P \cap N| \not\equiv 0 \pmod p$, namely, $P \cap N$ is a Sylow p-subgroup of N. The proof that PN/N is a Sylow p-subgroup of G/N is similar and will be left to the reader.

(4) For $\sigma \in \mathcal{N}_G(H)$ we have $H = H^\sigma \geq P^\sigma$, hence P and P^σ are Sylow p-subgroups of H. Therefore there exists an element τ of H such that $(P^\sigma)^\tau = P$, i.e. $\sigma\tau \in \mathcal{N}_G(P)$. Thus

$$\sigma \in \mathcal{N}_G(P)\tau^{-1} \subseteq \mathcal{N}_G(P)H = H, \text{ i.e. } \mathcal{N}_G(H) = H. \qquad \blacksquare$$

Theorem 2.1.3 *If $N \trianglelefteq G$ and if P is a Sylow p-subgroup of N, then $G = N \cdot \mathcal{N}_G(P)$.*

Proof For $\sigma \in G$, $P^\sigma (\le N^\sigma = N)$ is a Sylow p-subgroup of N. Therefore, there exists an element τ of N such that $P^\sigma = P^\tau$, i.e. $P^{\sigma\tau^{-1}} = P$. Thus $\sigma\tau^{-1} \in \mathcal{N}_G(P)$ and the desired result follows. ■

2.2 Direct products and semi-direct products

Let G_1 and G_2 be groups. We make $G = G_1 \times G_2$ into a group by defining

$$(\sigma_1, \sigma_2)(\tau_1, \tau_2) = (\sigma_1\tau_1, \sigma_2\tau_2) \quad ((\sigma_1, \sigma_2) \text{ and } (\tau_1, \tau_2) \in G).$$

The identity of G is the element $(1, 1)$; the inverse of an element $(\sigma, \tau) \in G$ is the element (σ^{-1}, τ^{-1}). The group G is called the direct product of the groups G_1 and G_2 and is denoted by $G = G_1 \times G_2$. Putting

$$H_1 = \{(\sigma, 1) \mid \sigma \in G_1\} \text{ and } H_2 = \{(1, \sigma) \mid \sigma \in G_2\}$$

we have $H_i \simeq G_i$, H_i is a normal subgroup of $G (i = 1, 2)$, $G = H_1 H_2$ and $H_1 \cap H_2 = 1$.

Conversely, if G is a group with normal subgroups H_1 and H_2 satisfying:

$$G = H_1 H_2, \quad H_1 \cap H_2 = 1, \tag{2.2.1}$$

then every element σ of G can be represented in exactly one way as $\sigma = \sigma_1 \sigma_2$ $(\sigma_i \in H_i)$. Hence G is isomorphic to the direct product $H_1 \times H_2$. We say that G is the direct product of (normal) subgroups H_1 and H_2 and we denote this by $G = H_1 \times H_2$.

The above discussion can easily be generalized to the case of n groups or n subgroups. Let $G_1, \ldots, G_n$ be groups. Again, $G = G_1 \times \ldots \times G_n$ is made into a group by defining:

$$(\sigma_1, \ldots, \sigma_n) \cdot (\tau_1, \ldots, \tau_n) = (\sigma_1\tau_1, \ldots, \sigma_n\tau_n) \quad ((\sigma_1, \ldots, \sigma_n) \text{ and } (\tau_1, \ldots, \tau_n) \in G).$$

G is called the direct product of $G_1, \ldots, G_n$. Putting:

$$H_i = \{(\sigma_1, \ldots, \sigma_n) \in G \mid \sigma_j = 1, \forall j \neq i\}$$

we have that $H_1, \ldots, H_n$ are normal subgroups of G, $H_i \simeq G_i$ $(i = 1, \ldots, n)$, $G = H_1 \ldots H_n$ and $H_1 \ldots H_i \cap H_{i+1} = 1$ $(i = 1, \ldots, n-1)$.

Conversely, if G is a group with normal subgroups $H_1, \dots, H_n$, satisfying:

$$G = H_1 \dots H_n, \; H_1 \dots H_i \cap H_{i+1} = 1 \quad (i = 1, \dots, n-1)$$

then G is called the direct product of the (normal) subgroups $H_1, \dots, H_n$ and denoted by: $G = H_1 \times \dots \times H_n$.

The following theorem is an easy consequence of the definitions.

Theorem 2.2.1 *Let $H_1, \dots, H_n$ be normal subgroups of G. Then:*

(1) $G = H_1 \times \dots \times H_n \Leftrightarrow$ *every element σ of G can be represented in exactly one way as $\sigma = \sigma_1 \dots \sigma_n$ $(\sigma_i \in H_i)$.*

(2) *If $G = H_1 \times \dots \times H_n$,* then $[H_i, H_j] = 1, \forall i, j (\neq)$. *Further, if N is a normal subgroup of H_i for some i, then N is a normal subgroup of G.* ■

Let G be a finite group, let N be a minimal normal subgroup of G and let N_0 be a minimal normal subgroup of N. Then N is the direct product of a certain number of normal subgroups $N_1, \dots, N_r$ that are conjugate to N_0 in G. (Proof: Let $\tilde{N}$ be a maximal element of the set of all subgroups of N that can be represented as the direct product of a certain number of normal subgroups of N, that are conjugate to N_0 in G. So, $\tilde{N} = N_1 \times \dots \times N_r$ with normal subgroups N_i of N that are conjugate to N_0. Suppose $N \gneq \tilde{N}$. Since N is a minimal normal subgroup of G, there exists an element $\sigma \in G$ such that $\tilde{N}^\sigma \neq \tilde{N}$. Hence, we must have for at least one $i: \tilde{N} \ngeq N_i^\sigma$. From $\tilde{N} \trianglelefteq N, N_i^\sigma \trianglelefteq N$ and $\tilde{N} \cap N_i^\sigma = 1$ we conclude:

$$N \geq \tilde{N} N_i^\sigma = \tilde{N} \times N_i^\sigma = N_1 \times \dots \times N_r \times N_i^\sigma$$

contrary to our choice of N. Hence $N = \tilde{N}$.) According to theorem 2.2.1(2) normal subgroups of N_i are normal subgroups of N, from which we conclude that the N_i are simple groups. So we have proved:

Theorem 2.2.2 *Let G be a finite group and let N be a minimal normal subgroup of G, then N is the direct product of a certain number of isomorphic simple groups.*

Theorem 2.2.3 *Let G_1 and G_2 be groups and let H be a subgroup of $G_1 \times G_2$, satisfying: for every $\sigma \in G_1$ there exists a $\tau \in G_2$ such that $(\sigma, \tau) \in H$ and for every $\tau \in G_2$ there exists a $\sigma \in G_1$ such that $(\sigma, \tau) \in H$. Putting*

$$N_1 = \{a \in G_1 \mid (a, 1) \in H\}, \quad N_2 = \{b \in G_2 \mid (1, b) \in H\}$$

we have:

(1) $N_i \trianglelefteq G_i \quad (i = 1, 2)$

(2) $G_1/N_1 \simeq G_2/N_2$.

Proof (1) Trivial.

(2) We define a map from G_1 to G_2/N_2 as follows: for $\sigma \in G_1$ there exists an element $\tau \in G_2$ such that $(\sigma, \tau) \in H$. If τ' is another element of G_2 such that $(\sigma, \tau') \in H$, then $(1, \tau\tau'^{-1}) \in H$, i.e. $\tau\tau'^{-1} \in N_2$ and $\tau N_2 = \tau' N_2$. Therefore the map $\sigma \to \tau N_2$ is well defined and it is easy to check that the kernel of this map is N_1. ■

If $N \trianglelefteq G$ and $K \leq G$, then $NK = KN$ is a subgroup of G. In particular, if $G = NK$ and $K \cap N = 1$, then G is called the semi-direct product of (a normal subgroup) N and (a subgroup) K of G. In this case, K operates on N via the conjugacy maps defined on N for all elements of K and these maps are automorphisms of N. Conversely, let $\tilde{K}$ and $\tilde{N}$ be groups, such that $\tilde{K}$ operates on $\tilde{N}$ and such that all permutations of this operation are automorphisms of $\tilde{N}$. Then it is easy to see that $\tilde{K} \times \tilde{N}$ provided with a multiplication defined by:

$$(\tau_1, \sigma_1)(\tau_2, \sigma_2) = (\tau_1\tau_2, \sigma_1^{\tau_2}\sigma_2) \quad ((\tau_1, \sigma_1), (\tau_2, \sigma_2) \in \tilde{K} \times \tilde{N})$$

is a group. This group is called the semi-direct product of $\tilde{K}$ and $\tilde{N}$ and is denoted by $\tilde{K}\tilde{N}$ (or $\tilde{N}\tilde{K}$). Putting

$$K = \{(\tau, 1) \mid \tau \in \tilde{K}\} \text{ and } N = \{(1, \sigma) \mid \sigma \in \tilde{N}\},$$

we have:

$$K \simeq \tilde{K}, N \simeq \tilde{N}, N \lhd \tilde{K}\tilde{N}, \tilde{K}\tilde{N} = KN, K \cap N = 1.$$

Thus, $\tilde{K}\tilde{N}$ is the semi-direct product of the subgroup K, which is isomorphic with $\tilde{K}$, and the normal subgroup N, which is isomorphic with $\tilde{N}$. If all isomorphisms of N (or $\tilde{N}$) induced by K (or $\tilde{K}$) are identity maps, then the semi-direct product becomes a direct product, i.e. direct products are special cases of semi-direct products.

If G is the semi-direct product of the subgroup K and the normal subgroup N, then we have for $\sigma \in G : G = G^\sigma = (KN)^\sigma = K^\sigma N$ and $K^\sigma \cap N = (K \cap N)^\sigma = \{1\}$ i.e. G is the semi-direct product of K^σ and N. However, there may exist a subgroup H that is not conjugate with K such that G is the semi-direct product of H and N. We will study this situation more closely under the assumption that N is commutative.

So, let N be a commutative group and let K be a group operating on N, such that all permutations of N induced by the elements of K are automorphisms. A map $f : K \to N$, satisfying:

$$f(\sigma\tau) = f(\sigma)^\tau f(\tau) \quad \sigma, \tau \in K \tag{2.2.2}$$

is called a derivation of K with coefficients in N. The collection of all derivations of K with coefficients in N is denoted by $\mathrm{Der}(K, N)$.

Let $x_0 \in N$, then the map $f_{x_0} : K \to N$ defined by

$$f_{x_0}(\sigma) = x_0^\sigma x_0^{-1} \quad \forall \sigma \in K$$

satisfies condition (2.2.2). The maps f_{x_0} are called the inner derivations of K with coefficients in N and the collection of all these inner derivations is denoted by $\mathrm{Inn}(K, N)$.

For two elements f and g of $\mathrm{Der}(K, N)$ the product $fg : K \to N$ defined as usual by

$$(fg)(\sigma) = f(\sigma)g(\sigma) \quad \forall \sigma \in K$$

also satisfies condition (2.2.2), i.e. $\mathrm{Der}(K, N)$ provided with this multiplication is a commutative group and $\mathrm{Inn}(K, N)$ is easily seen to be a subgroup of $\mathrm{Der}(K, N)$. The identity element of $\mathrm{Der}(K, N)$ is the map $f_0 : K \to N$ defined by $f_0(\sigma) = 1$ for all $\sigma \in K$. The inverse of an arbitrary element $f \in \mathrm{Der}(K, N)$ is given by

$$f^{-1}(\sigma) = f(\sigma)^{-1} \quad \forall \sigma \in K.$$

The quotient group $\mathrm{Der}(K, N)/\mathrm{Inn}(K, N)$ is denoted by $H^1(K, N)$ and is called the first cohomology group of K with coefficients in N.

Let G be the semi-direct product of a commutative, normal subgroup N and a subgroup K. Let K_1 be another subgroup, such that G is the semi-direct product of N and K_1, then both K and K_1 are representative systems of G/N, i.e. the map $\varphi: K \to K_1$ defined by:

for $\sigma \in K$, $\varphi(\sigma)$ is the unique element of K_1 satisfying $\sigma N = \varphi(\sigma)N$,

is a bijection.

Next, we define $f: K \to N$ as follows:

for $\sigma \in K$, $f(\sigma)$ is the unique element of N satisfying $\varphi(\sigma) = \sigma f(\sigma)$.

From:

$$\begin{array}{c} \varphi(\sigma)\varphi(\tau) = \varphi(\sigma\tau) = \sigma\tau f(\sigma\tau) \\ \| \\ \sigma f(\sigma)\tau f(\tau) = \sigma\tau f(\sigma)^{\tau} f(\tau) \quad \sigma, \tau \in K. \end{array}$$

We conclude:

$$f(\sigma\tau) = f(\sigma)^{\tau} f(\tau) \quad \forall \sigma, \tau \in K,$$

i.e. $f \in \mathrm{Der}(K, N)$. Conversely, if $f \in \mathrm{Der}(K, N)$, then it is easy to check that $K_1 = \{\sigma f(\sigma) | \sigma \in K\}$ is a subgroup of G such that G is the semi-direct product of N and K_1. Hence there exists a one-to-one correspondence between the collection $\mathfrak{M} = \{\tilde{K} | \tilde{K} \leq G$, G is the semi-direct product of N and $\tilde{K}\}$ and $\mathrm{Der}(K, N)$.

Let $K_1, K_2 \in \mathfrak{M}$ and let f_1 and $f_2 \in \mathrm{Der}(K, N)$ be the elements corresponding to M_1 and M_2 respectively, i.e. $K_i = \{\sigma f_i(\sigma) | \sigma \in K\}$ $(i = 1, 2)$. If K_1 and K_2 are conjugate in G, then there exists an element $a \in N$ such that $K_2 = aK_1a^{-1}$. For f_1 and f_2 we have in this case:

$$f_2(\sigma) = a^{\sigma} a^{-1} f_1(\sigma) \quad \forall \sigma \in K.$$

(Since $a\sigma f_1(\sigma)a^{-1} = \sigma' f_2(\sigma')$ for some $\sigma' \in K$. From:

$$a\sigma f_1(\sigma)a^{-1} = \sigma\sigma^{-1}a\sigma f_1(\sigma)a^{-1} \text{ and } \sigma^{-1}a\sigma f_1(\sigma)a^{-1} \in N$$

we conclude $\sigma' = \sigma$ and $f_2(\sigma) = a^{\sigma}a^{-1}f_1(\sigma)$.) Hence $f_2 f_1^{-1} \in \mathrm{Inn}(K, N)$. Conversely, if there exists an element $a \in N$, such that $(f_2 f_1^{-1})(\sigma) =$

$a^{\sigma}a^{-1}$ ($\forall\sigma\in K$), then obviously $K_2 = aK_1a^{-1}$. Therefore K_1 and K_2 are conjugate in $G \Leftrightarrow f_1f_2^{-1} \in \mathrm{Inn}(K, N)$. Hence, there exists a one-to-one correspondence between $H^1(K, N)$ and the set of equivalence classes in $\mathfrak{M}$ under the relation of being conjugate in G. We sum up our results in the following theorem.

Theorem 2.2.4 *Let G be the semi-direct product of a commutative normal subgroup N and a subgroup K. Then the number of equivalence classes in*

$$\mathfrak{M} = \{K_1 \leq G \mid G \textit{ is semi-direct product of } K_1 \text{ and } N\}$$

under the relation of being conjugate in G equals the order of the group $H^1(K, N)$.

Let N be a group and let K be a permutation group on $\Omega = \{1, 2, \dots, n\}$. Let $N_1, \dots, N_n$ be groups that are isomorphic to N and let $\varphi_i : N \to N_i$ be isomorphisms ($i = 1, \dots, n$). Putting $H = N_1 \times \dots \times N_n$, we have for every $\sigma \in K$ a map $\tilde{\sigma} : H \to H$ defined by

$$(\varphi_1(x_1), \dots, \varphi_n(x_n))^{\tilde{\sigma}} = (\varphi_1(x_{\sigma^{-1}(1)}), \dots, \varphi_n(x_{\sigma^{-1}(n)})) \quad (x_1, \dots, x_n \in N).$$

Then $\tilde{\sigma}$ is an automorphism of H, i.e. K operates on H. The semi-direct product KH is called the wreath product of N and K and is denoted by $N \wr K$. It is easy to prove that the order of $N \wr K$ is $|N|^n |K|$.

2.3 Normal series

Let $G_0, G_1, \dots, G_n$ be subgroups of the group G, satisfying:

$$G_i \trianglerighteq G_{i+1} \quad (i = 0, \dots, n-1),$$
$$G_0 = G \text{ and } G_n = 1.$$

Then $\{G_i \mid i = 0, \dots, n\}$ is called a normal series of G and the integer n is called the length of the series. Further, the quotient groups G_i/G_{i+1} ($i = 0, \dots, n-1$) are called the factors and the collection of factors $\{G_i/G_{i+1} \mid i = 0, \dots, n-1\}$ is called the factor series of the normal series.

A normal series $\{G_i | i=0,\ldots,n\}$ of G such that all factors are simple groups not equal to $\{1\}$ is called a composition series of G. Again, the integer n is called the length of the composition series and the factors of the composition series are called composition factors.

If $\{G_i | i=0,\ldots,n\}$ and $\{H_j | j=0,\ldots,m\}$ are two normal series of G satisfying:

(i) $n \geq m$

(ii) $\forall j\,(0 \leq j \leq m)\ \exists i_j\,(0 \leq i_j \leq n): H_j = G_{i_j}$,

then $\{G_i | i=0,\ldots,n\}$ is called a refinement of $\{H_j | j=0,\ldots,m\}$.

Theorem 2.3.1 *Let G be a finite group.*

(1) *If $\{G_i | i=0,\ldots,n\}$ is a normal series of G satisfying $G_i \neq G_{i+1}$ $(i=0,\ldots,n-1)$, then there exists a composition series, that is a refinement of $\{G_i | i=0,\ldots,n\}$.*

(2) *All composition series of G have the same length and their composition factors are uniquely determined by G apart from the order of the terms.*

Proof Since G is finite (1) is obvious. Let n represent the length of the shortest composition. We prove (2) by induction on n. For $n=1$, the statement is obvious. For $n>1$, let $\{G_i | i=0,\ldots,n\}$ represent a composition series of length n and let $\{H_j | j=0,\ldots,m\}$ represent an arbitrary composition series. If $G_1 = H_1$, then we apply the induction hypothesis to $G_1 = H_1$ and we are finished. If $G_1 \neq H_1$, then $G = G_1 H_1$ and $G/H_1 \simeq G_1/H_1 \cap G_1$ by the isomorphism theorem. Since all composition series of G_1 have length $n-1$ by the induction hypothesis and since $G_1/G_1 \cap H_1$ is simple, $G_1 \cap H_1$ has a composition series of length $n-2$. Since $H_1/G_1 \cap H_1$ is simple, H_1 has a composition series of length $n-1$, and then all composition series of H_1 have length $n-1$ by the induction hypothesis. On the other hand $\{H_j | j=1,\ldots,m\}$ is a composition series of H_1 of length $m-1$. Hence we have $m=n$. The rest of the assertion is almost obvious from the above argument. ■

If $\{G_i | i=0,\ldots,n\}$ is a normal series of maximal length G satisfying $G \trianglerighteq G_i\,(i=0,\ldots,n)$ and $G_i \neq G_{i+1}\,(i=0,\ldots,n-1)$, then

$\{G_i | i=0, \ldots, n\}$ is called a principal composition series of G. If instead of $G \trianglerighteq G_i$ the condition $G \trianglerighteq G_i$ is satisfied, then $\{G_i | i=0, \ldots, n\}$ is called a characteristic composition series of G.

If there exists a normal series of G, the factors of which are all commutative, then G is called a solvable group. It is easy to verify that subgroups and quotient groups of a solvable group are also solvable and that the direct product, semi-direct product and wreath product of two solvable groups are also solvable groups. The following theorem is a straightforward consequence of the definitions.

Theorem 2.3.2 *Let G be a finite group. Then the following conditions are equivalent:*

(1) *G is solvable.*

(2) *There is an integer n such that $D^{(n)}(G)=1$.*

(3) *The composition factors of G are all simple groups of prime order.*

If all composition factors of a group are p-groups or p'-groups (for some prime number p), then G is called p-solvable. Putting (with respect to G)

$$H_1 = O^p(G), \qquad K_1 = O_p(G),$$
$$H_2 = O^{p,p'}(G), \quad K_2 = O_{p,p'}(G),$$

and inductively

$$H_{2i-1} = O^p(H_{2i-2}), \quad K_{2i-1} = O_p(K_{2i-2}),$$
$$H_{2i} = O^{p'}(H_{2i-1}), \quad K_{2i} = O_{p'}(K_{2i-1}),$$

the following conditions are easily seen to be equivalent:

(i) G is p-solvable.

(ii) There is an integer n such that $H_n = 1$.

(iii) There is an integer m such that $K_m = G$.

Again, it is easy to prove that subgroups and quotient groups of a p-solvable group are also p-solvable and that the direct product, semi-direct product and wreath product of two p-solvable groups are p-solvable groups.

If G has a normal series $\{G_i | i=0,\ldots,n\}$ satisfying:

(i) $G \trianglerighteq G_i$

(ii) $Z(G/G_i) \geq G_i/G_{i+1} \quad (i=0,\ldots,n-1)$,

then G is called a nilpotent group. Since in this case the G_i/G_{i+1} are commutative, nilpotent groups are special examples of solvable groups. Putting

$$D^{[0]}(G) = G,$$
$$D^{[1]}(G) = [G, G], \ldots, D^{[i+1]}(G) = [D^{[i]}(G), G]$$

the following assertions are easy to prove:

(i) $G \trianglerighteq D^{[i]}(G)$.

(ii) $D^{[i]}(G) \geq D^{[i+1]}(G) \quad (i=1,2,\ldots)$.

(iii) $Z(G/D^{[i+1]}(G)) \geq D^{[i]}(G)/D^{[i+1]}(G) \quad (i=0,1,\ldots)$.

The series: $D^{[0]}(G), D^{[1]}(G), \ldots, D^{[i]}(G), \ldots$ is called the lower central series of G. The reader may easily prove the following theorem:

Theorem 2.3.3 *For a group G the following conditions are equivalent:*

(1) *G is a nilpotent group.*

(2) *There is an integer n such that $Z^{(n)}(G) = G$.*

(3) *There is an integer m such that $D^{[m]}(G) = 1$.*

The following theorem tells something about the structure of nilpotent groups.

Theorem 2.3.4 *For a finite group G the following conditions are equivalent:*

(1) *G is nilpotent.*

(2) *For every subgroup H of G ($H \neq G$), we have $\mathcal{N}_G(H) \gneqq H$.*

(3) *All Sylow subgroups of G are normal.*

(4) *G is the direct product of its Sylow subgroups.*

Proof (1)$\Rightarrow$(2): Since G is nilpotent, there exists a normal series $\{G_i | i=0,\ldots,n\}$ satisfying $Z(G/G_{i+1}) \geq G_i/G_{i+1}$. There exists an i such that $G_i \leq H$ and $G_{i-1} \nleq H$. From $[G, G_{i-1}] \leq G_i$, we conclude $[H, G_{i-1}] \leq G_i \leq H$, i.e. $\sigma^{-1}H\sigma = H$ ($\forall \sigma \in G_{i-1}$). Therefore $\mathcal{N}_G(H) \geq G_{i-1}$, and $\mathcal{N}_G(H) \gneqq H$.

(2) ⇒ (3): Let P be a Sylow p-subgroup of G. Putting $H = \mathcal{N}_G(P)$, we have $\mathcal{N}_G(H) = H$ according to theorem 2.1.2(4). Hence $H = G$, i.e. P is normal.

(3) ⇒ (4): Let $p_1, \ldots, p_r$ be the prime factors of the order of G and let P_i represent the Sylow p_i-subgroup of G. From $P_i \trianglelefteq G$, $G = P_1 \ldots P_r$ and $P_i \cap P_1 \ldots P_{i-1} = 1$, we conclude that G is the direct product of $P_1, \ldots, P_r$.

(4) ⇒ (1): Since it is easy to prove that the direct product of nilpotent groups is a nilpotent group again, it suffices to prove that p-groups are nilpotent. Let P be a p-group ($\neq 1$). P can be considered to operate on itself via the inner automorphisms. Since the length of an orbit of a permutation representation (P, P) is a divisor of $|P|$, the length of an orbit equals 1 or is a multiple of p. Since the identity element of P constitutes an orbit of length 1 and since $|P|$ is a power of p, we conclude that (P, P) has at least p fixed points. Since the collection of fixed points is just the centre of P, we have: $Z(P) \neq 1$. So the centre of a non-trivial p-group is non-trivial. Application of theorem 2.3.3 gives the desired result. ∎

Theorem 2.3.5 *If $N \neq 1$ is a normal subgroup of the nilpotent group G, then $N \cap Z(G) \neq 1$.*

Proof Putting

$$N^{(0)} = N, \quad N^{(1)} = [N, G], \ldots, \quad N^{(i)} = [N^{(i-1)}, G], \ldots$$

we have:

$$N^{(i)} \leq N \text{ and } N^{(i)} \leq D^{[i]}(G).$$

Since G is nilpotent, $D^{[m]}(G) = 1$ for some m, hence there is an n such that: $N^{(n)} \neq 1$, $N^{(n+1)} = 1$. From $[N^{(n)}, G] = 1$ we conclude $N^{(n)} \leq N \cap Z(G)$, i.e. $N \cap Z(G) \neq 1$. ∎

Subgroups and quotient groups of nilpotent groups are nilpotent. The direct product of two nilpotent groups is a nilpotent group again. Therefore, if N_1 and N_2 are normal nilpotent subgroups of G, then also $N_1 N_2$ is a nilpotent subgroup. Hence, if G is finite, there exists the unique maximal subgroup among the normal nilpotent subgroups of G. This subgroup is called the Fitting subgroup of G and is denoted by $\mathrm{Fit}(G)$.

2.4 Finite Abelian groups

Theorem 2.4.1 (Fundamental theorem of finite Abelian groups) *Let G be a finite Abelian group, then:*

(1) *G can be represented as the direct product of cyclic groups.*

(2) *Among the various representations of G as the direct product of cyclic groups there is a representation $G = G_1 \times \ldots \times G_r$ such that $|G_{i+1}| \,\|\, |G_i|$ $(i = 1, \ldots, r-1)$.*

(3) *Such a representation in (2) is not necessarily unique, but the orders of the cyclic groups G_i are uniquely determined.*

Proof (1) Since G is obviously nilpotent, G can be represented as the direct product of Sylow subgroups. Hence it suffices to prove that every commutative p-group can be written as the direct product of cyclic groups. Let G be a commutative p-group and let us prove the assertion by induction on $|G|$. Let σ be an element of G with maximal order. Select from $\{K \leq G \mid K \cap \langle \sigma \rangle = 1\}$ a maximal element H. We will prove $G = \langle \sigma \rangle \times H$. Suppose $\langle \sigma \rangle \times H \lneqq G$, then there is a $\tau \in G$ such that $\tau \notin \langle \sigma \rangle \times H$ and $\tau^p \in \langle \sigma \rangle \times H$. Writing $\tau^p = \tau_1 \tau_2$ with $\tau_1 \in \langle \sigma \rangle$ and $\tau_2 \in H$, we have $\langle \tau_1 \rangle \lneqq \langle \sigma \rangle$ since $|\tau| \leq |\sigma|$. Therefore, $\tau_1 = \sigma^{p^e}$ for some $e \geq 1$. Putting $\rho = \tau \sigma^{-p^{e-1}}$, we have $\rho \notin \langle \sigma \rangle \times H$ and $\rho^p \in H$. Hence $\langle \rho, H \rangle \cap \langle \sigma \rangle = 1$. Contradiction. Hence we have $G = \langle \sigma \rangle \times H$. Since $|H| < |G|$, we can apply the induction hypothesis to H, whereby the proof is completed.

(2) This is obvious for commutative p-groups. For an arbitrary commutative group G, write G as the product of its Sylow subgroups $G = H_1 \times \ldots \times H_t$ and write each Sylow group H_i as a direct product of cyclic subgroups $H_i = G_{i,1} \times G_{i,2} \times \ldots$ such that $|G_{i,1}| \geq |G_{i,2}| \geq \ldots$. The groups $G_1 = G_{1,1} \times \ldots \times G_{r,1}, G_2 = G_{2,1} \times \ldots \times G_{2,r}, \ldots$ satisfy the conditions of the assertion.

(3) It is easy to see that it suffices to prove this result for commutative p-groups. Let G be a commutative p-group, and let

$G = G_1 \times \dots \times G_r$ be a representation of G as a product of cyclic subgroups satisfying $|G_{i+1}| \,\big|\, |G_i|$. Let H_i be the subgroup of G generated by all elements of G the order of which is equal to or less than p^i. Putting $|H_i/H_{i-1}| = p^{e_i}$, the number of subgroups among $G_1, \dots, G_r$ the order of which equals p^i is equal to $e_i - e_{i-1}$, which proves the assertion. ■

The set of natural numbers $|G_1|, \dots, |G_r|$ uniquely determined by G is called the type of G. An Abelian group of type $(p, \dots, p)$ for some prime number p is called an elementary Abelian (p-) group. It is easy to verify that the subgroup of a commutative p-group G generated by all elements of order p is an elementary Abelian group. This group is written as $\Omega_1(G)$. Obviously, $\Omega_1(G) \trianglelefteq G$. Let P be an Abelian group consisting of elements of order p. Then P is isomorphic to the additive group underlying the finite field $GF(p) = F_p$ of p elements since all Abelian groups of order p are isomorphic. Choosing an isomorphism, P can be regarded as a finite field and hence an elementary Abelian p-group can be regarded as a vector space over F_p. For example, it is easy to verify that the automorphism group Aut G of an elementary Abelian p-group G corresponds with $GL(G)$, where G is considered as a vector space over F_p.

2.5 p-groups

According to theorem 2.3.4, p-groups are nilpotent. Using this we prove the following theorem.

Theorem 2.5.1 *If P is a p-group, then:*

(1) $Z(P) \neq 1$.

(2) $H \lneqq P \Rightarrow H \lneqq \mathcal{N}_P(H)$.

(3) *If H is a maximal subgroup of P, then $H \ntrianglelefteq P$ and $[P:H] = p$.*

(4) *If P is non-commutative, then $|P:Z(P)| \geq p^2$.*

(5) *If $|P| = p^2$, then P is commutative.*

(6) *If $|P| \geq p^2$, then $|P:D(P)| \geq p^2$.*

Proof (1) Follows from the definition of nilpotency.

(2) Theorem 2.3.4 (2).

(3) Follows from (2).

(4) Since P is not commutative, we have $|P:Z(P)| \geq p$. Suppose $|P:Z(P)| = p$, then $P = \langle a \rangle Z(P)$ for any element $a \in P$ such that $a \notin Z(P)$, i.e. every element of P can be written as $a^r b (r \in \mathbb{Z}, b \in Z(P))$. Since all elements of this form commute with each other, we have reached a contradiction.

(5) Follows from (1) and (4).

(6) Follows from (3) and (5). ■

The intersection of all the maximal subgroups of an arbitrary group G is called the Frattini subgroup of G and denoted by $\Phi(G)$. It follows from the definition that $\Phi(G)$ is a characteristic subgroup of G. The following two theorems are concerned with the properties of Frattini subgroups.

Theorem 2.5.2 *Let S be a subset of the finite group G. If $G = \langle S, \Phi(G) \rangle$, then $G = \langle S \rangle$.*

Proof Suppose $G \neq \langle S \rangle$. Let H be a maximal subgroup of G containing $\langle S \rangle$, then $\Phi(G) \leq H$. Hence

$$G \gneq H = \langle H, \Phi(G) \rangle \geq \langle S, \Phi(G) \rangle = G.$$

Contradiction. ■

Theorem 2.5.3 *Let P be a p-group of order p^n and let $|P:\Phi(P)| = p^d$. Then:*

(1) *$P/\Phi(P)$ is an elementary Abelian group.*

(2) $|\mathrm{Aut}\, P/\Phi(P)| = p^{(1/2)d(d-1)} \prod_{i=1}^{d} (p^i - 1)$.

(3) $|\mathrm{Aut}\, P|$ *is a divisor of* $p^{d(n-d)+(1/2)d(d-1)} \prod_{i=1}^{d} (p^i - 1)$.

Proof (1) If H is a maximal subgroup of P, then $|P:H| = p$, hence P/H is Abelian and $\Phi(P) \geq D(P)$ according to theorem 1.2.7. Therefore, $P/\Phi(P)$ is an elementary Abelian p-group.

(2) Since $P/\Phi(P)$ is an elementary Abelian p-group of order p^d, $P/\Phi(P)$ is written as a direct product of d cyclic groups T_i, $1 \leq i \leq d$,

of order p, and $P/\Phi(P)$ is generated by $\{u_1,\ldots,u_d\}$ where $\langle u_i\rangle = T_i$. Conversely if $P/\Phi(P)$ is generated by some d elements $v_1,\ldots,v_d$, then it is easily seen that $P/\Phi(P)$ is the direct product of d cyclic subgroups $\langle v_1\rangle,\ldots,\langle v_d\rangle$.

Since $P/\Phi(P)$ is generated by d elements, P is also generated by d elements. Conversely, if $P=\langle a_1,\ldots,a_d\rangle$ for certain d elements $a_1,\ldots,a_d\in P$, then $P/\Phi(P)=\langle\bar{a}_1\rangle\times\ldots\times\langle\bar{a}_d\rangle$. (For any element $a\in P$ let $\bar{a}=a\Phi(P)\in P/\Phi(P)$.) We will call an ordered set $(a_1,\ldots,a_d)$ which generates P (or $(\bar{a}_1,\ldots,\bar{a}_d)$ which generates $P/\Phi(P)$) an ordered system of generators of P (or of $P/\Phi(P)$). We denote the collection of all ordered systems of generators of P by Ω_1 and the collection of all ordered systems of generators of $P/\Phi(P)$ by Ω_2.

Let $(\bar{a}_1,\ldots,\bar{a}_d)$ be an element of Ω_2. Since every automorphism of $P/\Phi(P)$ must map a set of generators onto a set of generators, $(\bar{a}_1^\sigma,\ldots,\bar{a}_d^\sigma)$ is also an element of Ω_2 for any $\sigma\in\operatorname{Aut}P/\Phi(P)$. Let $(\bar{a}_1,\ldots,\bar{a}_d)$ and $(\bar{b}_1,\ldots,\bar{b}_d)$ be two elements of Ω_2. Since

$$P/\Phi(P)=\langle\bar{a}_1\rangle\times\ldots\times\langle\bar{a}_d\rangle=\langle\bar{b}_1\rangle\times\ldots\times\langle\bar{b}_d\rangle$$

it is easily seen that a mapping σ from $P/\Phi(P)$ to $P/\Phi(P)$ can be defined by

$$\bar{a}_1^{i_1}\ldots\bar{a}_d^{i_d}\to\bar{b}_1^{i_1}\ldots\bar{b}_d^{i_d}$$

and that σ is an automorphism of $P/\Phi(P)$. Furthermore if $(\bar{a}_1,\ldots,\bar{a}_d)=(\bar{b}_1,\ldots,\bar{b}_d)$, then σ is the identity automorphism. Hence we have

$$|\operatorname{Aut}P/\Phi(P)|=|\Omega_2|.$$

All elements of $P/\Phi(P)$ except the identity element have order p. Choose an element $\bar{a}_1\in P/\Phi(P)\,(\bar{a}_1\neq 1)$, then, choose an element $\bar{a}_2$ such that $\bar{a}_2\notin\langle\bar{a}_1\rangle$. Since $\langle\bar{a}_1\rangle\cap\langle\bar{a}_2\rangle=1$, we have $\langle\bar{a}_1,\bar{a}_2\rangle=\langle\bar{a}_1\rangle\times\langle\bar{a}_2\rangle$. Next, choosing an element $\bar{a}_3$ such that $\bar{a}_3\notin\langle\bar{a}_1,\bar{a}_2\rangle$, we have $\langle\bar{a}_1,\bar{a}_2,\bar{a}_3\rangle=\langle\bar{a}_1\rangle\times\langle\bar{a}_2\rangle\times\langle\bar{a}_3\rangle$. Repeating this procedure we arrive at

$$P/\Phi(P)=\langle\bar{a}_1\rangle\times\ldots\times\langle\bar{a}_d\rangle$$

for certain elements $\bar{a}_1,\ldots,\bar{a}_d\in P/\Phi(P)$. Since there are p^d-1 choices for $\bar{a}_1$, p^d-p choices for $\bar{a}_2$, p^d-p^2 choices for $\bar{a}_3,\ldots,$

$p^d - p^{d-1}$ choices for $\bar{a}_d$, we conclude that

$$|\text{Aut } P/\Phi(P)| = |\Omega_2| = \prod_{i=0}^{d-1} (p^d - p^i) = p^{(1/2)d(d-1)} \prod_{i=1}^{d} (p^i - 1).$$

(3) Since for any two elements $a, b \in P$ we have $\bar{a} = \bar{b} \Leftrightarrow ab^{-1} \in \Phi(P)$, it follows that $|\Omega_1| = p^{d(n-d)}|\Omega_2|$, from which the desired result follows easily. ■

Since an elementary Abelian p-group G is considered as a vector space over F_p as we noticed at the end of §2.4, subgroups of G become subspaces. Therefore, if P is an elementary Abelian group of order p^n, the number of its subgroups of order p^e is

$$N_e(n, p) = \prod_{i=1}^{e} \frac{p^{n-i+1} - 1}{p^i - 1} = \frac{\prod_{i=1}^{n} (p^i - 1)}{\prod_{i=1}^{e} (p^i - 1) \prod_{i=1}^{n-e} (p^i - 1)}$$

(according to §1.5).

For example, the number of subgroups of order p of an Abelian group of (p, p) type equals $p + 1$.

2.6 Groups with operators

Let G and H be groups. If G operates on the set H such that all permutations of H defined by this operation are automorphisms of the group H then the group G is said to operate on the group H. We also say, that H is a G-group. In other words, G operates on H if the permutation representation $G \to S^H$ is a homomorphism from G into Aut $H \leq S^H$. The kernel of the permutation representation (G, H) is called the kernel of the operation of G on H. If the kernel is 1, then H is called a faithful G-group. In this case, we also say that G operates faithfully on H. If the kernel is G, then H is called a trivial G-group or G is said to operate trivially on H. If K is a subgroup of H, such that K is invariant under the operation of G, K is called a G-invariant subgroup or simply a G-subgroup of H. Obviously, H and 1 are G-subgroups of H. If these are the only G-subgroups of H, H is called an irreducible (or minimal) G-group.

For a G-group H, the subgroup $\{u \in H \mid u^\sigma = u, \forall \sigma \in G\}$ is called the G-centre of H and is denoted by $\mathscr{C}_H(G)$. The G-centre of H is a G-subgroup of H and G acts trivially on it.

If H is a normal subgroup of G, G operates on H by conjugation and H becomes a G-group. In this case H is called a G-group by conjugation.

For the rest of this section, let G be a finite group and let H be a commutative G-group. Since H is commutative $\mathrm{Hom}(H, H)$ carries the structure of a ring (see §1.3). For $\sigma \in G$ let $\tilde{\sigma}$ represent the automorphism of H defined by σ and let $\alpha(G) \in \mathrm{Hom}(H, H)$ be defined by

$$\alpha(G) = \sum_{\sigma \in G} \tilde{\sigma}.$$

Theorem 2.6.1 *If G is a finite group, H a commutative G-group and if $G_0, G_1, \ldots, G_t$ are $t+1$ subgroups of G such that: $\bigcup_{i=0}^{t} G_i = G$ and $G_i \cap G_j = 1$ $(i \neq j)$, then at least one of the following statements is true:*

(1) $\mathscr{C}_H(G_i) \neq 1$ *for some i.*

(2) $v^t = 1$ *for all elements $v \in H$.*

Proof Since

$$\alpha(G) = \left(\sum_{i=0}^{t} \alpha(G_i) \right) - t\tilde{1}$$

(where $\tilde{1}$ represents the identity map from H to itself), we have

$$v^{\alpha(G)} = \prod_{i=0}^{t} v^{\alpha(G_i)} v^{-t}.$$

For an arbitrary i we have

$$v^{\alpha(G_i)\tilde{\sigma}} = v^{\alpha(G_i)}, \quad v^{\alpha(G)\tilde{\sigma}} = v^{\alpha(G)}, \quad \forall \sigma \in G_i.$$

Therefore if it is not the case (1), then

$$v^{\alpha(G_i)} = v^{\alpha(G)} = 1 \quad (\forall i),$$

hence $v^t = 1$. ■

Corollary 2.6.2 *Let H be a faithful, irreducible and commutative G-group. If $(|G|,|H|)=1$ then $Z(G)$ is cyclic.*

Proof Suppose that $Z(G)$ is not cyclic. Then $Z(G)$ contains an Abelian group K of (p,p)-type. K contains $p+1$ subgroups of order p (cf. end of §2.5), say $K_0,\ldots,K_p$. Since $K=\bigcup_{i=0}^{p}K_i$ and $K_i\cap K_j=1$ $(i\neq j)$, we can apply the previous theorem. From $(|H|,p)=1$, we conclude $\mathscr{C}_H(K_i)\neq 1$ for some i. For $\sigma\in G$ we have $\mathscr{C}_H(K_i)^\sigma=\mathscr{C}_{H^\sigma}(K_i^\sigma)=\mathscr{C}_H(K_i)$, hence $\mathscr{C}_H(K_i)=H$ since H is irreducible. Therefore K_i is contained in the kernel of (G,H). Contradiction, because H is faithful. ∎

Theorem 2.6.3 *Let H be a finite, commutative G-group such that $(|G|,|H|)=1$. Then we have:*

(1) *$H_1=\operatorname{Ker}\alpha(G)$ and $H_2=\operatorname{Im}\alpha(G)$ are G-subgroups of H and H is the direct product of H_1 and H_2. Moreover, $H_2=\mathscr{C}_H(G)$.*

(2) *If H is a non-trivial G-group, then $H_1\neq 1$ and G operates non-trivially on every G-subgroup of H_1 that is not equal to 1.*

Proof (1) It is obvious that H_1 and H_2 are G-subgroups. For an arbitrary $v\in H$, we have

$$v^{\alpha(G)\tilde{\sigma}}=v^{\alpha(G)},\quad \forall\sigma\in G,$$

hence $H_2\leq\mathscr{C}_H(G)$. Now, take $u\in\mathscr{C}_H(G)\cap H_1$, then: $u^{\alpha(G)}=u^{|G|}$ (since $u\in\mathscr{C}_H(G)$) and $u^{\alpha(G)}=1$ (since $u\in H_1$). Therefore $u^{|G|}=1$. From $(|G|,|H|)=1$ we conclude $u=1$, i.e. $\mathscr{C}_H(G)\cap H_1=1$. According to the homomorphism theorem $|H|=|H_1||H_2|$, hence $H_2=\mathscr{C}_H(G)$ and $H=H_1\times H_2$.

(2) According to (1) we have $\mathscr{C}_{H_1}(G)=1$, which proves (2). ∎

Corollary 2.6.4 *Let G be a p'-group and let H be a commutative p-group. If G operates non-trivially on H, G operates non-trivially on $\Omega_1(H)=\{u\in H\mid u^p=1\}$.*

Proof Obviously, $\Omega_1(H)$ is a G-subgroup of H. According to (2) of the previous theorem, $H_1=\operatorname{Ker}\alpha(G)\neq 1$ and the subgroup $\Omega_1(H_1)\neq 1$ of H_1 is a non-trivial G-subgroup. Therefore, $\Omega_1(H)(\geq\Omega_1(H_1))$ is also a non-trivial G-subgroup. ∎

2.7 Group extensions and the theorem of Schur–Zassenhaus

Let G, A and B be groups such that G contains (an isomorphic image of) A as a normal subgroup and such that G/A is isomorphic to B. Then G is called an extension of A by B. (Strictly speaking, an extension of A by B obtained from the given isomorphism $G/A \simeq B$).

Now, let A be an Abelian group. For $b \in B$ let $\sigma_b \in G$ be such that $\sigma_b A$ corresponds with b under the isomorphism $G/A \simeq B$. The element σ_b induces an automorphism of A by conjugation and since A is commutative, this automorphism is only dependent on b and independent of the choice of σ_b. (Therefore, instead of u^{σ_b} we can write simply u^b for $u \in A$.) So, a map from B into Aut A is given and this map is easily seen to be a homomorphism, i.e. A becomes a B-group. Now, fix a representative system $\{\sigma_b | b \in B\}$ of G/A. For $b_1, b_2 \in B$ we have

$$\sigma_{b_1}\sigma_{b_2}A = \sigma_{b_1}A \cdot \sigma_{b_2}A = \sigma_{b_1 b_2}A,$$

hence

$$\sigma_{b_1}\sigma_{b_2} = \sigma_{b_1 b_2} f(b_1, b_2) \text{ for some } f(b_1, b_2) \in A.$$

If b_3 is a third element from B we have

$$(\sigma_{b_1}\sigma_{b_2})\sigma_{b_3} = \sigma_{b_1}(\sigma_{b_2}\sigma_{b_3}),$$

hence

$$f(b_1 b_2, b_3) f(b_1, b_2)^{b_3} = f(b_1, b_2 b_3) f(b_2, b_3) \quad \forall b_1, b_2, b_3 \in B. \tag{2.7.1}$$

Generally, if A is a B-group and if $f: B \times B \to A$ satisfies condition (2.7.1), then f is called a factor set of the B-group A or a 2-dimensional cocycle. So, if G is an extension of the Abelian group A by B, then A is a B-group and a representative system $\{\sigma_b | b \in B\}$ of G/A determines a factor set. This factor set depends on the choice of the representative system in the following way. Let $\{\tau_b | b \in B\}$ be another representative system of G/A, determining the factor set g. Then $\tau_b A = \sigma_b A$, hence $\tau_b = \sigma_b a_b$ for some $a_b \in A$. So

$$\begin{aligned}\tau_{b_1}\tau_{b_2} &= \sigma_{b_1}a_{b_1}\sigma_{b_2}a_{b_2} = \sigma_{b_1 b_2} f(b_1, b_2) a_{b_1}^{b_2} a_{b_2} \\ &= \tau_{b_1 b_2} f(b_1, b_2) a_{b_1 b_2}^{-1} a_{b_1}^{b_2} a_{b_2}.\end{aligned}$$

Therefore

$$g(b_1, b_2) = f(b_1, b_2)a_{b_1b_2}^{-1}a_{b_1}^{b_2}a_{b_2}.$$

Let A be a commutative B-group, and let $F(B, A)$ represent the collection of all factor sets of A. If $f_1, f_2 \in F(B, A)$ we define $f_1 f_2 \in F(B, A)$ by

$$(f_1 f_2)(b_1, b_2) = f_1(b_1, b_2)f_2(b_1, b_2)$$

making $F(B, A)$ a commutative group. (The identity element of this group is the map $f_0 : B \times B \to A$ defined by $f_0(b_1, b_2) = 1, \forall b_1, b_2 \in B$.) For any map $b \to a_b$ from B into A the map $f : B \times B \to A$ defined by

$$f(b_1, b_2) = a_{b_1b_2}^{-1}a_{b_1}^{b_2}a_{b_2}, \quad \forall b_1, b_2 \in B \tag{2.7.2}$$

is easily seen to be an element of $F(B, A)$. These elements of $F(B, A)$ are called 2-dimensional coboundaries and the collection of all 2-dimensional coboundaries is denoted by $F_0(B, A)$. The coboundary $F_0(B, A)$ is a subgroup of $F(B, A)$, and the quotient group $F(B, A)/F_0(B, A)$ is called the 2-dimensional (or second) cohomology group of A with coefficients in B and denoted by $H^2(B, A)$. So, the discussion above shows that an extension G of A by B unambiguously determines:

(1) the structure of a B-group on A, and

(2) an element of the second cohomology group $H^2(B, A)$.

Conversely, let us start with a commutative B-group A and an element from $H^2(B, A)$. Let $f \in F(B, A)$ represent this element. We define the set G by $G = B \times A$ (as sets) and we define a multiplication on G by

$$(b_1, a_1)(b_2, a_2) = (b_1b_2, f(b_1, b_2)a_1^{b_2}a_2).$$

We may easily show that G together with this multiplication is a group. In fact

$$\begin{aligned}((b_1, a_1)(b_2, a_2))(b_3, a_3) &= (b_1b_2, f(b_1, b_2)a_1^{b_2}a_2)(b_3, a_3)\\ &= (b_1b_2b_3, f(b_1b_2, b_3)f(b_1, b_2)^{b_3}a_1^{b_2b_3}a_2^{b_3}a_3)\\ &= (b_1b_2b_3, f(b_1, b_2b_3)f(b_2, b_3)a_1^{b_2b_3}a_2^{b_3}a_3)\\ &= (b_1, a_1)(b_2b_3, f(b_2, b_3)a_2^{b_3}a_3)\\ &= (b_1, a_1)((b_2, a_2)(b_3, a_3))\end{aligned}$$

gives the associativity.

$$(1, f(1,1)^{-1})(b, a) = (b, f(1,b)f(1,1)^{-b}a) = (b, a)$$

since $f(1,1)^b = f(1,b)$ by (2.7.1), i.e. $(1, f(1,1)^{-1})$ is the identity element of G. The inverse of (b, a) is easily seen to be

$$(b^{-1}, f(1,1)^{-b^{-1}} f(b^{-1}, b)^{-b^{-1}} a^{-b}).$$

The map from G onto B defined by $(b, a) \to b$ is a homomorphism with kernel $\tilde{A} = \{(1, a) | a \in A\}$, and hence $G/\tilde{A} \simeq B$. The kernel $\tilde{A}$ is isomorphic to A via the isomorphism $(1, a) \to f(1,1)a (a \in A)$. Consequently, G is an extension of A by B. It is easy to check that the B-group structure on A and the element of $H^2(B, A)$ determined by this extension are identical with the B-group structure on A and the element of $H^2(B, A)$, which we started with. We have proved the following theorem:

Theorem 2.7.1 *Let A be a commutative group and let B be a group. An extension of A by B unambiguously determines a B-group structure on A and an element of $H^2(B, A)$. Conversely, a B-group structure on A and an element of $H^2(B, A)$ unambiguously determine an extension of A by B. These correspondences are each other's inverse.*

If G is an extension of A by B, such that there exists a subgroup $\tilde{B}$ of G satisfying $G = \tilde{B}A$ and $\tilde{B} \cap A = 1$ (i.e. G is the semi-direct product of $\tilde{B}$ and A), then we say that the extension splits. The extension splits if and only if it is possible to find a subgroup of G that is a representative system of G/A and isomorphic to B. Hence (if A is commutative) the element of $H^2(B, A)$ that corresponds with this extension is the identity element. Especially, if A is a commutative B-group, such that $F_0(B, A) = F(B, A)$, then all the extensions corresponding with the given B-group structure of A split.

The following theorem gives a sufficient condition for $F(B, A) = F_0(B, A)$ to be satisfied:

Theorem 2.7.2 *If A is a finite Abelian group and if B is a finite* *group, such that* $(|A|, |B|) = 1$, *then* $H^2(B, A) = 0$, *i.e.* $F(B, A) = F_0(B, A)$ *and therefore all extensions of A by B split.*

Proof For $f \in F(B, A)$

$$f^{|A|}(b_1, b_2) = (f(b_1, b_2))^{|A|} = 1 \quad \forall b_1, b_2 \in B,$$

since $f(b_1, b_2) \in A$, and so $f^{|A|}$ is the identity element of $F(B, A)$. Hence the order of each element of $H^2(B, A)$ is a divisor of $|A|$. On the other hand, using (2.7.1) we get

$$\prod_{b_1 \in B} f(b_1 b_2, b_3) f(b_1, b_2)^{b_3} = \prod_{b_1 \in B} f(b_1, b_2 b_3) f(b_2, b_3) \quad (b_1, b_2, b_3 \in B).$$

(Since A is commutative, the order of the factors is irrelevant.)

Putting $\alpha(b) = \prod_{c \in B} f(c, b)$, we get

$$\alpha(b_3)\alpha(b_2)^{b_3} = \alpha(b_2 b_3) f(b_2, b_3)^n, \text{ where } n = |B|$$

or:

$$f(b_2, b_3)^n = \alpha(b_2 b_3)^{-1} \alpha(b_2)^{b_3} \alpha(b_3).$$

Therefore, $f(b_2, b_3)^n \in F_0(B, A)$, from which we conclude that the orders of all elements of $H^2(B, A)$ are divisors of $|B|$. Combining this with the result of the first half of the proof we arrive at the desired result. ■

The following theorem is an immediate consequence of the previous theorem.

Theorem 2.7.3 *Let G be a finite group and let A be a commutative, normal subgroup of G. If* $(|G/A|, |A|) = 1$, *then G contains a subgroup B, that is isomorphic with G/A and G is the semi-direct product of A and B.* ■

The conclusion of this theorem is also true even in the case that A is not commutative, as shown by the following theorem.

Theorem 2.7.4 (Schur–Zassenhaus) *Let G be a finite group and let H be a normal subgroup of G. If* $(|G/H|, |H|) = 1$, *then G contains* *a subgroup K, that is isomorphic with G/H and G is the semi-direct product of K and H.*

Proof The proof is by induction on $|G|$. If there exists a normal subgroup N of G such that $1 \lneq N \lneq H$, then we can apply the induction hypothesis to $\bar{G} = G/N$ and conclude that there exists a subgroup T of G containing N such that: $\bar{G} = \bar{T}\bar{H}$ and $\bar{T} \cap \bar{H} = 1$. Since $T \lneq G$ and $(|T/N|, |N|) = 1$, we can apply the induction hypothesis to conclude that there exists a subgroup K of T such that $T = KN$ and $K \cap N = 1$. Therefore $G = TH = KNH = KH$ and $K \cap H = K \cap T \cap H = K \cap N = 1$, and the proof for this case is finished.

So, we may assume that H is a minimal normal subgroup of G. If H is a p-group, then H is Abelian and we can apply theorem 2.7.3. Therefore, we may assume that H is not a p-group. Let P be a Sylow p-subgroup of H, where p is a prime factor of $|H|$. According to theorem 2.1.3 we have $G = H\mathcal{N}_G(P)$, hence $G/H \simeq \mathcal{N}_G(P)/H \cap \mathcal{N}_G(P)$. Since $G \gneq \mathcal{N}_G(P)$ we can apply the induction hypothesis, giving the existence of a subgroup K such that $\mathcal{N}_G(P) = K(H \cap \mathcal{N}_G(P))$ and $K \cap (H \cap \mathcal{N}_G(P)) = 1$. Therefore

$$G = H\mathcal{N}_G(P) = HK(H \cap \mathcal{N}_G(P)) = HK$$

and

$$K \cap H = K \cap \mathcal{N}_G(P) \cap H = 1. \qquad \blacksquare$$

The following theorem is an application of the previous theorem:

Theorem 2.7.5 (Hall–Higman) *Let G be a p-solvable group such that $O_{p'}(G) = 1$, then $\mathscr{C}_G(O_p(G)) \subseteq O_p(G)$.*

Proof. It follows from the assumptions, that $O_p(G) \neq 1$. Putting $H = O_p(G)\mathscr{C}_G(O_p(G))$, we conclude from $H \lhd G$ that $O_p(G) = O_p(H)$. Suppose $H \gneq O_p(G)$, then $H \geq O_{p,p'}(H) \gneq O_p(H)$ because H is p-solvable. Hence, by theorem 2.7.4 there exists a p'-subgroup $K \neq 1$ satisfying $O_{pp'}(H) = O_p(H)K$ and $K \cap O_p(H) = 1$. Let $z \in K$ $(z \neq 1)$, then there exist elements $z_1 \in O_p(H)$ and $z_2 \in \mathscr{C}_G(O_p(G))$ such that $z = z_1 z_2$. Therefore the conjugacy operation defined by

z on $O_p(H)$ is the same as that defined by z_1 on $O_p(H)$. Since the orders of z and z_1 are relatively prime, the operation defined by z on $O_p(H)$ is the identity operation, therefore $z \in \mathscr{C}_G(O_p(G))$. Hence $O_{p,p'}(H) = O_p(H) \times K$. From $G \trianglerighteq H \trianglerighteq O_{p,p'}(H) \trianglerighteq K$ we conclude $G \trianglerighteq K$. Contradiction to the assumption $O_{p'}(G) = 1$. ■

2.8 Normal π-complements

If there is a normal Hall π-subgroup of a group G, then by the theorem of Schur–Zassenhaus (theorem 2.7.4) there exists a Hall π'-subgroup of G. In this section and the next, we consider conditions which assert the existence of normal Hall π'-groups (i.e. normal π-complements) from the existence of a Hall π-group.

Let H be a subgroup of G, then $G = \sum_{i=1}^{n} Hx_i$ for certain elements $x_1, \dots, x_n \in G$. Since $\{Hx_1, \dots, Hx_n\}$ are all left cosets of H, for $\sigma \in G$, $Hx_i\sigma$ is a member of $\{Hx_1, \dots, Hx_n\}$, and so we denote $Hx_i\sigma$ by $Hx_{i\sigma}$ where $x_{i\sigma} \in \{x_1, \dots, x_n\}$. Then there exists a unique $h_i \in H$ $(i = 1, \dots, n)$ such that $x_i\sigma = h_i x_{i\sigma}$. The product $h_1 h_2 \dots h_n \in H$ is dependent on the order of the factors. However, if we choose a normal subgroup H_0 of H such that H/H_0 is Abelian, the product $\prod_{i=1}^{n} h_i$ is unambiguously (i.e. independent from the order of the factors) defined mod H_0. Representing hH_0 by $\bar{h}\,(h \in H)$ we get a map $V: G \to H/H_0$ defined by

$$V(\sigma) = \prod_{i=1}^{n} \bar{h}_i.$$

This map is called the transfer of G into H/H_0 and is denoted by $V_{G \to H/H_0}$. In particular, if $H_0 = [H, H]$ we write $V_{G \to H}$ rather than $V_{G \to H/[H,H]}$ and we call $V_{G \to H}$ the transfer of G into H.

Theorem 2.8.1 (1) *The transfer $V_{G \to H/H_0}$ is a homomorphism.*

(2) *The transfer is independent from the choice of the representative system $x_1, \dots, x_n$ of $H \backslash G$.*

Proof (1) For $\sigma, \tau \in G$, let

$$x_i\sigma = h_i x_{i\sigma} \text{ and } x_i\tau = k_i x_{i\tau} \quad h_i, k_i \in H.$$

Noting that $\{x_{1\sigma}, \ldots, x_{n\sigma}\} = \{x_1, \ldots, x_n\}$ we conclude from $x_i(\sigma\tau) = h_i k_{i\sigma} x_{i\sigma\tau}$ that

$$V_{G\to H/H_0}(\sigma\tau) = \prod_i \overline{h_i k_{i\sigma}} = \prod_i \bar{h}_i \prod_i k_{i\sigma}$$
$$= \prod_i \bar{h}_i \prod_i \bar{k}_i = V_{G\to H/H_0}(\sigma) V_{G\to H/H_0}(\tau).$$

(2) Choose $y_1, \ldots, y_n \in G$ such that $Hx_i = Hy_i$ $(i = 1, \ldots, n)$. Then there exist $t_1, \ldots, t_n \in H$ such that $y_i = t_i x_i$ $(i = 1, \ldots, n)$. From $y_i\sigma = t_i h_i x_{i\sigma} = t_i h_i t_{i\sigma}^{-1} y_{i\sigma}$ and

$$\prod_i \overline{t_i h_i t_{i\sigma}^{-1}} = \prod_i \bar{h}_i,$$

we conclude that (2) is true. ■

For $H \leq G$ we define the focal subgroup $\mathrm{Foc}_G(H)$ of H in G by

$$\mathrm{Foc}_G(H) = \langle [\sigma, \tau] \mid \sigma \in H, \tau \in G, [\sigma, \tau] \in H \rangle.$$

Since $[H, H] \leq \mathrm{Foc}_G(H) \trianglelefteq H$ the quotient group $H/\mathrm{Foc}_G(H)$ is Abelian. We define groups $H_0, H_1, \ldots$ inductively by

$$H_0 = H, \quad H_1 = \mathrm{Foc}_G(H), \ldots, \quad H_{i+1} = \mathrm{Foc}_G(H_i), \ldots.$$

If $H_n = 1$ for some n, then H is called hyperfocal. It is almost obvious that subgroups of a hyperfocal subgroup H of G are hyperfocal subgroups of G and that for an arbitrary subgroup G_0 of G, the intersection $H \cap G_0$ is a hyperfocal subgroup of G_0.

Lemma 2.8.2 *Let H be a nilpotent subgroup of G. If every two elements of H that are conjugate in G are also conjugate in H, then H is a hyperfocal subgroup of G.*

Proof Let the lower central series of H be denoted by $H = L_0, L_1, \ldots, L_i, \ldots$. It suffices to prove that $H_i = L_i$ for all i. Let us prove this by induction on n. Suppose that $H_i = L_i$. From

$$L_{i+1} = \langle [\sigma, \tau] \mid \sigma \in L_i, \tau \in H \rangle$$
$$H_{i+1} = \langle [\sigma, \tau] \mid \sigma \in H_i, \tau \in G, [\sigma, \tau] \in H_i \rangle$$

we conclude: $L_{i+1} \leq H_{i+1}$. On the other hand, if $[\sigma, \tau] \in H_{i+1}$ with

$\sigma \in H_i$ and $\tau \in G$, then $\sigma^\tau \in H_i \leq H$ since $[\sigma, \tau] = \sigma^{-1}\sigma^\tau$. By the assumptions of the theorem, there exists a $\tau' \in H$ such that $\sigma^\tau = \sigma^{\tau'}$. Therefore $[\sigma, \tau] = [\sigma, \tau'] \in L_{i+1}$. Hence $H_{i+1} \leq L_{i+1}$. ∎

Theorem 2.8.3 *Let H be a hyperfocal Hall π-subgroup of G, then G contains a normal Hall π'-subgroup K, i.e. a subgroup K satisfying $K \trianglelefteq G$, $G = HK$ and $H \cap K = 1$.*

Proof The proof is by induction on $|G|$. We may assume that $|H| \neq 1$. Since H is a hyperfocal subgroup of G, we have $\mathrm{Foc}_G(H) \lneqq H$. Now we can prove that if $\sigma \in H$ but $\sigma \notin \mathrm{Foc}_G(H)$, then $\sigma \notin \mathrm{Ker}\, V_{G \to H/\mathrm{Foc}_G(H)}$. (In fact, consider $\langle \sigma \rangle$ operating on $H \backslash G$ and denote the orbits of $H \backslash G$ by $\Delta_1, \ldots, \Delta_r$. Pick one element Ha_i from each Δ_i. Putting $|\Delta_i| = n_i$, we have:

$$\Delta_i = \{Ha_i, Ha_i\sigma, \ldots, Ha_i\sigma^{n_i - 1}\} \text{ and } Ha_i\sigma^{n_i} = Ha_i.$$

Taking $\{a_i\sigma^j \mid i = 1, \ldots, r, j = 1, \ldots, n_i\}$ as a representative system of $H \backslash G$, we find:

$$V_{G \to H/\mathrm{Foc}_G(H)}(\sigma) = \prod_{i=1}^{r} \overline{a_i\sigma^{n_i}a_i^{-1}}.$$

Since $a_i\sigma^{n_i}a_i^{-1} \in H$, the commutator $[a_i^{-1}, \sigma^{-n_i}] \in H$. Hence $[a_i^{-1}, \sigma^{-n_i}] \in \mathrm{Foc}_G(H)$. Therefore: $\overline{a_i\sigma^{n_i}a_i^{-1}} = \overline{\sigma^{n_i}}$ and $V_{G \to H/\mathrm{Foc}_G(H)}(\sigma) = \overline{\sigma^n}$, where $n = |G : H|$. Now suppose $V_{G \to H/\mathrm{Foc}_G(H)}(\sigma) = 1$, then $\sigma^n \in \mathrm{Foc}_G(H)$ and therefore $\sigma \in \mathrm{Foc}_G(H)$, since $(n, |H|) = 1$. Contradiction.) Putting $G_0 = \mathrm{Ker}\, V_{G \to H/\mathrm{Foc}_G(H)}$, we have $G_0 \ntrianglelefteq G$. Since $H_0 = G_0 \cap H$ is a hyperfocal Hall π-subgroup of G_0, G_0 contains a normal Hall π'-subgroup K by the induction hypothesis. Since $K \lneqq G_0$ and $|G : G_0|$ is a π-number, the subgroup K is a normal Hall π'-subgroup of G. ∎

Corollary 2.8.4 *If H is a nilpotent Hall π-subgroup of G and if every two elements of H that are conjugate in G are also conjugate in H, then G contains a normal Hall π'-subgroup.*

Proof This follows immediately from theorems 2.8.2 and 2.8.3. ∎

Corollary 2.8.5 (Burnside) *Let P be a Sylow p-subgroup of G, such that $\mathscr{C}_G(P) = \mathscr{N}_G(P)$, then G contains a normal p-complement.*

Proof It follows from the assumptions, that P is Abelian. Let $\sigma, \tau \in P$ that are conjugate in G, say $\sigma = \tau^\rho$ $(\rho \in G)$. Since $\sigma \in P \cap P^\rho$, P and P^σ are subgroups of $\mathscr{C}_G(\sigma)$. According to Sylow's theorem there is a $\rho_1 \in \mathscr{C}_G(\sigma)$ such that $P^{\rho_1} = P^\rho$. Therefore, $\rho_1 \rho^{-1} \in \mathscr{N}_G(P) = \mathscr{C}_G(P)$, thus $\rho^{-1} \in \mathscr{C}_G(\sigma)$. Hence $\tau = \sigma^{\rho^{-1}} = \sigma$, and τ and σ are certainly conjugate in H, so that we can apply the previous corollary. ■

2.9 Normal p-complements

Sometimes it is possible to draw important conclusions regarding the structure of a group from the existence of normal p-complements. For example, if a group G has normal p-complements for all primes p, then all the Sylow subgroups of G are normal and G is nilpotent. The main objective of this section is to prove Thompson's theorem (theorem 2.9.3) concerning the existence of normal p-complements for odd primes p. We start with a lemma.

Lemma 2.9.1 *Assume that G has a normal p-complement N, then:*

(1) *If H is a subgroup of G, then $H \cap N$ is a normal p-complement in H. If K is a normal subgroup of G, then KN/K is a normal p-complement in G/K.*

(2) *For every p-subgroup X of G, the quotient group $\mathscr{N}_G(X)/\mathscr{C}_G(X)$ is a p-group.*

(3) *If P_1 and P_2 are two Sylow p-subgroups of G, then there exists an element $\sigma \in \mathscr{C}_G(P_1 \cap P_2)$ such that $P_1^\sigma = P_2$.*

Proof (1) Obvious.

(2) $\mathscr{N}_G(X)$ has a normal p-complement, say K. From $K \trianglelefteq \mathscr{N}_G(X)$ and $X \triangleleft \mathscr{N}_G(X)$ and $(|K|, |X|) = 1$, we conclude: $[X, K] = 1$. Therefore, $K \subseteq \mathscr{C}_G(X)$. Hence $\mathscr{N}_G(X)/\mathscr{C}_G(X)$ is a p-group.

(3) According to Sylow's theorem $P_1^\rho = P_2$ for some $\rho \in G$. Since $G = P_1 N$, ρ can be written as $\rho = \tau\sigma$ $(\tau \in P_1, \sigma \in N)$. Now, $P_1^\sigma = P_2$.

For $\xi \in P_1 \cap P_2$ we have

$$\sigma^{-1}\xi\sigma\xi^{-1} = (\sigma^{-1}\xi\sigma)\xi^{-1} = \sigma^{-1}(\xi\sigma\xi^{-1}) \in P_2 \cap N = 1,$$

i.e. $\sigma \in \mathscr{C}_G(P_1 \cap P_2)$. ■

The converse of (2) of this lemma is also true, as is shown by the following theorem.

Theorem 2.9.2 (Frobenius) *If for every p-subgroup X of G the quotient group $\mathscr{N}_G(X)/\mathscr{C}_G(X)$ is a p-group, then G has a normal p-complement.*

Proof Suppose that the theorem is not true and let G be a group of minimal order for which the theorem is false.

(*a*) Since every subgroup of G satisfies the conditions of the theorem, every proper subgroup of G contains a normal p-complement.

(*b*) The Sylow p-subgroups of G are not normal. (Proof: Suppose P is a normal Sylow p-subgroup of G, then G contains a p-complement H by the theorem of Schur–Zassenhaus. Since $G/\mathscr{C}_G(P)$ is a p-group, we have $H \leq \mathscr{C}_G(P)$. Hence $H \lhd P \times H = G$, contradiction.)

(*c*) $O_p(G) = 1$. (Proof: p-subgroups $\bar{X}$ of $\bar{G} = G/O_p(G)$ can be written as $\bar{X} = X/O_p(G)$ where X is a p-subgroup of G containing $O_p(G)$. From

$$\mathscr{N}_{\bar{G}}(\bar{X}) = \mathscr{N}_G(X)/O_p(G)$$

and

$$\mathscr{C}_{\bar{G}}(\bar{X}) \geq \mathscr{C}_G(X)O_p(G)/O_p(G)$$

we conclude that $\bar{G}$ satisfies the same conditions as G. Therefore, if $O_p(G) \neq 1$, $\bar{G}$ has a normal p-complement $\bar{M} = M/O_p(G)$, where M is a subgroup of G such that $O_p(G) \leq M \trianglelefteq G$. By (*b*) $\bar{G} \neq \bar{M}$, hence $M \underset{\neq}{\lhd} G$ and M has a normal p-complement N. Since $N \lessdot M$, we have $N \lhd G$ and G/N is a p-group. Therefore N is a normal p-complement of G. Contradiction.)

(*d*) If P, P_1 are two Sylow p-subgroups of G, there exists an element $\sigma \in \mathscr{C}_G(P \cap P_1)$ such that $P_1^\sigma = P$. (Proof: by induction on

$t=|P:P\cap P_1|=|P_1:P\cap P_1|$. For $t=1$ this is obvious. If $P\cap P_1=1$, then the assertion follows since $G=\mathscr{C}_G(P\cap P_1)$ in this case. So we may assume: $1\lneqq P\cap P_1\lneqq P$. Putting $H=\mathscr{N}_G(P\cap P_1)$ we have $G\neq H$ by (*c*), hence H has a normal p-complement. Let R denote a Sylow p-subgroup of H containing $P\cap H$ and let S denote a Sylow p-subgroup of G containing R. Further, let R_1 denote a Sylow p-subgroup of H containing $P_1\cap H$ and let S_1 denote a Sylow p-subgroup of G containing R_1. From $P\cap P_1\lneqq P$ we get $P\cap P_1\lneqq P\cap H$ by theorem 2.5.1, and so we get $P\cap P_1\lneqq P\cap S$. Hence there exists an element $\sigma_1\in\mathscr{C}_G(P\cap S)$ such that $P^{\sigma_1}=S$ by the induction hypothesis. In exactly the same way there exists an element $\sigma_2\in\mathscr{C}_G(P_1\cap S_1)$ such that $P_1=S_1^{\sigma_2}$. Since H has a normal p-complement, there exists a $\sigma_3\in\mathscr{C}_H(R\cap R_1)$ such that $R^{\sigma_3}=R_1$ by lemma 2.9.11(3). Since $S^{\sigma_3}\cap S_1\geq R_1\gneqq P\cap P_1$, there exists an element $\sigma_4\in\mathscr{C}_G(S^{\sigma_3}\cap S_1)$ Such that $S^{\sigma_3\sigma_4}=S_1$ by the induction hypothesis. Since $\sigma_1,\sigma_2,\sigma_3,\sigma_4\in\mathscr{C}_G(P\cap P_1)$, we have $\sigma=\sigma_1\sigma_3\sigma_4\sigma_2\in\mathscr{C}_G(P\cap P_1)$ and $P^\sigma=P_1$.)

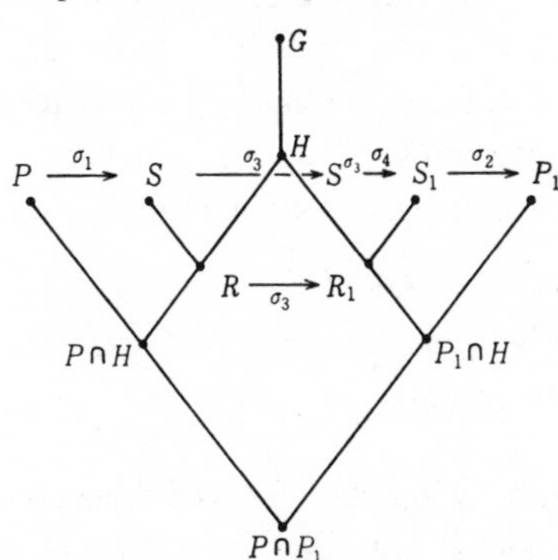

(*e*) If P is a Sylow p-subgroup of G, then every two elements of P that are conjugate in G are conjugate in P. (Proof: Let $\sigma_1,\sigma_2\in P$ be conjugate in G, say $\sigma_1^\tau=\sigma_2$ $(\tau\in G)$. By (*d*) there is a $\tau_1\in\mathscr{C}_G(P\cap P^\tau)$ such that $P^{\tau\tau_1}=P$ and $\sigma_1^{\tau\tau_1}=\sigma_2$. From $\mathscr{N}_G(P)=\mathscr{C}_G(P)P$ we conclude that there exist elements $\rho\in\mathscr{C}_G(P)$ and $\rho_1\in P$ such that $\tau\tau_1=\rho\rho_1$. Therefore,

$$\sigma_1^{\rho_1}=\sigma_1^{\rho\rho_1}=\sigma_1^{\tau\tau_1}=\sigma_2.)$$

Hence, G contains a normal p-complement by corollary 2.8.4. Contradiction. ■

Let A be an Abelian group, and let $m(A)$ denote the minimal number of generators of A. For a p-group P we put

$$d(P) = \max\{m(A) \mid A \text{ is an Abelian subgroup of } P\}.$$

We define the subgroup $J(P)$ of P to be generated by all Abelian subgroups A of P with $m(A) = m(P)$, i.e.

$$J(P) = \langle A \mid A \text{ is an Abelian subgroup of } P \text{ and } m(A) = d(P) \rangle.$$

$J(P)$ is called the Thompson subgroup of P. Obviously, $J(P) \trianglelefteq P$.

Theorem 2.9.3 (Thompson) *Let p be an odd prime and let G_p be a Sylow p-subgroup of a group G. If both $\mathscr{C}_G(Z(G_p))$ and $\mathscr{N}_G(J(G_p))$ have normal p-complements, then G has a normal p-complement.*

Proof Since all Sylow p-subgroups are conjugate, the assumptions of the theorem are true for all Sylow p-subgroups of G. Suppose the theorem is not true and let G be a group of minimal order for which the theorem is false. So, since $G_p, \mathscr{C}_G(Z(G_p))$ and $\mathscr{N}_G(J(G_p))$ are all proper subgroups of G, we conclude that every proper subgroup of G containing G_p has a normal p-complement.

(*a*) We put

$$\Lambda = \{H \mid H \text{ is a } p\text{-subgroup of } G, N_G(H) \text{ has no normal } p\text{-complements}\}.$$

Obviously, $1 \in \Lambda$ and by lemma 2.9.1 and theorem 2.9.2 Λ contains a p-subgroup of G that is not equal to 1. For $H, K \in \Lambda$, we will say, that K is bigger than H (notation: $H \prec K$) if one of the following conditions is satisfied:

(i) $|\mathscr{N}_G(H)|_p \lneqq |\mathscr{N}_G(K)|_p$

(ii) $|\mathscr{N}_G(H)|_p = |\mathscr{N}_G(K)|_p$ and $|H| \lneqq |K|$

(iii) $H = K$.

The relation $\prec$ is an order relation on Λ. Let H represent a maximal element of $\Lambda - \{1\}$ with respect to this order. Putting $N = \mathscr{N}_G(H)$, let P be a Sylow p-subgroup of $N (\geq H)$ and let G_p be a Sylow p-subgroup of G containing P. If $H = G_p$, then $\mathscr{N}_G(H) \leq \mathscr{N}_G(J(G_p))$

since $J(G_p) \lessdot H$. Since $\mathcal{N}_G(J(G_p))$ has a normal p-complement, $\mathcal{N}_G(H)$ also has a normal p-complement, contrary to $H \in \Lambda$. Therefore: $H \lneqq G_p$.

(*b*) We prove next that $G = N$. First, we show that N satisfies the conditions of the theorem. From $H \lneqq P \leq G_p$ we get $Z(G_p) \leq \mathcal{N}_G(H) = N$, hence $PZ(G_p) \leq N \cap G_p = P$. Therefore $Z(G_p) \leq P$ and thus $Z(G_p) \leq Z(P)$. Thus $\mathcal{C}_N(Z(P)) \leq \mathcal{C}_G(Z(P)) \leq \mathcal{C}_G(Z(G_p))$ and we conclude that $\mathcal{C}_N(Z(P))$ has a normal p-complement. Now, consider $\mathcal{N}_N(J(P))$. If $P = G_p$, then $\mathcal{N}_N(J(P)) \leq \mathcal{N}_G(J(G_p))$, hence $\mathcal{N}_N(J(P))$ has a normal p-complement. If $P \lneqq G_p$, then $P \lneqq \mathcal{N}_{G_p}(P) \leq G_p$, therefore $P \lneqq \mathcal{N}_G(P) \leq \mathcal{N}_G(J(P))$ since $J(P) \lessdot P$. Hence $|\mathcal{N}_G(H)|_p < |\mathcal{N}_G(J(P))|_p$ and $J(P) \notin \Lambda$ because of the maximality of H. Therefore, $\mathcal{N}_G(J(P))$ has a normal p-complement. So, N satisfies the conditions of the theorem in both cases. Suppose $G \gneqq N$, then N has a normal complement by the assumption (i.e. G is a group of minimal order for which the theorem is false), contrary to $H \in \Lambda$.

(*c*) We prove next, that $O_{p'}(G) = 1$. Put $\bar{G} = G/O_{p'}(G)$ and put $\bar{X} = XO_{p'}(G)/O_{p'}(G)$ for a subgroup X of G. It is easy to verify that, for every p-subgroup K of G_p, $\overline{\mathcal{N}_G(K)} = \mathcal{N}_{\bar{G}}(\bar{K})$ and $\overline{\mathcal{C}_G(K)} = \mathcal{C}_{\bar{G}}(\bar{K})$ from $(|K|, |O_{p'}(G)|) = 1$. Since $J(\bar{G}_p) = \overline{J(G_p)}$ and $Z(\bar{G}_p) = \overline{Z(G_p)}$ we have: $\mathcal{N}_{\bar{G}}(J(\bar{G}_p)) = \overline{\mathcal{N}_G(J(G_p))}$ and $\mathcal{C}_{\bar{G}}(Z(\bar{G}_p)) = \overline{\mathcal{C}_G(Z(G_p))}$. Therefore, $\bar{G}$ satisfies the conditions of the theorem. Now, suppose $O_{p'}(G) \neq 1$, then $|\bar{G}| < |G|$, hence $\bar{G}$ has a normal p-complement by the assumption. The inverse image of this normal p-complement in G is a normal p-complement in G. Contradiction.

(*d*) $H = O_p(G)$ and $\bar{G} = G/H$ has a normal p-complement and therefore $O_{p,p',p}(G) = G$ and G is p-solvable. (Proof: $O_p(G) \in \Lambda$ and $H \leq O_p(G)$ by (*b*), hence $H = O_p(G)$ because of the maximality of H in Λ. Suppose $\bar{G}$ has no normal p-complement. Since $|\bar{G}| < |G|$, we conclude that at least one of the groups $\mathcal{C}_{\bar{G}}(Z(\bar{G}_p))$ and $\mathcal{N}_{\bar{G}}(J(\bar{G}_p))$ does not have a normal p-complement by the assumption. Suppose first that $\mathcal{C}_{\bar{G}}(Z(\bar{G}_p))$ does not have a normal p-complement. Putting K the inverse image of $Z(\bar{G}_p)$ in G, K is a p-group and $K \gneqq H$. From $\overline{\mathcal{N}_G(K)} = \mathcal{N}_{\bar{G}}(Z(\bar{G}_p)) \geq \mathcal{C}_{\bar{G}}(Z(\bar{G}_p))$ we conclude that $\overline{\mathcal{N}_G(K)}$, and therefore also $\mathcal{N}_G(K)$, does not have a normal p-complement. Since $|\mathcal{N}_G(K)|_p = |G_p|$ and $H \lneqq K$, we have $H \prec K$, contrary to the maximality of H in Λ. Suppose next that $\mathcal{N}_{\bar{G}}(J(\bar{G}_p))$ does not have a normal p-complement. Putting K the inverse image of $J(\bar{G}_p)$ in G, we find in the same way that $H \prec K$ and $H \lneqq K$, hence that $\mathcal{N}_G(K)$ does not have a normal p-complement. Contradiction.) Hence $\bar{G}$ has a normal p-complement, from which $O_{p,p',p}(G) = G$ follows obviously.

(*e*) Since $H \lneqq G_p$ we have $O_{p,p'}(G) \lneqq G$. Putting $M = O_{p,p'}(G)$, then $\bar{M} = M/H$ is a $\bar{G}_p$-group under conjugation and we will prove that $\bar{M}$ is minimal as a $\bar{G}_p$-group. Suppose $\bar{M}$ is not minimal as a $\bar{G}_p$-group, and let $\bar{M}_0$ be a $\bar{G}_p$-group such that $1 \leq \bar{M}_0 \leq \bar{M}$. Let M_0 be the inverse image of $\bar{M}_0$ in G and put $G_0 = G_p M_0$. Clearly, $G_0 \lneqq G$ and G_0 has a normal p-complement $K \, (\neq 1)$. Since $K \lhd M_0$, $H \lhd M_0$ and $(|K|, |H|) = 1$, we have $[K, H] = 1$, i.e. $K \leq \mathscr{C}_G(H)$. On the other hand, according to the theorem of Hall–Higman (theorem 2.7.5) $\mathscr{C}_G(H) \leq H$. Contradiction.

(*f*) $\bar{M}$ is an elementary Abelian q-group for some prime q. (Proof: Let q be a prime divisor of $|\bar{M}|$. Then $\bar{G}_p$ operates by conjugation on the set consisting of the Sylow q-subgroups of $\bar{M}$. Since the number of Sylow q-subgroups of $\bar{M}$ is not divisible by p, $\bar{G}_p$ leaves a certain Sylow q-subgroup $\bar{M}_0$ of $\bar{M}$ fixed. Therefore $\bar{M} = \bar{M}_0$ by (*e*). Since $Z(\bar{M}) \trianglelefteq \bar{M}$, we conclude that $\bar{M}$ is Abelian. Since also $\Omega_1(\bar{M}) \trianglelefteq \bar{M}$, we have $\Omega_1(\bar{M}) = \bar{M}$. Therefore $\bar{M}$ is an elementary Abelian group.)

(*g*) G_p is a maximal subgroup of G. (Proof: Suppose there is a subgroup L, such that $G_p \lneqq L \lneqq G$, then $\bar{L} = \bar{G}_p(\overline{L \cap M}) \rhd \overline{L \cap M} \neq 1$. Since $G \gneqq L$, we have $\bar{M} \gneqq \overline{L \cap M}$, contrary to (*e*).)

(*h*) There exists an Abelian subgroup X of G_p, such that $d(G_p) = m(X)$ and $X \nleq H$. Picking an A of minimal order from the Xs satisfying these conditions, we have $\bar{G} = \bar{A}\bar{M}$ and $|\bar{A}| = p$. (Proof: Suppose $J(G_p) \leq H$. From $J(G_p) = J(H) \trianglelefteq H$, we have $G = \mathcal{N}(J(G_p))$. Contradiction. Therefore, there exists an Abelian subgroup X of G such that $m(X) = d(G_p)$ and $X \nleq H$. Let A be a subgroup of minimal order among the Xs. Since $A \cap H \lneqq A$, we have $\Omega_1(A/A \cap H) \neq 1$ and there exists a subgroup A_1 of A satisfying $\Omega_1(A/A \cap H) = A_1/A \cap H$ and $A \geq A_1 \gneqq A \cap H$. Since all elements of order p of A are elements of A_1, we have $m(A) = m(A_1)$. Since $A_1 \nleq H$, we have $A = A_1$ i.e. $\bar{A} = AH/H \simeq A/A \cap H$ is an elementary Abelian group.

Let $\bar{K}$ be the kernel of the operation defined by $\bar{G}_p$ on $\bar{M}$ and let K be the inverse image of $\bar{K}$ in G_p. Since $K \lhd G_p$ and $[K, M] \leq H \leq K$ we have $K \lhd G$. Since $H \leq K \leq O_p(G) = H$, we have $H = K$. Therefore, $\bar{G}_p$ operates faithfully on $\bar{M}$. Hence $\bar{A}(1 \neq \bar{A} \leq \bar{G}_p)$ operates non-trivially on $\bar{M}$. Therefore by theorem 2.6.3 there exists a minimal $\bar{A}$-subgroup $\bar{Q}$ of $\bar{M}$, on which $\bar{A}$ operates non-trivially. Let the inverse image of $\bar{A}\bar{Q}$ in G be denoted by G_1 and let P_1 denote a Sylow p-subgroup of G_1 containing A and let again G_p be a Sylow p-subgroup of G containing P_1. From

$$Z(G_p) \leq \mathscr{C}_G(G_p) \leq \mathscr{C}_G(H)(\leq H) \leq P_1$$

we conclude: $Z(G_p) \leq Z(P_1)$. Therefore,

$$\mathscr{C}_{G_1}(Z(P_1)) \leq \mathscr{C}_G(Z(P_1)) \leq \mathscr{C}_G(Z(G_p)),$$

hence $\mathscr{C}_{G_1}(Z(P_1))$ has a normal p-complement. If Q_1 represents a Sylow q-subgroup of $\mathscr{N}_{G_1}(J(P_1))$, we have $[A, Q_1] \leq [J(P_1), Q_1] \leq J(P_1)$ since $J(P_1) \geq A$. Therefore $[\bar{A}, \bar{Q}_1]$ is a p-group. On the other hand, we conclude from $[\bar{A}, \bar{Q}_1] \leq [\bar{A}, \bar{Q}] \leq \bar{Q}$, that $[\bar{A}, \bar{Q}_1]$ is a q-group. Therefore, $[\bar{A}, \bar{Q}_1] = 1$. Therefore $\bar{Q}_1$ is an $\bar{A}$-subgroup of $\bar{Q}$ and $\bar{A}$ operates trivially on $\bar{Q}_1$. Therefore, $\bar{Q}_1 = 1$, and $\mathscr{N}_{G_1}(J(P_1)) = P_1$. Hence $\mathscr{C}_{G_1}(Z(P_1))$ and $\mathscr{N}_{G_1}(J(P_1))$ both have normal p-complements. Suppose $G_1 < G$, then by the assumption on G, G_1 has a normal p-complement Q and Q satisfies $Q \leq \mathscr{C}_G(H) \leq H$. Contradiction. Therefore $G = G_1$. Since $\bar{M} \lhd \bar{A}\bar{M} = \bar{G}$ and $\bar{M}$ is an Abelian, faithful and irreducible $\bar{A}$-group, satisfying $(|\bar{M}|, |\bar{A}|) = 1$, we have $|\bar{A}| = p$ by corollary 2.6.2.)

(*i*) If Q is a Sylow q-subgroup of G, then $[Q, \Omega_1(Z(H))] \neq 1$. (Proof: Since $H \lhd G$, we have $\Omega_1(Z(H)) \lhd G$. Since $H \leq G_p$ we have $\mathscr{C}_G(G_p) \leq \mathscr{C}_G(H) \leq H$, hence $Z(G_p) \leq Z(H)$. Since $\mathscr{C}_G(Z(G_p)) \lneq G = G_pQ$, we have: $[Z(G_p), Q] \neq 1$. Therefore, $[Z(H), Q] \neq 1$, hence $[Q, \Omega_1(Z(H))] \neq 1$ by corollary 2.6.4.)

(*j*) Since $\Omega_1(Z(H)) \lhd G$ and $[H, \Omega_1(Z(H))] = 1$, the group $\bar{G}$ operates on $\Omega_1(Z(H))$. We know that $\bar{Q}$ is a unique minimal normal subgroup of $\bar{G}$ by (*e*) and $\bar{Q}$ operates non-trivially on $\Omega_1(Z(H))$ by (*i*). Therefore, $\bar{G}$ operates faithfully on $\Omega_1(Z(H))$. Regarding $\Omega_1(Z(H))$ as a $\bar{Q}$-group and applying theorem 2.6.3, we get that $\Omega_1(Z(H)) = V \times W$, i.e. $\Omega(Z(H))$ is the direct product of $\bar{Q}$-groups

V and W where $V = \text{Ker}\,\alpha(\bar{Q})$ and $W = \Omega_1(Z(H))^{\alpha(\bar{Q})}$. Since $\bar{Q} \lhd \bar{G}$, V is a $\bar{G}$-group. Since $\bar{Q}$ operates non-trivially on any $\bar{Q}$-subgroup $(\neq 1)$ of V and since $\bar{Q}$ is a unique normal subgroup of G, $\bar{G}$ operates faithfully on V and non-trivially on any $\bar{G}$-subgroup $(\neq 1)$ of V.

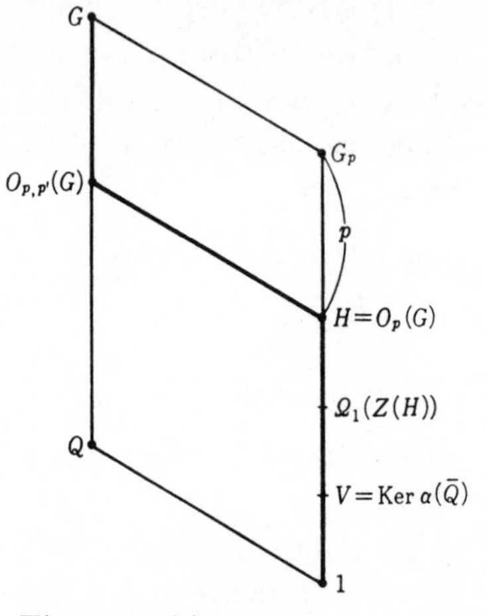

$O_{p'}(G)=1$, $\bar{G}=G/H$
$Q\simeq\overline{O_{p,p'}(G)}$ (an elementary Abelian q-group)
G_p is a maximal subgroup of G
$\overline{O_{p,p'}(G)}$ is a minimal normal subgroup of $\bar{G}$

The next thing we want to prove is that $|V| = p^2$ and $\bar{G} \leq SL(2, p)$.

Put $A \cap H = A_0$ and $V \cap A = V_0$. Since $|\bar{A}| = p$, we have $m(A_0) \geq d(G_p) - 1$. Since $V \leq Z(H)$, the group $\langle V, A_0 \rangle = VA_0$ is Abelian. Noticing that V is an elementary Abelian group and that $V \cap A_0 = V \cap A \cap H = V \cap A$, we get

$$\begin{aligned} d(G_p) \geq m(\langle V, A_0 \rangle) &= m(V) + m(A_0) - m(V \cap A_0) \\ &= m(V) + m(A_0) - m(V \cap A) \\ &= m(A_0) + m(V | V_0) \geq d(G_p) - 1 + m(V | V_0). \end{aligned}$$

Therefore, $V = V_0$ or V_0 is a maximal subgroup of V. Since $[\bar{A}, \bar{Q}] \neq 1$, there are $\bar{a} \in \bar{A}$ $(\bar{a} \neq 1)$ and $\bar{b} \in \bar{Q}$ $(\bar{b} \neq 1)$, such that $[\bar{a}, \bar{b}] \neq 1$. Therefore $\langle \bar{a} \rangle \neq \langle \bar{a}^{\bar{b}} \rangle$. (For suppose $\langle \bar{a} \rangle = \langle \bar{a}^{\bar{b}} \rangle$, then $[\bar{a}, \bar{b}] \in \bar{A}$. On the other hand, $\bar{Q} \trianglelefteq \bar{G}$, therefore $[\bar{a}, \bar{b}] \in \bar{Q}$. Hence $[\bar{a}, \bar{b}] = 1$.) Since $\bar{A}$ is a maximal subgroup of $\bar{G}$, we have $\langle \bar{a}, \bar{a}^{\bar{b}} \rangle = \bar{G}$. From $[V_0, \bar{a}] = 1$, we get that $[V_0^{\bar{b}}, \bar{a}^{\bar{b}}] = 1$ and $\langle \bar{a}, \bar{a}^{\bar{b}} \rangle$ operates trivially on $V_0 \cap V_0^{\bar{b}}$. Therefore, $V_0 \cap V_0^{\bar{b}} = 1$ and

$$\begin{aligned} |V| = |V : V_0 \cap V_0^{\bar{b}}| &= |V : V_0|\,|V_0 : V_0 \cap V_0^{\bar{b}}| \\ &\leq |V : V_0|\,|V : V_0^{\bar{b}}| \leq p^2. \end{aligned}$$

Since $\bar{G}$ is non-Abelian and operates faithfully on V, we conclude $|V| = p^2$ (by theorem 1.2.10) and $\bar{G} \leq GL(2, p)$. Since $\bar{G}$ is generated by two elements $\bar{a}$ and $\bar{a}^{\bar{b}}$ of order p and all elements of order p of $GL(2, p)$ are elements of $SL(2, p)$, we conclude $\bar{G} \leq SL(2, p)$.

Thus we have the following situation: $SL(2, p)$, where p is an odd prime, contains an element $\bar{a}$ of order p and an Abelian subgroup $\bar{Q}$ such that $\bar{Q}^{\bar{a}} = \bar{Q}$ and $[\bar{a}, \bar{Q}] \neq 1$. On the other hand we can show that such a situation does not occur. To show this we need the following lemma.

Lemma 2.9.4 *Let p be a prime.*

(1) $P_1 = \left\{ \begin{pmatrix} 1 & 0 \\ a & 1 \end{pmatrix} \middle| a \in F_p \right\}$ and $P_2 = \left\{ \begin{pmatrix} 1 & a \\ 0 & 1 \end{pmatrix} \middle| a \in F_p \right\}$

are Sylow p-subgroups of $SL(2, p)$ *and* $SL(2, p) = \langle P_1, P_2 \rangle$.

(2) *If the subgroup H of* $SL(2, p)$ *contains 2 or more Sylow p-subgroups of* $SL(2, p)$, *then* $H = SL(2, p)$.

(3) *The Sylow 2-subgroups of* $SL(2, 3)$ *are non-Abelian.*

Proof (1) According to theorem 2.5.3 and §1.4 we have $|SL(2, p)| = p(p+1)(p-1)$, therefore the order of the Sylow p-subgroups of $SL(2, p)$ equals p. Since both P_1 and P_2 are subgroups of order p of $SL(2, p)$, they are Sylow p-subgroups. Let $\begin{pmatrix} a & b \\ c & d \end{pmatrix}$ be an arbitrary element of $SL(2, p)$. We distinguish three cases:

(i) $c = 0$, then:

$$\begin{pmatrix} a & b \\ 0 & d \end{pmatrix} = \begin{pmatrix} a & b \\ 0 & a^{-1} \end{pmatrix}$$

$$= \begin{pmatrix} 1 & 0 \\ a^{-1}(a^{-1}-1) & 1 \end{pmatrix} \begin{pmatrix} 1 & a \\ 0 & 1 \end{pmatrix} \begin{pmatrix} 1 & 0 \\ 1-a^{-1} & 1 \end{pmatrix} \begin{pmatrix} 1 & ba^{-1}-1 \\ 0 & 1 \end{pmatrix}$$

$$\in \langle P_1, P_2 \rangle.$$

(ii) $c \neq 0$, $a \neq 0$, then:

$$\begin{pmatrix} a & b \\ c & d \end{pmatrix} = \begin{pmatrix} 1 & 0 \\ a^{-1}c & 1 \end{pmatrix} \begin{pmatrix} a & b \\ 0 & d - a^{-1}cb \end{pmatrix} \in \langle P_1, P_2 \rangle \text{ by case (i).}$$

(iii) $c \neq 0, a = 0$, then:

$$\begin{pmatrix} 0 & b \\ c & d \end{pmatrix} = \begin{pmatrix} 1 & -1 \\ 0 & 1 \end{pmatrix} \begin{pmatrix} c & b+d \\ c & d \end{pmatrix} \in \langle P_1, P_2 \rangle \text{ by case (ii)}$$

Therefore, $SL(2,p) = \langle P_1, P_2 \rangle$.

(2) Since

$$\mathcal{N}_{SL(2,p)}(P_1) = \left\{ \begin{pmatrix} a & 0 \\ b & a^{-1} \end{pmatrix} \middle| a, b \in F_p, a \neq 0 \right\},$$

the number of Sylow p-subgroups of $SL(2,p)$ equals $|SL(2,p) : \mathcal{N}_{SL(2,p)}(P_1)| = p+1$. Therefore, if a subgroup of $SL(2,p)$ contains two Sylow p-groups, it contains all Sylow p-groups of $SL(2,p)$ by Sylow's theorem, hence it is equal to $SL(2,p)$ by (1).

(3) The elements $x = \begin{pmatrix} 0 & 1 \\ -1 & 0 \end{pmatrix}$ and $y = \begin{pmatrix} 1 & 1 \\ 1 & -1 \end{pmatrix}$ of $SL(2,3)$ satisfy: $x^4 = y^4 = 1$ and $x^{-1}yx = y^{-1}$. Therefore, $\langle x, y \rangle$ is a non-Abelian 2-subgroup of $SL(2,3)$. ■

Now we can finish the proof of theorem 2.9.3. We put $\bar{P} = \langle \bar{a} \rangle$, then $\bar{P}$ is a Sylow p-subgroup of $SL(2,p)$. Let us consider the semi-direct product $\bar{P}\bar{Q}$. If $\bar{P} \lhd \bar{P}\bar{Q}$, then $\bar{P}\bar{Q} = \bar{P} \times \bar{Q}$ since $\bar{P} \cap \bar{Q} = 1$, contrary to $[\bar{a}, \bar{Q}] \neq 1$. Therefore, $\bar{P}\bar{Q}$ contains 2 or more Sylow p-subgroups, hence $\bar{P}\bar{Q} = SL(2,p)$ by (2) of the previous lemma. Therefore, $SL(2,p)$ has a normal p-complement. Since $|\mathcal{N}_{SL(2,p)}(\bar{P})| = p(p-1)$ and $|\mathscr{C}_{SL(2,p)}(\bar{P})| = 2p$, we conclude from lemma 2.9.1, that $p - 1 = 2$, i.e. $p = 3$. Therefore, $\bar{Q}$ is a Sylow 2-group of $SL(2,3)$ and $\bar{Q}$ is non-Abelian by (3) of the previous lemma, contrary to the commutativity of $\bar{Q}$. ■

2.10 Representation of finite groups

2.10.1 Representations

Let $\mathbb{C}$ represent the field of complex numbers, let V be an n-dimensional vector-space over $\mathbb{C}$ and let G be a finite group. A homomorphism $\rho : G \to GL(V)$ is called a representation of G on V. V is called the representation module of G or the G-module corresponding with ρ (or simply G-module). Once ρ is fixed, we write v^σ

rather than $v^{\rho(\sigma)}$ for $v \in V$ and $\sigma \in G$. The natural number n is called the degree of ρ. Notation: $n = \deg \rho$. If ρ is a monomorphism, i.e. if $\operatorname{Ker} \rho = 1$, ρ is called a faithful representation. Choice of a basis for V determines an isomorphism between $GL(V)$ and $GL(n, \mathbb{C})$ (§1.4). If we fix a basis for V, we get a homomorphism $P: G \to GL(n, \mathbb{C})$ by assigning to $\sigma \in G$ the matrix representation of $\rho(\sigma)$ (with regard to the chosen basis). The homomorphism P is called the matrix representation of ρ (with regard to the given basis). Conversely, if a basis for V and a homomorphism $P: G \to GL(n, \mathbb{C})$ are given, we get a representation $\rho: G \xrightarrow{P} GL(n, \mathbb{C}) \simeq GL(V)$, where the last isomorphism is determined by the given basis, and the matrix representation of ρ with regard to the given basis is P. A homomorphism $P: G \to GL(n, \mathbb{C})$ is called a matrix representation over $\mathbb{C}$.

Let ρ be a representation of G on V. If the subspace U of V is invariant under G, i.e. if $U^{\rho(\sigma)} = U$ for all $\sigma \in G$, then $\rho(\sigma)$ induces a non-singular linear transformation $\tilde{\tilde{\rho}}(\sigma): U \to U$. The assignment $\sigma \to \tilde{\rho}(\sigma)$ defines a representation of G on U, called the restriction of ρ to U and denoted by $\rho_{|U}$. In this case U is called a G-submodule of V. For G-submodules of V the assignment $\tilde{\tilde{\rho}}(\sigma): v + U \to v^{\rho(\sigma)} + U$ defines a representation of G on V/U, called the quotient representation of G on V/U induced by ρ and denoted by $\rho_{|V/U}$. In this case, V/U is called the quotient G-module of V with respect to U.

If $H \leq G$, then the representation ρ of G can be restricted to H to give a representation of H. This representation is called the restriction of ρ to H and denoted by $\rho_{|H}$.

Let ρ_1, ρ_2 be representations of G on V_1 and V_2 respectively. A $\mathbb{C}$-module homomorphism $f: V_1 \to V_2$, satisfying

$$f(v^{\rho_1(\sigma)}) = f(v)^{\rho_2(\sigma)}, \quad \forall \sigma \in G, \forall v \in V_1$$

is called a G-module homomorphism or simply a G-homomorphism. In particular, if f is an isomorphism from V_1 onto V_2 then f is called a G-isomorphism and V_1 and V_2 are called G-isomorphic and ρ_1 and ρ_2 are called equivalent representations of G. In this case, by choosing suitable bases for V_1 and V_2, the matrix representations of ρ_1 and ρ_2 can be made exactly the same. (In fact, given an arbitrary basis for V_1, it suffices to take as a basis for V_2 the image of that base under f.) Conversely, it is obvious that if ρ_1, ρ_2 are two representations of G, such that with regard to suitably chosen bases for V_1 and V_2 their matrix representations are exactly the same, then ρ_1 and ρ_2 are equivalent representations. If $f: V_1 \to V_2$ is a G-homomorphism, then $\operatorname{Im} f$ and $\operatorname{Ker} f$ are G-submodules of V_2 and V_1 respectively, and the $\mathbb{C}$-module isomorphism $V_1/\operatorname{Ker} f \to \operatorname{Im} f$ is easily seen to be a G-isomorphism.

A representation ρ such that $\deg \rho = 1$ is called linear, and the linear representation ρ of G such that $\operatorname{Ker} \rho = G$ is called the identity representation on G and denoted by 1_G. If the G-module $V(\neq 0)$ has no other G-submodules except V and $\{0\}$ we call V an irreducible G-module (or simple G-module) and the corresponding representation ρ is called an irreducible representation. G-modules (or representations) that are not irreducible are called reducible. If V is a reducible G-module, there exists a G-submodule U of V such that $\{0\} \subsetneq U \subsetneq V$. Choosing a basis $\omega_1, \ldots, \omega_n$ for V such that $\omega_1, \ldots, \omega_m$ $(m < n)$ is a basis for U (where $m = \dim U$), the matrix representation of ρ becomes:

$$P(\sigma) = \left[\begin{array}{c|c} P_1(\sigma) & 0 \\ \hline * & P_2(\sigma) \end{array}\right]$$

where P_1 and P_2 are the matrix representations of the representations induced on the G-modules U and V/U respectively. Let V be a reducible G-module, let V_1 be a maximal G-submodule of V, V_2 be a maximal G-submodule of V_1 and so on. In this way we get a sequence $V_1, V_2, \ldots$ of G-submodules of V such that $V = V_0 \supsetneq V_1 \supsetneq V_2 \supsetneq \cdots \supsetneq V_m = \{0\}$ (the sequence is finite, because $\dim V < \infty$) and such that V_i/V_{i+1} is an irreducible G-module for $i = 0, \ldots, m-1$. If we pick a basis for V_{m-1}, then extend that basis to a basis for V_{m-2}, then extend that basis to a basis for V_{m-3}, and so on till we have arrived at a basis for V, the matrix representation of ρ is given by:

$$P(\sigma) = \begin{bmatrix} P_{m-1}(\sigma) & & & \\ & P_{m-2}(\sigma) & 0 & \\ & & & \\ & * & & \\ & & & P_0(\sigma) \end{bmatrix} \qquad (\sigma \in G)$$

where P_i is the matrix representation of the irreducible representation induced by ρ on V_i/V_{i+1}.

Let ρ_1 and ρ_2 be representations of G on V_1 and V_2 respectively. We define a representation ρ of G on $V = V_1 \oplus V_2$ (the direct sum as $\mathbb{C}$-modules) by

$$(v_1, v_2)^{\rho(\sigma)} = (v_1^{\rho_1(\sigma)}, v_2^{\rho_2(\sigma)}), \quad \forall v_i \in V_i, \forall \sigma \in G.$$

Thus V is made into a G-module, called the direct sum of the G-modules V_1 and V_2. The representation ρ is called the direct sum of ρ_1 and ρ_2. Notation: $\rho = \rho_1 \oplus \rho_2$. Let $v_1, \ldots, v_n$ be a basis for V such that $v_1, \ldots, v_m$ is a basis for V_1 and $v_{m+1}, \ldots, v_n$ is a basis for V_2. With regard to this basis the matrix representation P of ρ is given by

$$P(\sigma) = \left[\begin{array}{c|c} P_1(\sigma) & 0 \\ \hline 0 & P_2(\sigma) \end{array}\right] \quad \forall \sigma \in G,$$

where P_1 and P_2 represent the matrix representations of ρ_1 and ρ_2 respectively. If the G-module V is the direct sum of a certain number of irreducible G-submodules, then V is called a completely reducible G-module and the corresponding representation is called a completely reducible presentation.

Let (G, Ω) be a permutation representation and let $\Omega = \{1, \ldots, n\}$. Then we can define a representation of degree n of G in the following way. For $\sigma \in G$ we define a matrix $P(\sigma) = (\lambda_{ij})$ of degree n by

$$\lambda_{ij} = \begin{cases} 1 & \text{if } i^\sigma = j \\ 0 & \text{if } i^\sigma \neq j. \end{cases}$$

It is easily seen that the assignment $\sigma \to P(\sigma)$ defines a homomorphism from G into $GL(n, \mathbb{C})$, and so this gives a representation of degree n of G. This representation is called the (permutation matrix) representation of G induced by (G, Ω). In particular, using the left (right) regular permutation representations of G (§1.2.3), we get the so-called left (right) regular representation of G.

Theorem 2.10.1 *The left and right regular presentations of a group G are equivalent.*

Proof Put $G=\{\sigma_1,\ldots,\sigma_n\}$, then a matrix representation $P(\sigma)=(\lambda_{ij})$ of the right regular representation of G is given by

$$\lambda_{ij}=\begin{cases}1 & \text{if } \sigma_i\sigma=\sigma_j\\ 0 & \text{if } \sigma_i\sigma\neq\sigma_j.\end{cases}$$

Since $G=\{\sigma_1^{-1},\ldots,\sigma_n^{-1}\}$ a matrix representation $Q(\sigma)=(\mu_{ij})$ of the left regular presentation of G is given by

$$\mu_{ij}=\begin{cases}1 & \text{if } \sigma^{-1}\sigma_i^{-1}=\sigma_j^{-1}\\ 0 & \text{if } \sigma^{-1}\sigma_i^{-1}\neq\sigma_j^{-1}.\end{cases}$$

Therefore $P=Q$. ■

The following theorem is one of the basic theorems in the theory of representations of finite groups.

Theorem 2.10.2 (Maschke) *Representations of finite groups are all completely reducible.*

Proof Let G be a finite group and let V be a G-module. It suffices to prove that for every G-submodule U of V there exists a G-submodule W such that $V=U\oplus W$. There exists a $\mathbb{C}$-module W' such that $V=U\oplus W'$ (as $\mathbb{C}$-modules). Since every $v\in V$ can be written in exactly one way as $v=u+w'$ $(u\in U, w'\in W')$, we can define a linear map $\theta: V\to W'$ by $v^\theta=\theta(v)=w'$ $(v\in V)$. Using θ, we define a linear map $\psi: V\to V$ by

$$v^\psi=\frac{1}{g}\sum_{\sigma\in G}((v^\sigma)^\theta)^{\sigma^{-1}}\quad (v\in V),$$

where $g=|G|$.

We put $V^\psi=W$, so W is a submodule of V (as $\mathbb{C}$-module). For $\tau\in G$ we have

$$(v^\psi)^\tau=\frac{1}{g}\sum_{\sigma\in G}((v^\sigma)^\theta)^{\sigma^{-1}\tau}=\frac{1}{g}\sum_{\sigma\in G}((v^{\tau\tau^{-1}\sigma})^\theta)^{\sigma^{-1}\tau}$$
$$=(v^\tau)^\psi\quad (v\in V).$$

Therefore $W^\tau=W$, i.e. W is a G-submodule of V. For $v\in V$ we can write $v = (v - v^{\psi}) + v^{\psi}$ with $v^{\psi} \in W$ and

$$v - v^{\psi} = v - \frac{1}{g} \sum_{\sigma \in G} ((v^{\sigma})^{\theta})^{\sigma^{-1}} = \frac{1}{g} \sum_{\sigma \in G} (v^{\sigma} - v^{\sigma\theta})^{\sigma^{-1}} \in U$$

since $v^{\sigma} - v^{\sigma\theta} \in U$. Therefore $V = U + W$. If $u \in U$, then $u^{\theta} = 0$, hence $u^{\psi} = 0$. Therefore, if $v \in V$, then $(v - v^{\psi})^{\psi} = 0$, i.e. $v^{\psi^2} = v^{\psi}$. Now, take $w \in U \cap W$, then $w^{\psi} = 0$ because $w \in U$ and $w = v^{\psi}$ for some $v \in V$ because $w \in W$. Therefore $w^{\psi} = v^{\psi^2} = v^{\psi} = w$, hence $w = 0$, i.e. $V = U \oplus W$. ■

Let ρ_1 and ρ_2 be representations of G on V_1 and V_2 respectively. The collection of all G-homomorphisms of V_1 into V_2 is denoted by $\mathrm{Hom}_G(V_1, V_2)$ or $\mathrm{Hom}_G(\rho_1, \rho_2)$, i.e.

$$\mathrm{Hom}_G(\rho_1, \rho_2) = \{f \in \mathrm{Hom}_{\mathbb{C}}(V_1, V_2) | f\rho_2(\sigma) = \rho_1(\sigma)f, \forall \sigma \in G\}.$$

It is easy to check that $\mathrm{Hom}_G(\rho_1, \rho_2)$ is a submodule of $\mathrm{Hom}_{\mathbb{C}}(V_1, V_2)$ (as $\mathbb{C}$-module). In particular, if $V_1 = V_2 = V$ and $\rho_1 = \rho_2 = \rho$, then $\mathrm{Hom}_{\mathbb{C}}(V, V)$ is a ring and $\mathrm{Hom}_G(\rho, \rho)$ is the commutator ring of $\rho(G)$ in $\mathrm{Hom}_{\mathbb{C}}(V, V)$. This is also called the commutator ring of ρ.

Theorem 2.10.3 (Schur's lemma) *Let ρ_1 and ρ_2 be irreducible representations of G on V_1 and V_2 respectively.*

(1) *If $f \in \mathrm{Hom}_G(\rho_1, \rho_2)$ and $f \neq 0$, then f is an isomorphism from V_1 onto V_2.*

(2) $\dim_{\mathbb{C}} \mathrm{Hom}_G(\rho_1, \rho_2) \leq 1$ *and* $\dim_{\mathbb{C}} \mathrm{Hom}_G(\rho_1, \rho_2) = 1 \Leftrightarrow \rho_1$ *and ρ_2 are equivalent.*

(3) *If $V_1 = V_2 = V$ and $\rho_1 = \rho_2 = \rho$, then* $\mathrm{Hom}_G(\rho, \rho) = \{\lambda 1_V | \lambda \in \mathbb{C}\}$ *where 1_V is the identity element in* $\mathrm{Hom}_G(\rho, \rho)$. *In other words,* $\mathrm{Hom}_G(\rho, \rho)$ *is the set of all scalar transformations of V.*

Proof (1) Since $f \neq 0$, we have $V_1^f \neq 0$, therefore $V_1^f = V_2$ because ρ_2 is irreducible. Since $\mathrm{Ker}\, f \neq V_1$, we have $\mathrm{Ker}\, f = 0$ because ρ_1 is irreducible. This proves that $f: V_1 \to V_2$ is a G-isomorphism.

(2) If ρ_1 and ρ_2 are not equivalent, then $\mathrm{Hom}_G(\rho_1, \rho_2) = 0$ by (1). If ρ_1 and ρ_2 are equivalent, then there exists a G-isomorphism

$f_0 : V_1 \to V_2$. For $f \in \mathrm{Hom}_G(\rho_1, \rho_2)$, we have $ff_0^{-1} \in \mathrm{Hom}_G(\rho_1, \rho_1)$. The assignment $f \to ff_0^{-1}$ sets up an isomorphism between the vector spaces $\mathrm{Hom}_G(\rho_1, \rho_2)$ and $\mathrm{Hom}_G(\rho_1, \rho_1)$. Therefore, it suffices to prove (3) in order to prove (2).

(3) Let $f \in \mathrm{Hom}_G(\rho, \rho)(f \neq 0)$, let λ be an eigenvalue of f and let $v_0 \in V$ be an eigenvector corresponding to λ. Putting $U = \{v \in V \mid v^f = \lambda v\}$ we have that U is a G-submodule of V and $U \neq 0$, since $v_0 \in U$. Therefore, $U = V$ and $f = \lambda 1_V$. ■

If ρ is a faithful and irreducible representation of G on V, then $\rho(Z(G)) \subseteq \mathrm{Hom}_G(\rho, \rho) = \{\lambda 1_V \mid \lambda \in \mathbb{C}\}$. Therefore, $Z(G)$ is isomorphic to a finite subgroup of $\mathbb{C} - \{0\}$. Using theorem 1.2.4 we get:

Corollary 2.10.4 *If G has a faithful and irreducible representation, then $Z(G)$ is cyclic.*

In particular,

Corollary 2.10.5 *Let G be Abelian.*

(1) *All irreducible presentations of G are linear.*

(2) *G has a faithful and irreducible presentation $\Leftrightarrow$ G is cyclic.*

Proof The only thing that remains to be proved is the $\Leftarrow$ part of (2). Let $|G| = m$ and let σ be a generating element of G and consider the linear representation of G defined by assigning a primitive mth root of unity to σ. ■

2.10.2 Characters

Let ρ be a representation of G on V and let P be the matrix representation of ρ with regard to some fixed basis of V. The map from G into $\mathbb{C}$ assigning to every $\sigma \in G$ the trace of $P(\sigma)$ is called the character of ρ. We also refer to ρ as a character of the G-module V or simply as a character of G. The character of ρ is independent from the choice of the basis for V. (Let $\{v_1, \ldots, v_n\}$ and $\{u_1, \ldots, u_n\}$ be two bases for V and let P_1 and P_2 be the matrix representations of ρ with regard to these bases. For $\sigma \in G$, put $P_1(\sigma) = (\lambda_{ij})$ and

$P_2(\sigma) = (\mu_{ij})$. Then

$$v_i^{\rho(\sigma)} = \sum_j v_{ij} v_j \text{ and } u_i^{\rho(\sigma)} = \sum_j \mu_{ij} u_j. \qquad (2.10.1)$$

Let $T = (v_{ij})$ and $S = (v'_{ij})$ be matrices such that

$$v_i = \sum v_{ij} u_j \text{ and } u_i = \sum v'_{ij} v_j, \qquad (2.10.2)$$

then T and S are non-singular matrices such that $T = S^{-1}$. It follows from (2.10.1) and (2.10.2) that

$$P_2(\sigma) = T^{-1} P_1(\sigma) T,$$

hence trace $P_1(\sigma) =$ trace $P_2(\sigma)$.)

Since equivalent representations have identical matrix representations, their characters are also the same. The character of an irreducible representation is called an irreducible character. The character of a linear representation ρ coincides with its matrix representation. The character χ of a representation of G induced by a permutation representation (G, Ω) is given by:

$$\chi(\sigma) = |\{i \in \Omega \,|\, i^\sigma = i\}| \quad (\sigma \in G).$$

In particular, the character π of a regular representation of G is given by

$$\pi(\sigma) = \begin{cases} |G| & \text{if } \sigma = 1 \\ 0 & \text{if } \sigma \neq 1. \end{cases}$$

If σ and τ are conjugate elements of G, say $\sigma = \xi^{-1} \tau \xi \; (\xi \in G)$, then the matrix representation P of a representation ρ of G satisfies

$$P(\sigma) = P(\xi^{-1}) P(\tau) P(\xi).$$

Hence trace $P(\sigma) =$ trace $P(\tau)$, i.e. the value of the character of ρ at two conjugate elements is the same. Generally, a mapping from G into $\mathbb{C}$ that assumes the same value at conjugate elements is called a class function of G. The collection of all class functions of G is denoted by $CF(G)$. For $f, g \in CF(G)$ and $\lambda \in \mathbb{C}$ we define

$$(f+g)(\sigma) = f(\sigma) + g(\sigma) \quad (\sigma \in G)$$
$$(\lambda f)(\sigma) = \lambda(f(\sigma)) \quad (\sigma \in G)$$

making $CF(G)$ into a $\mathbb{C}$-module.

For a conjugacy class K of G we define $f_K : G \to \mathbb{C}$ by

$$f_K(\sigma) = \begin{cases} 1 & \text{if } \sigma \in K \\ 0 & \text{if } \sigma \notin K. \end{cases}$$

The map f_K is a class function and the collection of all f_K forms a basis for $CF(G)$. Therefore

$$\dim CF(G) = \text{class number of } G.$$

Let V be a reducible G-module, U a G-submodule of G and V/U the quotient G-module of V with respect to U. If χ, χ_1 and χ_2 represent the characters corresponding to V, U and V/U respectively, then: $\chi = \chi_1 + \chi_2$. By considering a sequence

$$V = V_0 \supsetneqq V_1 \supsetneqq \dots \supsetneqq V_n = 0$$

of G-submodules such that V_i/V_{i+1} is irreducible for $i = 0, \dots, n-1$, it is clear that a character χ can be written as a sum of irreducible characters.

Theorem 2.10.6 *Let χ be the character of a representation ρ of G of degree n. For $\sigma \in G$, $\chi(\sigma)$ is the sum of n $|G|$th roots of unity. Furthermore, $\chi(\sigma^{-1}) = \overline{\chi(\sigma)}$.*

Proof Put $\langle \sigma \rangle = H$. The representation $\rho_{|H}$ can be written as the direct sum of irreducible representations of H (theorem 2.10.2). Since H is Abelian, all irreducible representations of H are linear, hence

$$\rho_{|H} = \rho_1 \oplus \dots \oplus \rho_n$$

where $\rho_1, \dots, \rho_n$ are linear representations of H. Denoting the order of σ by m, we have $\rho_i(\sigma)^m = \rho_i(\sigma^m) = 1$, i.e. $\rho_i(\sigma)$ is an mth root of unity. Therefore $\rho_i(\sigma)^{-1} = \overline{\rho_i(\sigma)}$. Finally, $\chi(\sigma) = \sum \rho_i(\sigma)$ proving the first assertion and

$$\chi(\sigma^{-1}) = \sum \rho_i(\sigma^{-1}) = \sum \rho_i(\sigma)^{-1} = \overline{\sum \rho_i(\sigma)} = \overline{\chi(\sigma)}. \qquad \blacksquare$$

Theorem 2.10.7 *Let ρ be a representation of G of degree n and χ be the character of ρ. Then we have:*

(1) $|\chi(\sigma)| \le n, \forall \sigma \in G$.

(2) $|\chi(\sigma)| = n \Leftrightarrow \rho(\sigma)$ *is a scalar product* (i.e. $\rho(\sigma) = \lambda 1_V$ *for some* λ). *In particular, if* ρ *is faithful, then:* $|\chi(\sigma)| = n \Leftrightarrow \sigma \in Z(G)$.

(3) $\chi(\sigma) = n \Leftrightarrow \sigma \in \operatorname{Ker} \rho$.

(4) *Putting* $H_1 = \{\sigma \in G \mid \chi(\sigma) = n\}$ *and* $H_2 = \{\sigma \in G \,\big|\, |\chi(\sigma)| = n\}$, *we have* $H_1 \leq H_2$ *and* $H_i \trianglelefteq G (i = 1, 2)$.

Proof Put $\chi(\sigma) = \sum_{i=1}^{n} \lambda_i$, where λ_i are roots of unity.

(1) $|\chi(\sigma)| \leq \sum |\lambda_i| = n$.

(2) $|\chi(\sigma)| = n \Leftrightarrow \chi(\sigma)$ lies on a circle with centre O and radius n in the complex plane $\Leftrightarrow \lambda_1 = \ldots = \lambda_n \Leftrightarrow \chi(\sigma) = n\lambda$ for some root of unity $\lambda \Leftrightarrow \rho(\sigma) = \lambda 1_V$.

(3) and (4) Left to the reader. ∎

Using Schur's lemma (theorem 2.10.3) we can derive a fundamental relationship among irreducible characters, the so called orthogonal relations.

Theorem 2.10.8 (First orthogonality relations) *Let G be a finite group of order g.*

(1) *If* χ *is an irreducible character of G then*

$$\sum_{\sigma \in G} \chi(\sigma)\chi(\sigma^{-1}) = g.$$

(2) *If* χ *and* χ' *are characters of two non-equivalent, irreducible characters, then*

$$\sum_{\sigma \in G} \chi(\sigma)\chi'(\sigma^{-1}) = 0.$$

Proof (1) Let V be a n-dimensional, irreducible G-module, ρ an irreducible representation of G on V and χ the character of ρ. For $f \in \operatorname{Hom}_{\mathbb{C}}(V, V)$ and $\tau \in G$, we have

$$\rho(\tau)\left(\sum_{\sigma \in G} \rho(\sigma) f \rho(\sigma^{-1}) \right) = \left(\sum_{\sigma \in G} \rho(\tau\sigma) f \rho(\sigma^{-1}\tau^{-1}) \right) \rho(\tau)$$

$$= \left(\sum_{\sigma \in G} \rho(\sigma) f \rho(\sigma^{-1}) \right) \rho(\tau).$$

Therefore, $\sum_{\sigma \in G} \rho(\sigma) f \rho(\sigma^{-1})$ is a G-homomorphism from V into V. According to Schur's lemma, this G-homomorphism has to be a scalar multiplication. Choosing a basis for V, let P be the matrix representation of ρ with regard to this basis. Then we have

$$\sum_{\sigma\in G} P(\sigma)AP(\sigma^{-1}) = \lambda_A E, \quad \forall A\in(\mathbb{C})_n. \tag{2.10.3}$$

Hence: g (trace A) $= n\lambda_A$. Applying this to the matrix A_{ij} with 1 on the (i,j)-entry and 0 elsewhere, we get $\lambda_{A_{ij}} = \delta_{ij}g/n$. Therefore, putting $P(\sigma) = (\lambda_{ij}(\sigma))$, and setting $A = A_{kl}$ in (2.10.3) we get

$$\sum_{\sigma\in G} \lambda_{ik}(\sigma)\lambda_{lj}(\sigma^{-1}) = \delta_{ij}\delta_{kl}g/n.$$

Therefore

$$\sum_{\sigma\in G} \chi(\sigma)\chi(\sigma^{-1}) = \sum_{\sigma\in G}\left(\sum_i \lambda_{ii}(\sigma)\right)\left(\sum_j \lambda_{jj}(\sigma^{-1})\right) = g.$$

(2) Let ρ be an irreducible representation of G on V and let ρ' be an irreducible representation of degree n' of G on V', that is not equivalent to ρ, and let χ' be the character of ρ'. In the same way as above, choosing bases for V and V' and representing ρ, ρ' and the elements of $\mathrm{Hom}_{\mathbb{C}}(V, V')$ by matrices, we get, from Schur's lemma,

$$\sum P(\sigma)AP'(\sigma^{-1}) = 0 \quad \forall A\in(K)_{(n,n')}, \tag{2.10.4}$$

where $P(\sigma)$, $P'(\sigma)$ are the matrix representations of ρ and ρ' respectively with regard to the given bases. Setting $A = A_{kl}$ in (2.10.4), we get

$$\sum_\sigma \lambda_{ik}(\sigma)\mu_{lj}(\sigma^{-1}) = 0 \quad (1 \le i, k \le n; 1 \le l, j \le n')$$

Therefore

$$\sum_\sigma \chi(\sigma)\chi'(\sigma^{-1}) = \sum_\sigma \left(\sum \lambda_{ii}(\sigma)\right)\left(\sum \mu_{jj}(\sigma^{-1})\right) = 0. \qquad ■$$

Let $\chi_1, \ldots, \chi_t$ be a finite number of irreducible characters of G and let $\lambda_1, \ldots, \lambda_t \in \mathbb{C}$ be such that $\sum_i \lambda_i \chi_i = 0$, then

$$\begin{aligned} 0 &= \sum_\sigma\left(\sum_i \lambda_i\chi_i(\sigma)\right)\chi_j(\sigma^{-1}) = \sum_i \lambda_i\left(\sum_\sigma \chi_i(\sigma)\chi_j(\sigma^{-1})\right) \\ &= |G|\lambda_j \quad (1 \le j \le t). \end{aligned}$$

Therefore $\lambda_1 = \ldots = \lambda_t = 0$, *i.e.* $\chi_1, \ldots, \chi_t$ are linearly independent in the $\mathbb{C}$-vector space $CF(G)$. Hence the number of irreducible characters of G is finite and less than or equal to the class number of $G(=\dim_{\mathbb{C}} CF(G))$. Now, let $\chi_1, \ldots, \chi_t$ represent all characters of G. Since every character can be written as a sum of irreducible characters, the character π of the regular representation of G can be written as

$$\pi = \sum_{i=1}^{t} m_i \chi_i,$$

where the m_i are non-negative integers. Hence by theorem 2.10.8:

$$\sum_{\sigma} \pi(\sigma)\chi_j(\sigma^{-1}) = \sum_{i} m_i \Big(\sum_{\sigma} \chi_i(\sigma)\chi_j(\sigma^{-1}) \Big) = |G| m_j.$$

On the other hand, from $\pi(1) = |G|$ and $\pi(\sigma) = 0$ $(\forall \sigma \in G - \{1\})$ we conclude:

$$\sum_{\sigma} \pi(\sigma)\chi_j(\sigma^{-1}) = |G| \chi_j(1).$$

Therefore $\chi_j(1) = m_j$, hence:

$$\pi = \sum_{i=1}^{t} \chi_i(1)\chi_i \tag{2.10.5}$$

Let ρ be an irreducible representation of G on V; and let χ be the character of ρ. The mapping ρ from G into $\mathrm{Hom}_{\mathbb{C}}(V, V)$ can be extended to a mapping from the group ring $\mathbb{C}(G)$ into $\mathrm{Hom}_{\mathbb{C}}(V, V)$ by putting

$$\rho(\sum \lambda_\sigma \sigma) = \sum \lambda_\sigma \rho(\sigma).$$

From the definition, the extended ρ is a linear mapping between the vector spaces $\mathbb{C}(G)$ and $\mathrm{Hom}_{\mathbb{C}}(V, V)$ and it is easy to see that ρ is also a homomorphism for the rings. Since, for every conjugacy class K of G, $\sum_{\sigma \in K} \sigma = \bar{K}$ is in the centre of $\mathbb{C}(G)$, $\rho(\bar{K})$ commutes with all $\rho(\sigma)$ $(\sigma \in G)$ and so $\rho(\bar{K}) \in \mathrm{Hom}_G(\rho, \rho) = \mathbb{C} 1_V$, i.e. $\rho(\bar{K})$ is a scalar multiplication on V. Now, let $K_1, \ldots, K_s$ represent all conjugacy classes of G and put $\rho(\bar{K}_i) = \omega_i 1_V$ $(\omega_i \in \mathbb{C})$. Since, as we saw in §1.3,

$$\bar{K}_i \bar{K}_j = \sum_{k} c_{ijk} \bar{K}_k \quad (c_{ijk} \text{ non-negative integers})$$

we get

$$\omega_i \omega_j = \sum_k c_{ijk} \omega_k. \tag{2.10.6}$$

From $\omega_i 1_V = \rho(\bar{K}_i) = \sum_{\sigma \in K_i} \rho(\sigma)$ we get, by taking their traces,

$$n\omega_i = |K_i| \chi(\sigma_i), \tag{2.10.7}$$

where σ_i represents an arbitrary element of K_i and $n = \dim V = \chi(1)$. Substitution in (2.10.6) gives

$$\chi(\sigma_i)\chi(\sigma_j) = \sum_k c_{ijk} \frac{|K_k|}{|K_i||K_j|} \chi(1)\chi(\sigma_k).$$

Applying this to $\chi_1, \ldots, \chi_t$ and adding them gives

$$\sum_{\lambda=1}^{t} \chi_\lambda(\sigma_i)\chi_\lambda(\sigma_j) = \sum_k c_{ijk} \frac{|K_k|}{|K_i||K_j|} \left(\sum_\lambda \chi_\lambda(1)\chi_\lambda(\sigma_k) \right)$$

$$= \sum_k c_{ijk} \frac{|K_k|}{|K_i||K_j|} \pi(\sigma_k) = \frac{c_{ij1}|G|}{|K_i||K_j|}.$$

Since $c_{ij1} = |\{(x, y) | x \in K_i, y \in K_j, xy = 1\}|$ (see §1.3), we get $c_{ij1} = |K_i| \delta_{K_i^{-1}, K_j}$ (where $K_i^{-1} = \{\sigma^{-1} | \sigma \in K_i\}$ and this is also a conjugacy class of G). Therefore we get the following theorem:

Theorem 2.10.9 (Second orthogonality relations) *Let G be a finite group of order g and let $\chi_1, \ldots, \chi_t$ represent all irreducible characters of G. If $\sigma, \tau \in G$, then:*

$$\sum_{i=1}^{t} \chi_i(\sigma)\chi_i(\tau) = \begin{cases} \dfrac{g}{|K|} & \text{if } \sigma \text{ and } \tau^{-1} \text{ belong to the same conjugacy class } K \\ 0 & \text{if } \sigma \text{ and } \tau^{-1} \text{ are not conjugate.} \end{cases}$$

Corollary 2.10.10 *The number of irreducible characters of G equals the class number of G. Therefore the irreducible characters form a basis for $CF(G)$.*

Proof Let $\chi_1, \ldots, \chi_t$ represent all irreducible characters of G, let $K_1, \ldots, K_s$ represent all conjugacy classes of G and put $|K_i| = h_i$.

Choose elements $\sigma_i \in K_i (i = 1, \ldots, s)$. Let A be the (t, s)-matrix with $\chi_i(\sigma_j)$ as (i, j)-th entry and let B be the (s, t)-matrix with $h_i \chi_j(\sigma_i^{-1})$ as (i, j)-th entry. Then by the first and second orthogonality relations

$$AB = \begin{bmatrix} |G| & & 0 \\ & \ddots & \\ 0 & & |G| \end{bmatrix} = BA.$$

Therefore AB and BA are both non-singular matrices, so $t \leq s$ and $s \leq t$, i.e. $s = t$. ■

Theorem 2.10.11 (1) *The values of a character are algebraic integers.*

(2) *If χ is an irreducible character, K a conjugacy class of G and $\sigma \in K$, then $|K|\chi(\sigma)/\chi(1)$ is an algebraic integer.*

Proof (1) According to theorem 2.10.6 the values of a character are sums of roots of unity and therefore algebraic integers by lemma 1.1.3.

(2) Noting that $c_{ijk} \in \mathbb{Z}$ we conclude from (2.10.6), (2.10.7) and lemma 1.1.2 that $|K|\chi(\sigma)/\chi(1)$ is an algebraic integer. ■

An element of $CF(G)$ that is a linear combination with rational integer coefficients of irreducible characters of G is called a generalized character of G. The characters of G are the generalized characters with non-negative coefficients. Since the irreducible characters form a basis for $CF(G)$ the coefficients of a generalized character are unambiguously determined. Since by Maschke's theorem representations of G are completely reducible, every representation can be written as a direct sum of irreducible representations and the irreducible representations appearing in this direct sum and their multiplicities are unambiguously determined. Hence:

Theorem 2.10.12 *A representation of G is completely determined by its character, that is: if ρ_1 and ρ_2 are representations of G with characters χ_1 and χ_2 respectively, then*

$$\rho_1 \textit{ and } \rho_2 \textit{ are equivalent} \Leftrightarrow \chi_1 = \chi_2.$$

The irreducible representations appearing in the direct-sum

representation of a representation are called the irreducible components of that representation.

For $f, g \in CF(G)$,

$$(f, g) = \frac{1}{|G|} \sum_{\sigma} f(\sigma)\overline{g(\sigma)}$$

is called the inner product of f and g in $CF(G)$. For clarity's sake we sometimes write $(f, g)_G$ instead of (f, g). According to the orthogonality relations $(\chi_i, \chi_j) = \delta_{ij}$, therefore we have for $f = \sum_{i=1}^{t} m_i \chi_i \in CF(G)$:

$$(f, \chi_j) = \left(\sum_{i=1}^{t} m_i \chi_i, \chi_j \right) = \sum_{i=1}^{t} m_i (\chi_i, \chi_j) = m_j.$$

Hence f can be represented as

$$f = \sum_{i=1}^{t} (f, \chi_i)\chi_i.$$

For $f \in CF(G)$ we put

$$\| f \| = \sqrt{(f, f)},$$

and we call $\| f \|$ the norm of f.

2.10.3 Induced characters

Let H be a subgroup of G and let $G = H\sigma_1 + \ldots + H\sigma_r$ be the left coset decomposition of G. For a representation ρ of H we will define a representation ρ^G of G in the following way. Let V be an H-module corresponding with ρ and let $V_1, \ldots, V_r$ be r H-modules that are all isomorphic (as H-modules) with V, and choose an isomorphism $f_i : V \to V_i$ for each i. Then every element of V_i can be represented as $f_i(v)$ for some $v \in V$. Put $V^G = V_1 \oplus \ldots \oplus V_r$. Let $\sigma \in G$, then for every i there is a unique j such that $\sigma_i \sigma \in H\sigma_j$, i.e. $\sigma_i \sigma \sigma_j^{-1} \in H$. By putting

$$f_i(v)^{\sigma} = f_j(v^{\sigma_i \sigma \sigma_j^{-1}}) \in V_j,$$

we get a mapping from V^G into V^G determined by σ, and it is easy to see that V^G becomes a G-module by this action. This representation of G on V^G is denoted by ρ^G and called the induced representation of G from ρ. The character χ^G of ρ^G is called the induced character of G from the character χ of ρ. For $\sigma \in G$, $\chi^G(\sigma)$ is given by

$$\chi^G(\sigma) = \sum_{i=1}^{r} \dot{\chi}(\sigma_i \sigma \sigma_i^{-1}) = \frac{1}{|H|} \sum_{\tau \in G} \dot{\chi}(\tau \sigma \tau^{-1})$$

where

$$\dot{\chi}(\tau) = \begin{cases} \chi(\tau) & \text{if } \tau \in H \\ 0 & \text{if } \tau \notin H. \end{cases}$$

Generally, if f is a class function of H, the mapping $f^G : G \to \mathbb{C}$ defined by:

$$f^G(\sigma) = \frac{1}{|H|} \sum_{\tau \in G} \dot{f}(\tau \sigma \tau^{-1}),$$

where

$$\dot{f}(\tau) = \begin{cases} f(\tau) & \text{if } \tau \in H \\ 0 & \text{if } \tau \notin H, \end{cases}$$

is a class function of G, called the induced class function of G from f. It is easy to check that if f is a generalized character, so is f^G.

A class function of G naturally induces a class function of $H (H \leq G)$, called the restriction of f to H and denoted by $f_{|H}$. If f is a character, so is $f_{|H}$; if f is a generalized character, so is $f_{|H}$.

Theorem 2.10.13 *If ψ is a class function of a finite group G and if θ is a class function of a subgroup H of G, then:*

$$(\theta^G, \psi)_G = (\theta, \psi_{|H})_H.$$

Proof

$$\begin{aligned}
(\theta^G, \psi)_G &= \frac{1}{|G|} \sum_{\tau \in G} \theta^G(\tau) \overline{\psi(\tau)} = \frac{1}{|G||H|} \sum_{\tau \in G} \sum_{\sigma \in G} \theta(\sigma \tau \sigma^{-1}) \overline{\psi(\tau)} \\
&= \frac{1}{|G||H|} \sum_{\substack{\tau, \sigma \in G \\ \sigma \tau \sigma^{-1} \in H}} \theta(\sigma \tau \sigma^{-1}) \overline{\psi(\tau)} \\
&= \frac{1}{|G||H|} \sum_{\substack{\tau \in H \\ \sigma \in G}} \theta(\tau) \overline{\psi(\sigma^{-1} \tau \sigma)}
\end{aligned}$$

$$= \frac{1}{|H|} \sum_{\tau \in H} \theta(\tau) \left(\frac{1}{|G|} \sum_{\sigma \in G} \overline{\psi(\sigma^{-1}\tau\sigma)} \right) = \frac{1}{|H|} \sum_{\tau \in H} \theta(\tau)\overline{\psi(\tau)}$$
$$= (\theta, \psi_{|H})_H. \qquad \blacksquare$$

Using this we get the following theorem:

Theorem 2.10.14 (Reciprocity theorem of Frobenius) *Let $\chi_1, \ldots, \chi_t$ represent all irreducible characters of a finite group G and let $\theta_1, \ldots, \theta_s$ represent all irreducible characters of a subgroup H of G. Then:*

$$\chi_{i|H} = \sum_{j=1}^{s} m_{ij}\theta_j \Leftrightarrow \theta_j^G = \sum_{i=1}^{r} m_{ij}\chi_i.$$

Proof According to the previous theorem, we have:

$$(\chi_i, \theta_j^G)_G = (\chi_{i|H}, \theta_j)_H \quad \forall i, j,$$

from which the assertion follows. $\blacksquare$

Let (G, Ω) be a permutation representation of G of degree n. The character χ of the permutation matrix representation of (G, Ω) is called a permutation character. Putting $\Omega = \{1, \ldots, n\}$ we have for $\sigma \in G$

$$\chi(\sigma) = |\{i \in \Omega \mid i^\sigma = i\}|.$$

Now, let (G, Ω) be transitive and let H be the stabilizer of $1 \in \Omega$, i.e. $H = \{\sigma \in G \mid 1^\sigma = 1\}$. Choosing $\sigma_i \in G$ for $i = 1, \ldots, n$ such that $1^{\sigma_i} = i$, we get the left coset decomposition of G with respect to H as

$$G = H\sigma_1 + \ldots + H\sigma_n.$$

Using

$$i^\sigma = i \Leftrightarrow 1^{\sigma_i\sigma\sigma_i^{-1}} = 1 \Leftrightarrow \sigma_i\sigma\sigma_i^{-1} \in H,$$

we find for the induced representation of G from the identity representation 1_H of H

$$(1_H)^G(\sigma) = \sum_i \dot{1}_H(\sigma_i\sigma\sigma_i^{-1}) = |\{i \mid \sigma_i\sigma\sigma_i^{-1} \in H\}| = \chi(\sigma).$$

Hence:

Theorem 2.10.15 *Let (G, Ω) be a transitive permutation representation of G and let χ be the character of its permutation matrix representation. Then χ equals the character induced on G from the identity character of the stabilizer of an arbitrary point of Ω.*

As an application of the theory of group characters, we shall prove that a group of order $p^a q^b$ (p and q prime) is solvable. We first need a lemma.

Lemma 2.10.16 *Let ρ be an irreducible representation of the finite group G, let χ be its character and put $n = \chi(1)$ ($= \deg \rho$). Let K be a conjugacy class of G and put $h = |K|$. If $(h, n) = 1$, then for $\sigma \in G$ we have*

$$\chi(\sigma) = 0 \text{ or } \rho(\sigma) \in Z(\rho(G)).$$

Proof Suppose $\rho(\sigma) \notin Z(\rho(G))$, then $|\chi(\sigma)| < n$ by theorem 2.10.7. Let the order of σ be m, then $\chi(\sigma)$ can be written as the sum of n mth roots of unity:

$$\chi(\sigma) = \sum_{i=1}^{n} \lambda_i,$$

according to theorem 2.10.6. Let ω be a primitive mth root of unity, then $\lambda_i = \omega^{n_i}$ for certain non-negative integers n_i. Therefore $\chi(\sigma)$ can be written as a polynomial expression in ω, say $\chi(\sigma) = f(\omega)$. Let $\omega = \omega_1, \omega_2, \ldots, \omega_{\varphi(m)}$ represent all primitive mth roots of unity, then $\prod_{i=1}^{\varphi(m)} f(\omega_i)$ is invariant under permutations of $\{\omega_1, \ldots, \omega_{\varphi(m)}\}$. Hence $\prod_{i=1}^{\varphi(m)} f(\omega_i)$ can be expressed as a polynomial expression with rational coefficients in the elementary symmetric polynomials $\alpha_1, \ldots, \alpha_{\varphi(m)}$ of $\omega_1, \ldots, \omega_{\varphi(m)}$ (lemma 1.1.4). Since, as we saw in chapter 1, the $\alpha_1, \ldots, \alpha_{\varphi(m)}$ are the coefficients of the mth cyclotomic polynomial $\Phi_m(X)$, they are rational integers, hence $\prod_{i=1}^{\varphi(m)} f(\omega_i)$ is a rational number. From $\omega_i = \omega^{m_i}$ for some m_i with $(m, m_i) = 1$ we get $f(\omega_i) = \chi(\sigma^{m_i})$. Therefore $\prod_{(m_i, m) = 1, 1 \le m_i < m} |h\chi(\sigma^{m_i})/n|$ is rational. Since this product is an algebraic integer by theorem 2.10.11, we conclude from lemma 1.1.1 that it is a rational integer. Hence, since $(h, n) = 1$, we infer that $\prod_{(m, m_i) = 1, 1 \le m_i < m} |\chi(\sigma^{m_i})/n|$ is a rational integer. According to theorem 2.10.7 $|\chi(\sigma^{m_i})| \le n$, while $|\chi(\sigma)| < n$,

hence

$$\prod_{\substack{(m_i,m)=1\\1\le m_i<m}} |\chi(\sigma^{m_i})/n| < 1,$$

i.e.

$$\prod_{\substack{(m_i,m)=1\\1\le m_i<m}} |\chi(\sigma^{m_i})/n| = 0.$$

Therefore $\chi(\sigma^{m_i})=0$ for some i, hence $f(\omega_i)=\chi(\sigma^{m_i})=0$. Since $\Phi_m(X)$ is an irreducible polynomial by theorem 1.3.14 and ω_i is a root of $\Phi_m(X), f(X)$ is a multiple of $\Phi_m(X)$ and hence $\chi(\sigma)= f(\omega)=0$. ■

Theorem 2.10.17 (1) *Let G be a non-Abelian group. If there exists a conjugacy class in G that is not equal to* $\{1\}$ *and the number of elements of which is a prime power, then G is not simple.*

(2) (Burnside) *A group G of order* $p^a q^b$ *(p, q prime) is solvable.*

Proof (1) Let us suppose that G is simple. Let $\chi_1 = 1_G, \chi_2, \ldots, \chi_t$ represent all irreducible representations of G, let $K_1 = \{1\}, \ldots, K_t$ represent all conjugacy classes of G and put $h_i = |K_i| (i = 1, \ldots, t)$. By the assumptions, there is an $i_0 \ge 2$ such that $h_{i_0} = p^a \neq 1$ (p prime). The character π of the regular representation of G can be written as $\pi = \sum_{i=1}^{t} \chi_i(1)\chi_i$ by (2.10.5). Therefore we have for $\sigma \in K_{i_0} (\sigma \neq 1)$ that

$$1 + \chi_2(1)\chi_2(\sigma) + \ldots + \chi_t(1)\chi_t(\sigma) = \pi(\sigma) = 0. \qquad (2.10.8)$$

If $p \nmid \chi_i(1)$, $\chi_i(\sigma)=0$ by theorem 2.10.16. Hence from (2.10.8) there exists an algebraic integer α satisfying:

$$1 + p\alpha = 0.$$

Hence: $\alpha = -1/p$ is a rational number, in contradiction to lemma 1.1.1.

(2) It suffices to prove that either G is Abelian or G is not simple. Assume G is not Abelian. Let $P \neq 1$ be a Sylow subgroup of G and let $\sigma \neq 1$ be an element of the centre of P. From $P \le \mathscr{C}_G(\sigma)$ we con-

clude that the number of elements of the conjugacy class of G containing σ is a prime power, hence G is not simple by (1). ■

2.11 Frobenius groups

A group G containing a proper subgroup H (i.e. $1 \subsetneq H \subsetneq G$) satisfying the following condition:

$$H \cap H^{\sigma} = \{1\} \quad \forall \sigma \in G - H \tag{2.11.1}$$

is called a Frobenius group. The subgroup H is called a Frobenius complement in G. Notice that since the structure of a Frobenius group depends upon a subgroup H satisfying (2.11.1), we should say a 'Frobenius group with respect to a subgroup H'. But later, in theorem 2.11.6, we will prove that if G contains subgroups H_1 and H_2 satisfying (2.11.1), then H_1 and H_2 are G-conjugate, so that the structure as a Frobenius group is uniquely determined. So we may say simply a 'Frobenius group' instead of a 'Frobenius group with respect to a subgroup'.

Now we give a necessary and sufficient condition for being a Frobenius group in terms of permutation groups.

Theorem 2.11.1 *G is a Frobenius group if and only if there exists a set Ω on which G operates transitively such that the following condition is satisfied:*

$$\begin{aligned} &G_a \neq 1 \quad \text{for any } a \in \Omega, \text{ and} \\ &G_{a,b} = 1 \quad \text{for any } a, b \in \Omega,\ a \neq b. \end{aligned} \tag{2.11.2}$$

Proof $\Rightarrow$: Let G be a Frobenius group and let H be a Frobenius complement in G. Put $\Omega = H\backslash G = \{Hx_1, \ldots, Hx_n\}$ for certain $x_1, \ldots, x_n \in G$ and let G operate on Ω in the obvious way. (G, Ω) is clearly transitive. Suppose that $\sigma \in G$ leaves two elements Hx_i and $Hx_j (Hx_i \neq Hx_j)$ fixed, i.e. $Hx_i\sigma = Hx_i$ and $Hx_j\sigma = Hx_j$. Then $\sigma \in H^{x_i} \cap H^{x_j}$ and since $H^{x_i} \cap H^{x_j} = 1$ (because $x_i x_j^{-1} \notin H$) we conclude $\sigma = 1$. Hence (G, Ω) satisfies the second condition of (2.11.2). Since the subgroup of G that leaves $H \in \Omega$ fixed is $H (\neq 1)$, the first condition of (2.11.2) is also satisfied.

$\Leftarrow$: Let (G, Ω) be a permutation representation satisfying (2.11.2).

Put $H = G_a$ for some $a \in \Omega$. Then for $\sigma \in G - H$ we have $H^\sigma = (G_a)^\sigma = G_{a^\sigma}$ and $a^\sigma \neq a$, hence $H \cap H^\sigma = G_a \cap G_{a^\sigma} = G_{a,a^\sigma} = 1$. Since H is clearly a proper subgroup of G, we conclude that G is a Frobenius group. ∎

When there is no danger of confusion, we will also call transitive permutation representations (G, Ω) satisfying (2.11.2) Frobenius groups. In this case, the groups G_a leaving one point a fixed are Frobenius complements. It is easy to verify that $|G_a| \mid (|\Omega| - 1)$. In particular, when $|G_a| = |\Omega| - 1$, then (G, Ω) is called a complete Frobenius group.

Let G be a Frobenius group, H a Frobenius complement in G. Since $H \cap H^\sigma = 1$, $\forall \sigma \in G - H$, we have: $\mathcal{N}_G(H) = H$. Hence the number of subgroups of G that are conjugate to H is $n = |G:H|$. Let $H_1, \ldots, H_n$ represent all subgroups conjugate to H, then $N_0 = G - \bigcup_{i=1}^n H_i$ is just the set of all elements of G that are not conjugate to any element of H. Putting $|H| = m$, we have $|N_0| = |G| - |\bigcup_{i=1}^n H_i| = nm - n(m-1) - 1 = n - 1$. We will prove that $N = N_0 \dot{\cup} \{1\}$ is a normal subgroup of G as an application of the theory of group characters.

Theorem 2.11.2 (Frobenius) *Let G be a Frobenius group, H a Frobenius complement in G. Then the subset*

$$N = \left(G - \bigcup_{\sigma \in G} H^\sigma \right) \cup \{1\}$$

is a normal subgroup of G. Therefore, $N \lhd NH = G$ and $N \cap H = \{1\}$, i.e. G is the semi-direct product of N and H.

Proof Let f be a generalized character of H such that $f(1) = 0$. From

$$f^G(\sigma) = \frac{1}{|H|} \sum_{\tau \in G} \dot{f}(\tau\sigma\tau^{-1})$$

we conclude $f^G(1) = 0$. Furthermore, it is easy to check that for $\sigma \in G - \bigcup_{\tau \in G} H^\tau$ we have $f^G(\sigma) = 0$ and that for $\sigma \in H$ we have $f^G(\sigma) = f(\sigma)$, namely $f^G{}_{|H} = f$. Therefore, letting f' represent another generalized character of H satisfying $f'(1)=0$, we have by theorem 2.10.13:

$$(f^G, f'^G)_G = (f^G{}_{|H}, f')_H = (f, f')_H.$$

For an arbitrary group X we put

$$I_0(X) = \{f \mid f \text{ is generalized character of } X \text{ and } f(1)=0\}.$$

Then the mapping from $I_0(H)$ into $I_0(G)$ defined by $f \to f^G$ preserves the inner products defined on $I_0(H)$ and $I_0(G)$. Let $\varphi_1 = 1_H$, $\varphi_2, \ldots, \varphi_k$ represent all irreducible characters of H and put $n_i = \deg \varphi_i (= \varphi_i(1))$. For $i = 2, \ldots, k$ define the generalized character α_i of H by

$$\alpha_i = n_i 1_H - \varphi_i$$

Then α_i satisfy $\alpha_i \in I_0(H)$ and $(\alpha_i, \alpha_i)_H = n_i^2 + 1$. From $(\alpha_i^G, 1_G) = (\alpha_i, 1_H) = n_i$ and $(\alpha_i^G, \alpha_i^G) = n_i^2 + 1$ and $\alpha_i^G \in I_0(G)$, we conclude that for every i there exists an irreducible character χ_i such that $\alpha_i^G = n_i 1_G - \chi_i$ $(2 \leq i \leq k)$. Putting $\tilde{N} = \bigcap_{i=2}^k \operatorname{Ker} \rho_i$, where ρ_i are irreducible representations of G having χ_i as characters, we have $\tilde{N} \trianglelefteq G$. Let $\sigma \in \tilde{N} \cap H$. Then from $\sigma \in \tilde{N}$, $\chi_i(\sigma) = \chi_i(1) = n_i$ $(i = 2, \ldots, k)$ by theorem 2.10.7 (3). On the other hand, we conclude from $\sigma \in H$ that $\alpha_i^G(\sigma) = \alpha_i(\sigma)$, i.e. $\varphi_i(\sigma) = \chi_i(\sigma) = n_i$ $(= \deg \varphi_i)$ $(2 \leq i \leq k)$. Hence the character π of the regular representation of H satisfies $\pi(\sigma) = |H|$, and therefore $\sigma = 1$, i.e. $\tilde{N} \cap H = 1$. Hence $\tilde{N} \subseteq N$. Now, if $\tau \in N$, then as we saw before, $\alpha_i^G(\tau) = 0$ and $\chi_i(\tau) = n_i$, i.e. $\tau \in \tilde{N}$ and $N = \tilde{N}$. The last part of the theorem follows from $|N| = n = |G:H|$. ■

The subgroup N of G in theorem 2.11.2 is called the Frobenius kernel of G.

Theorem 2.11.3 *For a group G the following conditions are equivalent.*

(1) *G is a Frobenius group and the order of one of its Frobenius complements equals m.*

(2) *G has a proper normal subgroup N (i.e. $1 \neq N \ntrianglelefteq G$) satisfying: $|G:N| = m$ and if $\sigma \in N (\sigma \neq 1)$ then $\mathscr{C}_G(\sigma) \leq N$.*

(3) *There are natural numbers m and n such that: $|G| = mn$, $(m, n) = 1$, the order of any element of G is a divisor of either n or m and $\{\sigma \mid \sigma^n = 1\}$ forms a non-trivial normal subgroup of G.*

Proof (1)$\Rightarrow$(2). Let N be the Frobenius kernel of G corresponding with the Frobenius complement of order m. Then obviously $1 \neq N \trianglelefteq G$ and $|G:N| = m$. Pick $\sigma \in N$ $(\sigma \neq 1)$ and suppose $\mathscr{C}_G(\sigma) \nleq N$. Then there exists a τ such that $\mathscr{C}_G(\sigma) \cap H^\tau \neq 1$. Therefore, $H^\tau \cap H^{\tau\sigma} \neq 1$, and so we have $\sigma \in \mathscr{N}(H^\tau) = H^\tau$. From $H^\tau \cap N = 1$ we conclude $\sigma = 1$. Contradiction.

(2)$\Rightarrow$(3). Put $|N| = n$, let p be a prime divisor of n and let P and Q be Sylow p-subgroups of G and N respectively satisfying $P \geq Q$. Since $\mathscr{C}_G(\alpha) \leq N$ for $\alpha(\neq 1) \in Q$, we have $Z(P) \leq N$. Hence again since $P \leq \mathscr{C}_G(\tau) \leq N$ for $\tau(\neq 1) \in Z(P)$, we conclude $P = Q$. Thus we have $|G| = mn$ and $(m, n) = 1$. By the theorem of Schur–Zassenhaus, there exists a subgroup H of G such that $G = HN$ and $H \cap N = 1$. Picking $\sigma \in G - N$ we have $\sigma^m \in N$ since $|G/N| = m$. If $\sigma^m \neq 1$, then $\sigma \in \mathscr{C}_G(\sigma^m) \leq N$, contradiction. Therefore $\sigma^m = 1$ and the last two conditions of (3) are also satisfied.

(3)$\Rightarrow$(1). Put $N = \{\sigma | \sigma^n = 1\}$. By the theorem of Schur–Zassenhaus (theorem 2.7.4) there exists a subgroup H of G such that $G = HN$, $H \cap N = 1$ and $|H| = m$. Suppose that $\sigma \in G$ is such that $H \cap H^\sigma \neq 1$. Writing $\sigma = \sigma_1 \sigma_2$ with $\sigma_1 \in H, \sigma_2 \in N$, we have $H^\sigma = H^{\sigma_1\sigma_2} = H^{\sigma_2}$, hence $H \cap H^{\sigma_2} \neq 1$. Let $\tau \neq 1$ be an element from $H \cap H^{\sigma_2}$. Since

$$\sigma_2 \tau \sigma_2^{-1} \tau^{-1} = (\sigma_2 \tau \sigma_2^{-1})\tau^{-1} = \sigma_2(\tau\sigma_2^{-1}\tau^{-1}) \in H \cap N = 1,$$

we conclude: $\sigma_2 \tau = \tau \sigma_2$. Since the order of $\sigma_2 \tau$ is a multiple of $|\tau|$, and since $|\tau| \mid m$, we have $|\sigma_2| \mid m$ from the condition (3), hence $\sigma_2 = 1$, i.e. $\sigma \in H$. Therefore $H \cap H^\sigma = 1$ for $\sigma \in G - H$. ■

The following lemma is an immediate consequence of the previous theorem:

Lemma 2.11.4 *Let G be a Frobenius group, N the Frobenius kernel in G and K a subgroup of G. Then:*

(1) *If $K \nleq N$ and $K \cap N \neq 1$, then K is a Frobenius group and $K \cap N$ is the Frobenius kernel of K.*

(2) *If $1 \lneq K \trianglelefteq G$ and $K \lneq NK \lneq G$, then G/K is a Frobenius group and NK/K is the Frobenius kernel of G/K.*

Theorem 2.11.5 (Thompson) *The Frobenius kernel of a Frobenius group is nilpotent.*

Proof Let G be a Frobenius group, H a Frobenius complement and N the Frobenius kernel corresponding with H. The proof is by induction on $|G|$. According to lemma 2.11.4 (1) we may assume that $|G:N| = p$ (p a prime). Suppose first that $Z(N) \neq 1$. If $Z(N) = N$, then N is Abelian. If $Z(N) \lneq N$, then we apply lemma 2.11.4 (2) and conclude from the induction hypothesis that $N/Z(N)$ is nilpotent. Hence N is nilpotent. So now we assume $Z(N) = 1$ and we will derive a contradiction from this assumption.

We first prove that there exists a prime q, such that the Sylow q-subgroups of N are normal. To prove this we distinguish two cases.

(i) Where there exists an odd prime r such that N has no normal r-complements. Let H operate on the set of Sylow r-subgroups of N by conjugation, then there is a Sylow r-subgroup R of N that is invariant under this operation. Since $Z(R) \leqq R$ and $J(R) \leqq R$, both $\mathscr{C}_N(Z(R))$ and $\mathscr{N}_N(J(R))$ are invariant under this operation, hence $H\mathscr{C}_N(Z(R))$ and $H\mathscr{N}_H(J(R))$ are Frobenius groups by lemma 2.11.4. Since $Z(N) = 1$, we have $\mathscr{C}_N(Z(R)) \neq N$, hence $H\mathscr{C}_N(Z(R)) \lneq G$. Therefore $\mathscr{C}_N(Z(R))$ has a nilpotent normal r-complement by the induction hypothesis. If $\mathscr{N}_N(J(R)) \neq N$, then it is shown in the same way that $\mathscr{N}_N(J(R))$ has a normal r-complement. Therefore N has a normal r-complement by Thompson's theorem (theorem 2.9.3), which is contrary to the assumption. Thus we have:

$$N = \mathscr{N}_N(J(R)) \text{ and } G = HN > J(R).$$

By the induction hypothesis, the Frobenius kernel $N/J(R)$ of the Frobenius group $G/J(R)$ is nilpotent. Hence $R \lhd N$ and R is the group we are looking for.

(ii) Where N has normal r-complements for all odd primes. In this case we put

$$Q = \bigcap_{r:\text{odd prime}} (\text{normal } r\text{-complement of } N)$$

then Q is a normal Sylow 2-subgroup of N.

So, let Q denote a normal Sylow subgroup of N. Since $Z(Q) \leqq N$ we have $Z(Q) \unlhd G$, $\mathscr{C}_G(Z(Q)) \leq G$ and since $Z(N) = 1$ we have $\mathscr{C}_G(Z(Q)) \unlhd N$. Therefore $\bar{G} = G/\mathscr{C}_G(Z(Q))$ is a Frobenius group with $\bar{N} = N/\mathscr{C}_G(Z(Q))$ as Frobenius kernel and

$\bar{H} = H\mathscr{C}_G(Z(Q))/\mathscr{C}_G(Z(Q))$ as Frobenius complement. Hence $\bar{G}$ can be written as the union of $1+|\bar{N}|$ of its subgroups such that the mutual intersections of these subgroups are $\{1\}$

$$\bar{G} = \bigcup_{\bar{\sigma}\in\bar{N}} \bar{H}^{\bar{\sigma}} \cup \bar{N}.$$

$\bar{G}$ operates faithfully on the Abelian q-group $Z(Q)$. Since $(q, |\bar{N}|) = 1$, we conclude from theorem 2.6.1 that the operation of $\bar{N}$ or $\bar{H}^{\bar{\sigma}}(\bar{\sigma}\in\bar{N})$ on $Z(Q)-\{1\}$ has a fixed point. If the operation of $\bar{N}$ on $Z(Q)-\{1\}$ has a fixed point a, then $a\in Z(N)$, contrary to $Z(N)=1$. If on the other hand the operation of $\bar{H}^{\bar{\sigma}}$ on $Z(Q)-\{1\}$ has a fixed point a, then $H^{\sigma}\subseteq\mathscr{C}_G(a)$, contrary to theorem 2.11.3. ∎

Theorem 2.11.6 *A Frobenius group G has exactly one Frobenius kernel N that coincides with the Fitting subgroup of G and so its structure as a Frobenius group is unambiguously determined.*

Proof Let N be a Frobenius kernel of G. Since N is nilpotent, $N \le \mathrm{Fit}\,(G)$. Since Fit (G) is nilpotent and $1 \ne N \trianglelefteq \mathrm{Fit}\,(G)$ we have $N\cap Z(\mathrm{Fit}(G)) \ne \{1\}$ (theorem 2.3.5). Hence we have $\mathrm{Fit}\,(G) \le \mathscr{C}_G(\sigma) \le N$ for $\sigma\in N\cap Z(\mathrm{Fit}(G))$ $(\sigma \ne 1)$, therefore $\mathrm{Fit}\,(G) = N$. From this, it is easily seen that the Frobenius complement of G is uniquely determined (apart from G-conjugacy).

Theorem 2.11.7 *The Frobenius kernel of a complete Frobenius group is an elementary Abelian group.*

Proof Let G be a complete Frobenius group, let H be its Frobenius complement and N its Frobenius kernel. Choose $a\in N(a \ne 1)$. Since $\mathscr{C}_G(a)\cap H = 1$, we have $a^x \ne a^y$ for any two elements x and $y(x \ne y)$ of H. Since $|H| = |N|-1$ we have $N-\{1\} = \{a^x | x\in H\}$, i.e. the orders of the elements of $N-\{1\}$ are all equal. The theorem is a straightforward consequence of this. ∎

3

Fundamental theory of permutation groups

3.1 Permutations

Let Ω be a finite set. An element $\sigma \in S^{\Omega}$, i.e. a permutation of Ω is often represented in the following way, where we have put $\Omega = \{1, \dots, n\}$:

$$\sigma = \begin{pmatrix} 1 & 2 \dots n \\ 1^{\sigma} & 2^{\sigma} \dots n^{\sigma} \end{pmatrix} \quad \text{or } \sigma = \begin{pmatrix} \alpha \\ \alpha^{\sigma} \end{pmatrix}.$$

The collection of all fixed points of σ is denoted by $F_{\Omega}(\sigma)$ or simply by $F(\sigma)$ and called the fixed point subset of σ, i.e.

$$F_{\Omega}(\sigma) = \{\alpha \in \Omega \mid \alpha^{\sigma} = \alpha\}.$$

The number $|\Omega - F(\sigma)|$ is called the degree of σ. Elements $\sigma, \tau \in S^{\Omega}$ are called independent if

$$(\Omega - F(\sigma)) \cap (\Omega - F(\tau)) = \varnothing.$$

It is easily seen that if σ and τ are independent, then $\sigma\tau = \tau\sigma$.

If $\Omega - F(\sigma) = \{\alpha_1, \dots, \alpha_r\}$ and if $\alpha_1^{\sigma} = \alpha_2, \alpha_2^{\sigma} = \alpha_3, \dots, \alpha_r^{\sigma} = \alpha_1$ then σ is called a cycle of length r and denoted by

$$\sigma = (\alpha_1, \alpha_2, \dots, \alpha_r).$$

Let $\Delta \subset \Omega$ be an orbit of length $r(\geq 2)$ of the group $\langle \sigma \rangle$ generated by $\sigma \in S^{\Omega}$. If α is an arbitrary element of Δ, then $\alpha, \alpha^{\sigma}, \dots, \alpha^{\sigma^{r-1}}$ are all different points and $\alpha^{\sigma^r} = \alpha$. Therefore, denoting the number of orbits of $\langle \sigma \rangle$ in Ω of length $i(\geq 1)$ by $t_i(\sigma)$, the permutation σ can be represented as the product of $t_2(\sigma)$ cycles of length 2, ... and $t_n(\sigma)$ cycles of length n, that are mutually independent. Such an expression of σ as the product of cycles is called the cycle decomposition of σ. The cycles occurring in the cycle decomposition of σ are called the cycle components of σ. If τ is a cycle component of σ, then

$\Omega - F(\tau)$ is an orbit of σ. Therefore the cycle decomposition of σ is uniquely determined. The set of natural numbers $(t_1(\sigma), t_2(\sigma), \ldots, t_n(\sigma))$ is called the type of σ. Summarizing the above, we have the following theorem:

Theorem 3.1.1 (1) *Let $\sigma \in S^{\Omega}$ be a permutation of type $(t_1, \ldots, t_n)$ then σ can be represented in a unique way as the product of t_2 cycles of length 2, ... and t_n cycles of length n, that are mutually independent.*

(2) *Conversely, if σ is written as the product of mutually independent cycles, then the type of σ is given by: $(n - \sum_{i=2}^{n} m_i i,\ m_2, \ldots, m_n)$ where m_i represents the number of cycles of length i $(2 \leq i \leq n)$ occurring in the product.*

Theorem 3.1.2 *For $\sigma, \tau \in S^{\Omega}$ we have*

$$\sigma \text{ and } \tau \text{ have the same type} \Leftrightarrow \sigma \underset{S^{\Omega}}{\sim} \tau.$$

Proof $\Leftarrow$: By the definition of multiplications in S^{Ω} we have

$$\rho^{-1}\tau\rho = \begin{pmatrix} 1^{\rho} & 2^{\rho} \ldots n^{\rho} \\ 1^{\tau\rho} & 2^{\tau\rho} \ldots n^{\tau\rho} \end{pmatrix} \quad \text{for } \tau, \rho \in S^{\Omega}. \tag{3.1.1}$$

Hence, if $\tau_i = (i_1, \ldots, i_r)$ is a cycle component of τ, then $\rho^{-1}\tau_i\rho = (i_1^{\rho}, \ldots, i_r^{\rho})$ is a cycle component of $\rho^{-1}\tau\rho$ and this sets up a one-to-one correspondence between the sets of components of the same length of τ and $\rho^{-1}\tau\rho$.

$\Rightarrow$: We have to find a permutation ρ such that $\rho^{-1}\sigma\rho = \tau$. Since $|F(\sigma)| = |F(\tau)|$, we can choose a bijection from $F(\sigma)$ onto $F(\tau)$ and define ρ on $F(\sigma)$ via this bijection. Generally for each r $(2 \leq r \leq n)$, choose a bijection from the set of cycles of length r of σ onto the set of cycles of length r of τ. Let $\sigma_i = (\alpha_1, \ldots, \alpha_r)$ and $\tau_i = (\beta_1, \ldots, \beta_r)$ be corresponding cycles of length r of σ and τ respectively and define ρ on the elements $\alpha_1, \ldots, \alpha_r$ by $\alpha_j^{\rho} = \beta_j$ $(j = 1, \ldots, r)$. It is easy to check from (3.1.1) that ρ is a permutation of Ω such that $\rho^{-1}\sigma\rho = \tau$. ∎

Cycles of length 2 are called transpositions. An arbitrary cycle $(\alpha_1, \alpha_2, \ldots, \alpha_r)$ can be represented as a product of transpositions in the following way:

$$(\alpha_1, \ldots, \alpha_r) = (\alpha_1, \alpha_2)(\alpha_1, \alpha_3) \ldots (\alpha_1, \alpha_r).$$

Therefore, by theorem 3.1.1 every permutation on Ω can be written as the product of (not necessarily independent) transpositions. The following theorem is of basic importance.

Theorem 3.1.3 *Let $\sigma \in S^\Omega$. If σ is written as the product of r transpositions and also as the product of s transpositions, then $r - s \equiv 0$ (mod 2).*

Proof For a polynomial $f = f(X_1, \ldots, X_n)$ of n variables $X_1, \ldots, X_n$ over $\mathbb{Z}$ and $\sigma \in S^\Omega$ we define f^σ by

$$f^\sigma(X_1, \ldots, X_n) = f(X_{1\sigma}, \ldots, X_{n\sigma}).$$

If $\sigma, \tau \in S^\Omega$, then $f^{\sigma\tau} = (f^\sigma)^\tau$. If we put $f(X_1, \ldots, X_n) = \prod_{i<j}(X_i - X_j)$ then $f^\tau = -f$ for any transposition τ. Therefore, if σ can be written as the product of r transpositions and also as the product of s transpositions, we have $f^\sigma = (-1)^r f$ and $f^\sigma = (-1)^s f$. Since $f \neq -f$, we conclude $(-1)^r = (-1)^s$, i.e. $r \equiv s$ (mod 2). ∎

If the number of transpositions in any representation of a permutation as a product of transpositions is even, the permutation is called an even permutation, otherwise it is called an odd permutation. The collection of all even permutations of Ω is denoted by A^Ω or A_n. Since the product of two even permutations as well as the product of two odd permutations is even, we get the following theorem:

Theorem 3.1.4 *$A^\Omega \lhd S^\Omega$ and $|S^\Omega : A^\Omega| = 2$.*
A^Ω is called the alternating group on Ω (or of degree n).

Corollary 3.1.5 *If G operates on Ω and if a certain element of G induces an odd permutation of Ω, then G contains a normal subgroup of index 2.*

Proof The subgroup of all elements of G that induces an even permutation on Ω satisfies the requirements. ∎

In §1.2 we defined the right permutation representation of a group G with regard to its subgroup H by assigning to $\sigma \in G$ the permutation of $H\backslash G$ given by

$$(H\xi)^\sigma = H\xi\sigma.$$

From the definition it is easily seen that $(G, H\backslash G)$ is transitive. Conversely, let (G, Ω) be a transitive permutation representation of G. Pick a point $\alpha \in \Omega$ and put $H = G_\alpha$. Then we have, for $\tau_1, \tau_2 \in G$

$$\alpha^{\tau_1} = \alpha^{\tau_2} \Leftrightarrow H\tau_1 = H\tau_2.$$

Therefore the correspondence $\alpha^\tau \leftrightarrow H\tau$ is a one-to-one correspondence between Ω and $H\backslash G$, and this correspondence preserves the action of G, i.e.

$$(\alpha^\tau)^\sigma \leftrightarrow (H\tau)\sigma \quad \forall \sigma, \tau \in G.$$

Hence we have

$$(G, \Omega) \simeq (G, H\backslash G).$$

Thus we have proved:

Theorem 3.1.6 *Let H be a subgroup of G. By defining a permutation of $H\backslash G$ (for $\sigma \in G$) by*

$$(H\xi)^\sigma = H\xi\sigma,$$

we arrive at a transitive permutation representation $(G, H\backslash G)$ of G. Conversely, if (G, Ω) is a transitive permutation representation of G, and if H is the stabilizer of some point $a \in \Omega$, then $(G, \Omega) \simeq (G, H\backslash G)$.

The left permutation representation $(G, G/H)$ of G with regard to H is also transitive. The one-to-one correspondence between $H\backslash G$ and G/H given by $H\xi \leftrightarrow \xi^{-1}H$ satisfies:

$$(H\xi)^\sigma = H\xi\sigma \leftrightarrow \sigma^{-1}\xi^{-1}H = (\xi^{-1}H)^\sigma.$$

Therefore:

Theorem 3.1.7 $(G, H\backslash G) \simeq (G, G/H)$.
In particular, putting $H = \{1\}$, we get $(G, \{1\}\backslash G) \simeq (G, G/\{1\})$.

Theorem 3.1.8 *The kernel of* $(G, H\backslash G)$ *is* $\bigcap_{\xi\in G}\xi^{-1}H\xi$. *Therefore,* $\left|G:\bigcap_{\xi\in G}\xi^{-1}H\xi\right| \,\Big|\, n!$, *where* $n=|G:H|$. *In other words, if* G *contains a subgroup* H *of index* n, *then* G *contains a normal subgroup contained in* H *the index of which is a factor of* n!

Proof

$$\begin{aligned}\sigma\in\text{kernel of }(G, H\backslash G) &\Leftrightarrow H\xi = H\xi\sigma \quad \forall\xi\in G\\ &\Leftrightarrow \sigma\in\xi^{-1}H\xi \quad \forall\xi\in G.\end{aligned}$$ ■

Theorem 3.1.9 *A group* G *of order* $2m$, *where* m *is an odd number, contains a normal subgroup of order* m.

Proof Let $\sigma\in G$ be an element of order 2 given by the Sylow theorem. Since $\tau\sigma \neq \tau$ $\forall\tau\in G$, σ has no fixed points on G if we consider σ as a permutation in $(G, \{1\}\backslash G)$, and hence σ is written as the product of m transpositions. The assertion follows from this and corollary 3.1.5. (This is also a very special case of the general results on the existence of the normal 2-complement, e.g. theorem 2.9.2.) ■

3.2 Transitivity and intransitivity

Let G operate on Ω. For a subset X of G let $F_\Omega(X)$ (or simply $F(X)$) represent the set of those elements of Ω that are left fixed by all elements of X, i.e.

$$F_\Omega(X) = \{\alpha\in\Omega \mid \alpha^\sigma = \alpha, \forall\sigma\in X\} = \bigcap_{\sigma\in X} F_\Omega(\sigma).$$

$F_\Omega(X)$ is called the fixed point subset under X. In §1.2, we defined the pointwise stabilizer of a subset Δ of Ω by

$$G_\Delta = \{\sigma\in G \mid \alpha^\sigma = \alpha, \forall\alpha\in\Delta\} = \bigcap_{\alpha\in\Delta} G_\alpha$$

and the setwise stabilizer by

$$G_{\langle\Delta\rangle} = \{\sigma\in G \mid \Delta^\sigma = \Delta\}.$$

When we consider $\sigma\in G_{\langle\Delta\rangle}$ to be an operation on Δ, we write $\sigma_{|\Delta}$.

Since G_Δ is the kernel of the permutation representation $(G_{\langle\Delta\rangle}, \Delta)$, we have that $G_\Delta \trianglelefteq G_{\langle\Delta\rangle}$ and $G_{\langle\Delta\rangle}/G_\Delta$ can be regarded as a permutation group on Δ, that will be denoted by $G^\Delta_{\langle\Delta\rangle}$ or simply by G^Δ. We often use simplified notation if it causes no confusion; for instance we will write $G_{\alpha,\beta,\ldots}$ instead of $G_{\{\alpha,\beta,\ldots\}}$, $G_{\Delta_1,\langle\Delta_2\rangle}$ instead of $(G_{\Delta_1})_{\langle\Delta_2\rangle}$, Δ^H instead of $\bigcup_{\sigma\in H}\Delta^\sigma$ for $H \subseteq G$, etc..

Some basic facts, either already explained or easily derivable from the definitions above, are collected in the following theorem:

Theorem 3.2.1 (1) $G_\varnothing = G, \quad G_{\alpha,\beta} = G_{\beta,\alpha} \quad \alpha, \beta \in \Omega.$

(2) $G_{\Delta\cup\Gamma} = G_\Delta \cap G_\Gamma = (G_\Gamma)_\Delta = (G_\Delta)_\Gamma, \quad \Delta, \Gamma \subseteq \Omega.$

(3) $\sigma^{-1} G_\Delta \sigma = G_{\Delta^\sigma}, \quad \sigma^{-1} G_{\langle\Delta\rangle} \sigma = G_{\langle\Delta^\sigma\rangle} \quad \sigma \in G, \Delta \subseteq \Omega.$

In particular, $G_\Delta \trianglelefteq G_{\langle\Delta\rangle}$.

(4) *For* $\alpha \in \Omega$ *and* $\sigma, \tau \in G$ *we have*

$$\alpha^\sigma = \alpha^\tau \Leftrightarrow G_\alpha \sigma = G_\alpha \tau.$$

This sets up a one-to-one correspondence between $\alpha^G = \Delta$ *(the orbit containing* α*) and* $G_\alpha \backslash G$. *Therefore,*

$$|\Delta| = |\alpha^G| = |G : G_\alpha|.$$

Hence the length of an orbit of (G, Ω) *is a divisor of* $|G|$.

(5) $F(X)^\sigma = F(X^\sigma)\ (X \subseteq G, \sigma \in G).$

In particular,

$$F(G_\Delta)^\sigma = F(G_\Delta^\sigma) = F(G_{\Delta^\sigma}).$$

Theorem 3.2.2 (1) $|G : G_{\alpha,\beta}| = |\alpha^G||\beta^{G_\alpha}| = |\beta^G||\alpha^{G_\beta}|.$

(2) *If* (G, Ω) *is transitive, then the length of the orbit of* (G_β, Ω) *containing* α *is the same as the length of the orbit of* (G_α, Ω) *containing* β $(\alpha, \beta \in \Omega)$.

(3) *If* P *is a Sylow p-subgroup of* G *and if* $\alpha \in \Omega$, *then:*

$$p^m \,\|\, |\alpha^G| \Rightarrow p^m \,\|\, |\alpha^P|.$$

Conversely letting ψ *denote an orbit of* (P, α^G) *of minimal length, we have* $|\psi| \,\|\, |\alpha^G|$.

Proof (1)

$$|\alpha^G||\beta^{G_\alpha}| = |G:G_\alpha||G_\alpha:G_{\alpha,\beta}| = |G:G_{\alpha,\beta}| = |G:G_\beta||G_\beta:G_{\alpha,\beta}|$$
$$= |\beta^G||\alpha^{G_\beta}|.$$

(2) If (G,Ω) is transitive, then $\alpha^G = \beta^G = \Omega$, therefore $|\alpha^{G_\beta}| = |\beta^{G_\alpha}|$ by (1).

(3)

$$|\alpha^G||G_\alpha:P_\alpha| = |G:G_\alpha||G_\alpha:P_\alpha| = |G:P_\alpha| = |G:P||P:P_\alpha|$$
$$= |G:P||\alpha^P|.$$

Since $p \nmid |G:P|$ we get the first assertion. The second assertion follows from the fact that the length of every orbit of (P, α^G) is a power of p. ■

Theorem 3.2.3 *Let (G,Ω) be a permutation group. The following two conditions are necessary and sufficient for the order of G to be odd.*

(1) *The length of every orbit of (G,Ω) is odd.*
(2) *For every $\alpha\in\Omega$ the length of every orbit of (G_α,Ω) is odd.*

Proof If $|G|$ is odd, then (1) and (2) are true by theorem 3.2.1 (4). Let $|G|$ be even. Let σ be an element of G of order 2, then there exist $\alpha, \beta\in\Omega$ such that:

$$\sigma = \begin{pmatrix} \alpha & \beta \ldots \\ \beta & \alpha \ldots \end{pmatrix}.$$

Since $\sigma \notin G_{\alpha,\beta}$ and $G^\sigma_{\alpha,\beta} = G_{\alpha\beta}$, we infer that $|G:G_{\alpha\beta}|$ is even. Therefore, using $|G:G_{\alpha\beta}| = |\alpha^G||\beta^{G_\alpha}|$, we conclude that $|\alpha^G|$ or $|\beta^{G_\alpha}|$ is even. ■

For $\sigma\in G$ let $\alpha_1(\sigma)$ denote the number of fixed points of σ, i.e. $\alpha_1(\sigma) = |F(\sigma)|$. We saw in §2.10 that α_1 is the character of the matrix representation of G corresponding with (G,Ω). We have:

Theorem 3.2.4 *Let (G,Ω) be a permutation representation. Then*

$$(\alpha_1, 1_G) = \frac{1}{|G|}\sum_{\sigma\in G} \alpha_1(\sigma) = \textit{number of orbits of } (G,\Omega).$$

Proof Let us evaluate the number m of elements of $M = \{(a, \sigma) | a \in \Omega, \sigma \in G, a^\sigma = a\}$ in two different ways. For a fixed $\sigma \in G$, there are $\alpha_1(\sigma)$ elements $(a, \sigma) \in M$, i.e.

$$m = \sum_{\sigma \in G} \alpha_1(\sigma).$$

On the other hand, for a fixed $a \in \Omega$, there are $|G_a|$ elements $(a, \sigma) \in M$, i.e.

$$m = \sum_{a \in \Omega} |G_a|.$$

Let (G, Ω) have r orbits, and let $\Omega = \Delta_1 + \ldots + \Delta_r$ be the orbit decomposition of Ω. Applying theorem 3.2.1 and, using the fact that $|G_a| = |G_b|$ for $a, b \in \Delta_i$, we get

$$\sum_{a \in \Delta_i} |G_a| = |\Delta_i||G_a| = |G|.$$

Hence:

$$\sum_{\sigma \in G} \alpha_1(\sigma) = \sum_{a \in \Omega} |G_a| = \sum_{i=1}^{r} \Big(\sum_{a \in \Delta_i} |G_a| \Big) = r|G|. \qquad ■$$

Corollary 3.2.5 *If (G, Ω) is transitive, then:*

$$\sum_{\sigma \in G} \alpha_1(\sigma) = |G|.$$

Therefore, since $\alpha_1(1) = n(= |\Omega|)$, there exists an element $\sigma \in G$ such that $\alpha_1(\sigma) = 0$, i.e. an element σ that has no fixed points (supposing $n > 1$).

Theorem 3.2.6 *Let (G, Ω) be transitive. If for $a \in \Omega$ the number of orbits of (G_a, Ω) is denoted by r, we have:*

$$(\alpha_1, \alpha_1) = \frac{1}{|G|} \sum_{\sigma \in G} \alpha_1(\sigma)^2 = r.$$

Proof Let us evaluate the number of elements m of the set $M = \{(a, b, \sigma) | a, b \in \Omega, \sigma \in G, a^\sigma = a, b^\sigma = b\}$ in two different ways. For a fixed $\sigma \in G$ there are $\alpha_1(\sigma)^2$ elements $(a, b, \sigma) \in M$, i.e. $m = \sum_{\sigma \in G} \alpha_1(\sigma)^2$. On the other hand, for a fixed $a \in \Omega$ there are $\sum_{\sigma \in G_a} \alpha_1(\sigma)$

elements $(a, b, \sigma) \in M$, and this number equals $r|G_a|$ by theorem 3.2.4. Therefore:

$$\sum_{\sigma \in G} \alpha_1(\sigma)^2 = \sum_{a \in \Omega}\left(\sum_{\sigma \in G_a} \alpha_1(\sigma) \right) = \sum_{a \in \Omega} r|G_a| = r|G|. \qquad \blacksquare$$

If the length of all orbits of (G, Ω) is the same and greater than or equal to 2, then (G, Ω) is called $\frac{1}{2}$-transitive. Transitive permutation representations with $n = |\Omega| > 1$ are special examples of $\frac{1}{2}$-transitive permutation representations.

Theorem 3.2.7 *Let (G, Ω) be a transitive permutation group, and let N be such that $1 \neq N \trianglelefteq G$, then (N, Ω) is $\frac{1}{2}$-transitive and $F(N) = \varnothing$.*

Proof Let Δ be an orbit of (N, Ω). Since for $\sigma \in G$ we have $\Delta^{\sigma N} = \Delta^{\sigma N \sigma^{-1} \sigma} = \Delta^{N\sigma} = \Delta^{\sigma}$, Δ^{σ} is also an orbit of (N, Ω). Since (G, Ω) is transitive, there exist elements $\sigma_1, \ldots, \sigma_r \in G$, such that $\Omega = \Delta^{\sigma_1} + \ldots + \Delta^{\sigma_r}$ is the orbit decomposition of Ω with regard to N. If $|\Delta| = 1$, then $|\Delta^{\sigma_i}| = 1$ and $N = \{1\}$. Therefore, $|\Delta^{\sigma_1}| = \ldots = |\Delta^{\sigma_r}| \geq 2$, and so we have our assertion. $\blacksquare$

Let (G, Ω) be a permutation representation. The minimal value of the set $\Lambda(G) = \{|\Omega - F_{\Omega}(\sigma)| \,|\, \Omega \neq F_{\Omega}(\sigma), \sigma \in G\}$ is called the minimal degree of (G, Ω). If $\Lambda(G) = \varnothing$ or if the minimal degree of (G, Ω) equals $n = |\Omega|$, then (G, Ω) is called semi-regular. If (G, Ω) is semi-regular and transitive, then it is called regular. The following theorem is a straightforward consequence of the definitions.

Theorem 3.2.8 *Let (G, Ω) be a permutation group.*

(1) *(G, Ω) is semi-regular $\Leftrightarrow G_{\alpha} = \{1\}$ $\forall \alpha \in \Omega$.*

(2) *(G, Ω) is regular $\Leftrightarrow (G, \Omega)$ is transitive and $|G| = |\Omega|$.*

(3) *If (G, Ω) is semi-regular, then the length of every orbit equals $|G|$. In particular, when $G \neq 1$, (G, Ω) is $\frac{1}{2}$-transitive.*

Theorem 3.2.9 *Let (G, Ω) be a permutation group. If $\mathscr{C}_{S^{\Omega}}(G)$ operates transitively on Ω, then (G, Ω) is semi-regular.*

Proof Suppose that $\sigma \in G$ has a fixed point α. Since $(\mathscr{C}_{S^{\Omega}}(G), \Omega)$

is transitive, there exists for every $\beta \in \Omega$ an element $\tau \in \mathscr{C}_{S\Omega}(G)$ such that $\alpha^\tau = \beta$. Therefore, $\beta^\sigma = \alpha^{\tau\sigma} = \alpha^{\sigma\tau} = \alpha^\tau = \beta$, i.e. $\sigma = 1$. ■

The converse of this theorem is also true. We first prove:

Theorem 3.2.10 *If (G, Ω) is a regular permutation group, then $\mathscr{C}_{S\Omega}(G)$ also operates regularly on Ω and $G \simeq \mathscr{C}_{S\Omega}(G)$.*

Proof Since (G, Ω) is regular, we have $(G, \Omega) \simeq (G, \{1\}\backslash G)$. Therefore it suffices to prove the theorem for $(G, \{1\}\backslash G)$ which is a subgroup of the symmetric group S^G. For the subgroup

$$G^* = \left\{ \sigma^* = \begin{pmatrix} \xi \\ \sigma^{-1}\xi \end{pmatrix} \middle| \sigma \in G \right\}$$

of the symmetric group S^G we have: $G^* \simeq G$, (G^*, G) is transitive and $\mathscr{C}_{SG}(G) \geq G^*$, hence $\mathscr{C}_{SG}(G)$ operates transitively on G. Since $G \leq \mathscr{C}_{SG}(\mathscr{C}_{SG}(G))$, we see that $\mathscr{C}_{SG}(\mathscr{C}_{SG}(G))$ operates transitively on G. Therefore $\mathscr{C}_{SG}(G)$ operates semi-regularly on G by theorem 3.2.9. Hence $\mathscr{C}_{SG}(G) = G^* \simeq G$. ■

Theorem 3.2.11 *If (G, Ω) is a semi-regular permutation group, then $\mathscr{C}_{S\Omega}(G)$ operates transitively on Ω.*

Proof If there exists a subgroup H of S^Ω containing G such that (H, Ω) is regular, then $\mathscr{C}_{S\Omega}(H) \leq \mathscr{C}_{S\Omega}(G)$ and the assertion of the theorem follows from theorem 3.2.10. So we will construct such a subgroup H of S^Ω. Let $\Omega = \Delta_1 + \ldots + \Delta_r$ be the G-orbit decomposition of Ω, and pick a point $\alpha_i \in \Delta_i$ for $i = 1, \ldots, r$.

Since every point of Δ_i can be represented in a unique way as a_i^σ for some $\sigma \in G$, we can set up a correspondence $\alpha_i^\sigma \to \alpha_j^\sigma$, whereby we have $(G, \Delta_i) \simeq (G, \Delta_j)$. Therefore, if we consider $\tau = \prod_{\sigma \in G}(\alpha_1^\sigma, \alpha_2^\sigma, \ldots, \alpha_r^\sigma) \in S^\Omega$, we have $\tau \in \mathscr{C}_{S\Omega}(G)$, and $H = G\langle\tau\rangle = G \times \langle\tau\rangle$ operates regularly on Ω. ■

The following corollary follows immediately from theorems 3.2.9 and 3.2.11.

Corollary 3.2.12 *Let (G, Ω) be a transitive permutation group, then:*

(1) *$(\mathscr{C}_{S^\Omega}(G), \Omega)$ is semi-regular.*

(2) *If G is Abelian, then $\mathscr{C}_{S^\Omega}(G) = G$ and (G, Ω) is regular.*

Theorem 3.2.13 *If (G, Ω) is a transitive permutation group, then:*

$$|\mathscr{C}_{S^\Omega}(G)| = |F(G_\alpha)| \quad \textit{for } \alpha \in \Omega.$$

Proof Since $G_\alpha = G_\alpha^\sigma = G_{\alpha^\sigma}$ for $\sigma \in \mathscr{C}_{S^\Omega}(G)$, the orbit Δ of $\mathscr{C}_{S^\Omega}(G)$ that contains α is contained in $F(G_\alpha)$. Since $\mathscr{C}_{S^\Omega}(G)$ is semi-regular, it suffices to prove that $F(G_\alpha) \subseteq \Delta$, i.e. that for every $\beta \in F(G_\alpha)$ there exists an element $\tau \in \mathscr{C}_{S^\Omega}(G)$ such that $\alpha^\tau = \beta$. We can choose points $\alpha_1 = \alpha, \alpha_2, \ldots, \alpha_r \in \Omega$ such that Ω is written as the direct sum of $F(G_{\alpha_i})$, i.e.

$$\Omega = \Delta_1 + \ldots + \Delta_r \quad \text{with } \Delta_i = F(G_{\alpha_i}).$$

Let $\sigma_i \in G$ be such that $\alpha^{\sigma_i} = \alpha_i$. Then $\Delta_1^{\sigma_i} = \Delta_i$. Since $G_{\langle \Delta_1 \rangle}$ operates regularly on Δ_1, there exists an element $\tau_1 \in \mathscr{C}_{S^{\Delta_1}}(G^{\Delta_1})$ such that $\alpha^{\tau_1} = \beta$ by theorem 3.2.10. Since $\Delta_1^{\tau_1} = \Delta_1$, we have $\Delta_i^{\sigma_i^{-1}\tau_1\sigma_i} = \Delta_i$. Defining $\tau \in S^\Omega$ by $\tau_{|\Delta_i} = \sigma_i^{-1}\tau_1\sigma_i (i = 1, \ldots, r)$, we have $\tau \in \mathscr{C}_{S^\Omega}(G)$. (For an element $\sigma \in G$ induces a permutation of the set $\{\Delta_1, \ldots, \Delta_r\}$, say $\Delta_i^\sigma = \Delta_j$. Then $\Delta_1^{\sigma_i\sigma\sigma_j^{-1}} = \Delta_1$, and $\tau_1\sigma_i\sigma\sigma_j^{-1}$ and $\sigma_i\sigma\sigma_j^{-1}\tau_1$ operate in the same way on Δ_1 since $\sigma_i\sigma\sigma_j^{-1}{}_{|\Delta_1} \in G^{\Delta_1}$ and $\tau_1 \in \mathscr{C}_{S^{\Delta_1}}(G^{\Delta_1})$. Therefore for $\gamma \in \Delta_i$

$$\begin{aligned} \gamma^{\tau\sigma} &= \gamma^{\sigma_i^{-1}\cdot\tau_1\cdot\sigma_i\sigma\sigma_j^{-1}\cdot\sigma_j} = \gamma^{\sigma_i^{-1}\cdot\sigma_i\sigma\sigma_j^{-1}\cdot\tau_1\cdot\sigma_j} \\ &= \gamma^{\sigma\sigma_j^{-1}\tau_1\sigma_j} = \gamma^{\sigma\tau}, \end{aligned}$$

i.e. we have $\tau \in \mathscr{C}_{S^\Omega}(G)$.) From:

$$\alpha^\tau = \alpha^{\sigma_1^{-1}\tau_1\sigma_1} = \alpha^{\tau_1} = \beta$$

we conclude: $F(G_\alpha) \subseteq \Delta$. ∎

Theorem 3.2.14 *Let (G, Ω) be transitive and let $\Delta, \Gamma \subseteq \Omega$. If $(G_\Delta, \Omega - \Delta)$ and $(G_\Gamma, \Omega - \Gamma)$ are both transitive and if $|\Gamma| \leq |\Delta|$, then there is an element $\sigma \in G$ such that $\Gamma \subseteq \Delta^\sigma$, i.e. such that $\sigma^{-1}G_\Delta\sigma \leq G_\Gamma$.*

Proof The proof is by induction on $|\Gamma| - |\Gamma \cap \Delta|$. If $|\Gamma \cap \Delta| = |\Gamma|$

we can choose $\sigma = 1$. So we may assume $|\Gamma \cap \Delta| < |\Gamma|$, especially $\Delta \neq \Omega$. If $\Gamma \cup \Delta = \Omega$, then $(\Omega - \Gamma) \cap (\Omega - \Delta) = \varnothing$. For arbitrary $\alpha \in \Omega - \Gamma$ and $\beta \in \Omega - \Delta$, pick an element $\sigma \in G$ such that $\beta^\sigma = \alpha$. Then $(\Omega - \Delta^\sigma) \cap (\Omega - \Gamma) \neq \varnothing$, hence we get the assertion of the theorem by the induction hypothesis since $|\Delta \cap \Gamma| < |\Delta^\sigma \cap \Gamma|$. So we may assume $\Gamma \cup \Delta \neq \Omega$. $G_\Gamma, G_\Delta \leq G_{\Gamma \cap \Delta}$, G_Γ and G_Δ operate transitively on $\Omega - \Gamma$ and $\Omega - \Delta$ respectively, and $(\Omega - \Gamma) \cap (\Omega - \Delta) \neq \varnothing$, therefore $G_{\Gamma \cap \Delta}$ operates transitively on $(\Omega - \Gamma) \cup (\Omega - \Delta) = \Omega - \Gamma \cap \Delta$. Therefore, choosing $\alpha \in \Delta - (\Gamma \cap \Delta)$, $\beta \in \Gamma - (\Gamma \cap \Delta)$ arbitrarily and $\sigma \in G_{\Gamma \cap \Delta}$ such that $\alpha^\sigma = \beta$, we have $\{\beta\}, \Gamma \cap \Delta \subseteq \Gamma \cap \Delta^\sigma$. Hence the theorem follows from the induction hypothesis. ■

3.3 Primitivity and imprimitivity

Let (G, Ω) be transitive. The subset Δ of Ω is called a set of imprimitivity of (G, Ω) if for every $\sigma \in G$ either $\Delta^\sigma = \Delta$ or $\Delta^\sigma \cap \Delta = \varnothing$. The empty set, the one-point subsets and Ω itself are sets of imprimitivity, called the trivial sets of imprimitivity. If (G, Ω) has a non-trivial set of imprimitivity, then (G, Ω) is called imprimitive, otherwise primitive. Let ψ denote a non-empty set of imprimitivity and let $\{\psi_1, \ldots, \psi_r\}$ denote the different elements of the set $\{\psi^\sigma | \sigma \in G\}$, then Ω can be written as a direct sum:

$$\Omega = \psi_1 + \ldots + \psi_r.$$

We call $\{\psi_1, \ldots, \psi_r\}$ a system of sets of imprimitivity of (G, Ω).

The following theorem gives some basic properties which are easily verified.

Theorem 3.3.1 *Let (G, Ω) be transitive:*

(1) *If ψ is a set of imprimitivity, then so is ψ^σ $(\sigma \in G)$.*

(2) *If ψ and ψ' are sets of imprimitivity, so is $\psi \cap \psi'$.*

(3) *If $\Delta \subseteq \Omega$ and $\alpha \in \Omega$, then $\psi = \bigcap_{\alpha \in \Delta^\sigma, \sigma \in G} \Delta^\sigma$ is a set of imprimitivity containing α.*

(4) *If ψ is a non-empty set of imprimitivity, then $|\psi| \,\big|\, n = |\Omega|$. In particular, if $|\Omega|$ is prime, then (G, Ω) is primitive.*

(5) *The number of different systems of imprimitivity sets of (G, Ω) equals the number of sets of imprimitivity containing a given point $\alpha \in \Omega$.*

Theorem 3.3.2 *Let (G, Ω) be transitive and $\alpha \in \Omega$.*

(1) *If Δ is a set of imprimitivity containing α, then $G_{\langle\Delta\rangle}$ is a subgroup of G containing G_α and $\alpha^{G_{\langle\Delta\rangle}} = \Delta$. Conversely, if H is a subgroup of G containing G_α, then $\Delta = \alpha^H$ is a set of imprimitivity and $G_{\langle\Delta\rangle} = H$. Therefore there exists a one-to-one correspondence between the collection of sets of imprimitivity containing α and the collection of subgroups of G containing G_α. This correspondence is given by*

$$\Delta \to G_{\langle\Delta\rangle} \text{ and conversely } \alpha^H \leftarrow H.$$

Moreover if H_1 and H_2 are two subgroups of G containing G_α, then $\alpha^{H_1} \cap \alpha^{H_2} = \alpha^{H_1 \cap H_2}$ and if Δ_1 and Δ_2 are two sets of imprimitivity containing α, then

$$G_{\langle\Delta_1 \cap \Delta_2\rangle} = G_{\langle\Delta_1\rangle} \cap G_{\langle\Delta_2\rangle}.$$

(2) *In particular, (G, Ω) is primitive if and only if for $\alpha \in \Omega$ the subgroup G_α is a maximal subgroup of G.*

Proof (1) As Δ is a set of imprimitivity containing α, $\Delta^{G_\alpha} = \Delta$, hence $G_\alpha \leq G_{\langle\Delta\rangle}$. For $\beta \in \Delta$ there exists a $\sigma \in G$ such that $\alpha^\sigma = \beta$ since (G, Ω) is transitive. Since Δ is a set of imprimitivity, we have $\Delta^\sigma = \Delta$, hence $\alpha^{G_{\langle\Delta\rangle}} = \Delta$. Conversely, let $H \geq G_\alpha$ and put $\Delta = \alpha^H$. Pick $\sigma \in G$ and suppose $\Delta \cap \Delta^\sigma \neq \varnothing$. For $\beta \in \Delta \cap \Delta^\sigma$ there are τ_1, $\tau_2 \in H$ such that $\beta = \alpha^{\tau_1} = \alpha^{\tau_2\sigma}$, hence $\tau_2\sigma\tau_1^{-1} \in G_\alpha$ and so $\sigma \in H$. Therefore $\Delta^\sigma = \Delta$. The rest of the proof of (1) and the proof of (2) are almost trivial. ■

Theorem 3.3.3 *Let (G, Ω) be a transitive permutation group, and let $\{\Delta_1, \ldots, \Delta_r\}$ represent a system of sets of imprimitivity, then G is isomorphic to a subgroup of $S_m \wr S_r$, where m is given by $|\Omega| = n = mr$. In particular, the order of G is a divisor of $(m!)^r r!$.*

Proof Let $N = S_m \times \ldots \times S_m$ denote the direct product of r copies of $S_m = S^{\Delta_1}$ with itself. We make N into an S_r-group by defining:

$$(a_1, \ldots, a_r)^\sigma = (a_{1\sigma^{-1}}, \ldots, a_{r\sigma^{-1}})(\sigma \in S_r, (a_1, \ldots, a_r) \in N).$$

Then $S_m \wr S_r$ is the semi-direct product of N and S_r by definition, i.e. the set $S_m \wr S_r$ is equal to $N \times S_r$ and the group operation on $S_m \wr S_r$ is defined by

$$(a, \sigma)(b, \rho) = (ab^{\sigma^{-1}}, \sigma\rho) \quad \text{for } (a, \sigma), (b, \rho) \in N \times S_r.$$

Let $\tau_i : \Delta_1 \to \Delta_i$ represent a one-to-one mapping for $i = 1, \ldots, r$. For every $\sigma \in G$ we define a permutation

$$\tilde{\sigma} = \begin{pmatrix} 1, \ldots, r \\ 1^{\tilde{\sigma}}, \ldots, r^{\tilde{\sigma}} \end{pmatrix} \in S_r$$

by $\Delta_i^{\sigma} = \Delta_{i\tilde{\sigma}}$ and it is easy to verify that the mapping $\sigma \to \tilde{\sigma}$ is a homomorphism from G into S_r. Now, for $\sigma \in G$, we put $\sigma_i = \tau_i \sigma \tau_{i\tilde{\sigma}}^{-1} \in S^{\Delta_1}$ $(1 \leq i \leq r)$ and $\sigma^* = (\sigma_1, \ldots, \sigma_r) \in N$. We define $f : G \to S_m \wr S_r$ by

$$f(\sigma) = (\sigma^*, \tilde{\sigma}) \quad \forall \sigma \in G.$$

For $\rho, \sigma \in G$:

$$\sigma_i \rho_{i\tilde{\sigma}} = \tau_i \sigma \tau_{i\tilde{\sigma}}^{-1} \tau_{i\tilde{\sigma}} \rho \tau_{(i\tilde{\sigma})\tilde{\rho}}^{-1} = \tau_i \sigma \rho \tau_{i\tilde{\sigma}\tilde{\rho}}^{-1} = (\sigma\rho)_i.$$

Hence:

$$\begin{aligned} \sigma^* \rho^{*\tilde{\sigma}^{-1}} &= (\sigma_1, \ldots, \sigma_m)(\rho_{1\tilde{\sigma}}, \ldots, \rho_{m\tilde{\sigma}}) = (\sigma_1 \rho_{1\tilde{\sigma}}, \ldots, \sigma_m \rho_{m\tilde{\sigma}}) \\ &= (\sigma\rho)^*. \end{aligned}$$

Therefore, we have $f(\sigma)f(\rho) = f(\sigma\rho)$. If $f(\sigma) = 1$, then $\tilde{\sigma} = 1$ and $\sigma^* = (\tau_1 \sigma \tau_1^{-1}, \ldots, \tau_m \sigma \tau_m^{-1}) = 1$, therefore each Δ_i is an invariant domain under σ and σ operates on each Δ_i as the identity map. Therefore $\sigma = 1$. ■

This theorem shows that an imprimitive transitive permutation group (G, Ω) can be 'analysed' into a permutation group that operates non-transitively on Ω, and a transitive permutation group that operates on the systems of sets of imprimitivity (and is therefore of lower degree). Since non-transitive permutation groups can be 'analysed' into transitive permutation groups of lower degree, we see that an imprimitive, transitive permutation group is 'composed' of a certain number of primitive permutation groups of lower degree. (The words 'analysed' and 'composed' have no strict mathematical meaning here, they are just used to help the imagination of the reader.)

Theorem 3.3.4 *Let (G, Ω) be transitive, then $F(G_\alpha)$ is a set of imprimitivity for $\alpha \in \Omega$.*

Proof Follows from: $G_\alpha = G_\beta \Leftrightarrow F(G_\alpha) = F(G_\beta)$. ■

Corollary 3.3.5 *If (G, Ω) is primitive and not regular, then $G_\alpha \neq G_\beta$ and $G = \langle G_\alpha, G_\beta \rangle$ for any two elements $\alpha, \beta \in \Omega$ $(\alpha \neq \beta)$.*

Theorem 3.3.6 *Let (G, Ω) be a transitive permutation group, and let N be such that $\{1\} \neq N \trianglelefteq G$. Then:*

(1) *(N, Ω) is $\frac{1}{2}$-transitive and the collection of all orbits of (N, Ω) is a system of sets of imprimitivity of (G, Ω).*

(2) *If (G, Ω) is primitive, then (N, Ω) is transitive. In particular, if in addition (N, Ω) is regular, then N is a minimal normal subgroup of G.*

(3) *If (N, Ω) is transitive, N a minimal, normal and Abelian subgroup of G, then (G, Ω) is primitive.*

Proof (1) (N, Ω) is $\frac{1}{2}$-transitive by theorem 3.2.7. If Δ is an orbit of (N, Ω) then Δ^σ is an orbit of $N^\sigma = N$ for all $\sigma \in G$. Therefore, the collection of all orbits of (N, Ω) forms a system of sets of imprimitivity of (G, Ω).

(2) Obvious from (1).

(3) Pick $\alpha \in \Omega$. Suppose that H satisfies $G_\alpha \lneqq H \leq G$. From $G = G_\alpha N$ we conclude $H = G_\alpha(H \cap N)$ and $H \cap N \neq \{1\}$. From $N \trianglelefteq G$ we get $H \cap N \lhd H$, and since N is Abelian, $H \cap N \trianglelefteq N$. Therefore, $H \cap N$ is a normal subgroup of $G = HN$. Since N is minimal, $H \cap N = N$, i.e. $H \geq N$ and $H = G$. ■

Corollary 3.3.7 *Let (G, Ω) be a primitive permutation group with $|\Omega| = 2m$, where m is an odd number greater than 1. Then $4 \mid |G|$.*

Proof If $4 \nmid |G|$, then $2 \Vert |G|$. Then there exists a normal subgroup

N of G such that $|G:N| = 2$ according to theorem 3.1.9. Since (N, Ω) is transitive by the previous theorem, $2 \mid |N|$. Therefore $4 \mid |G|$. Contradiction. ∎

Theorem 3.3.8 *Let (G, Ω) be primitive, $\Delta, \Gamma \subsetneqq \Omega$ such that $|\Delta| = |\Gamma| \geq 1$. Then for any two elements $\alpha, \beta \in \Omega\,(\alpha \neq \beta)$ there exists an element $\sigma \in G$ such that $\alpha \in \Delta^\sigma$ and $\beta \notin \Gamma^\sigma$.*

Proof Putting $T_\alpha = \{\sigma \in G \mid \alpha \in \Delta^\sigma\}$ we have $T_\alpha \neq G$. Since (G, Ω) is transitive, for each $\gamma \in \Delta$ there is an element $\sigma_\gamma \in G$ such that $\gamma^{\sigma_\gamma} = \alpha$ and T_α can be written as $T_\alpha = \bigcup_{\gamma \in \Delta} \sigma_\gamma G_\alpha$. Therefore $T_\alpha G_\alpha = T_\alpha$ and $|T_\alpha| = |G_\alpha|\,|\Delta|$. In the same way, putting $S_\beta = \{\sigma \in G \mid \beta \in \Gamma^\sigma\}$ we find $S_\beta G_\beta = S_\beta$ and $|S_\beta| = |G_\beta|\,|\Gamma|$. Now suppose that $\beta \in \Gamma^\sigma$ for each $\sigma \in T_\alpha$, then $T_\alpha \subseteq S_\beta$. Hence $T_\alpha = S_\beta$ since $|T_\alpha| = |S_\beta|$. Therefore $T_\alpha G_\beta = S_\beta G_\beta = S_\beta = T_\alpha$, hence $T_\alpha G = T_\alpha \langle G_\alpha, G_\beta \rangle = T_\alpha$. Therefore $T_\alpha = G$. Contradiction. ∎

Theorem 3.3.9 *Let (G, Ω) be transitive.*

(1) *Let $H < G$ be such that H operates primitively on some orbit Γ. If $|\Omega| \lneqq 2|\Gamma|$, then (G, Ω) is primitive.*

(2) *Let Γ_1, Γ_2 be subsets of Ω, and let H_i be subgroups of $G_{\Omega - \Gamma_i}$ $(i = 1, 2)$. If $G = \langle H_1, H_2 \rangle$, and (H_1, Γ_1) and (H_2, Γ_2) are primitive, then (G, Ω) is primitive.*

Proof (1) Let $\{\Delta_1, \ldots, \Delta_r\}$ be a system of sets of imprimitivity of (G, Ω). Since $\Gamma \cap \Delta_i$ is a set of imprimitivity of (H, Γ), we have $|\Gamma \cap \Delta_i| = 0, 1$ or $|\Gamma|$ for each i. If $|\Gamma \cap \Delta_i| = |\Gamma|$ for some i, then $\Gamma \subseteq \Delta_i$, and so $\Delta_i = \Omega$ since $|\Omega| \lneqq 2|\Gamma|$, i.e. $\{\Delta_1, \ldots, \Delta_r\} = \{\Omega\}$. If $|\Gamma \cap \Delta_i| = 0$ or 1 for each i, then $|\Gamma| = \sum_{i=1}^r |\Gamma \cap \Delta_i| \leq r$, hence $|\Omega| = |\Delta_1| r \geq |\Delta_1|\,|\Gamma|$. Therefore $|\Delta_1| = 1$ and $\{\Delta_1, \ldots, \Delta_r\}$ becomes trivial.

(2) $\Gamma_1 \cup \Gamma_2 = \Omega$ and $\Gamma_1 \cap \Gamma_2 \neq \varnothing$ since $G = \langle H_1, H_2 \rangle$. Therefore, (H_1, Γ_1) or (H_2, Γ_2) satisfies the conditions of (1). ∎

3.4 Multiple transitivity

Let G operate on Ω and let t be a natural number. If for every two sequences $(\alpha_1, \ldots, \alpha_t)$ and $(\beta_1, \ldots, \beta_t)$ of elements of Ω, there exists a $\sigma \in G$ such that $\alpha_i^\sigma = \beta_i$ $(i = 1, \ldots, t)$, then (G, Ω) is called t-transitive. So 1-transitive is just transitive. $t \geq 2$ transitive is also called multiply transitive. The following theorem is almost obvious:

Theorem 3.4.1 *Let G operate on Ω. The following three conditions are equivalent:*

(1) *(G, Ω) is t-transitive.*

(2) *Choose a sequence $(\alpha_1, \ldots, \alpha_t)$ of elements of Ω. Then for every sequence $(\beta_1, \ldots, \beta_t)$ of elements of Ω there exists an element $\sigma \in G$ satisfying $\alpha_i^\sigma = \beta_i$ $(i = 1, \ldots, t)$.*

(3) *(G, Ω) is transitive and for any point $\alpha \in \Omega$, $(G_\alpha, \Omega - \{\alpha\})$ is $(t-1)$-transitive.*

Let (G, Ω) be t-transitive. If there is a subset Δ of Ω consisting of $t-1$ elements such that $(G_\Delta, \Omega - \Delta)$ is primitive, then (G, Ω) is called t-primitive. If there is a subset Δ of Ω consisting of t elements such that $(G_\Delta, \Omega - \Delta)$ is $\frac{1}{2}$-transitive, then (G, Ω) is called $(t + \frac{1}{2})$-transitive. It is easy to check that if these conditions are satisfied for one subset of $(t-1)$, respectively t elements, then they are satisfied for all such subsets.

Theorem 3.4.2 *Let (G, Ω) be transitive and put $|\Omega| = n$.*

(1)

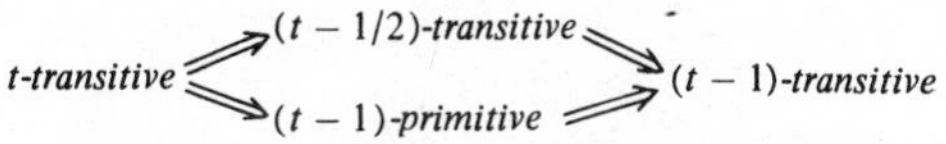

$$t\text{-transitive} \Rightarrow (t-1/2)\text{-transitive} \Rightarrow (t-1)\text{-transitive}$$
$$t\text{-transitive} \Rightarrow (t-1)\text{-primitive} \Rightarrow (t-1)\text{-transitive}$$

(2) *$|G : G_\alpha| = n$ for $\alpha \in \Omega$. Therefore, if (G, Ω) is t-transitive, then $|G : G_{\alpha_1, \ldots, \alpha_t}| = n(n-1)\ldots(n-t+1)$ for any t elements $\alpha_1, \ldots, \alpha_t \in \Omega$.*

(3) *S^Ω is n-transitive and A^Ω is $(n-2)$-primitive. Conversely, if a subgroup G of S^Ω is of order $\geq n!/2$ then G is either S^Ω or A^Ω.*

(4) *For $\alpha \in \Omega$ let $G = \sum_{i=1}^r G_\alpha \sigma_i G_\alpha$ be the double coset representation with regard to G_α. Then $\alpha^{G_\alpha \sigma_i G_\alpha} = \alpha^{\sigma_i G_\alpha}$ are the orbits of (G_α, Ω), hence the number of orbits of (G_α, Ω) equals r. In particular, (G, Ω) is 2-transitive if and only if $r = 2$.*

Proof (1) The only part that is not obvious is the implication: 't-transitive $\Rightarrow$ $(t-1)$-primitive'. It suffices to prove this for the case $t = 2$. So, let (G, Ω) be 2-transitive and let Δ be a set of imprimitivity with $|\Delta| > 1$. Let $\alpha, \beta \in \Delta$ such that $\alpha \neq \beta$. Since $(G_\alpha, \Omega - \{\alpha\})$ is transitive, for any $\gamma \in \Omega$ there exists an element $\sigma \in G_\alpha$ such that $\beta^\sigma = \gamma$. From $\alpha \in \Delta \cap \Delta^\sigma$ we have $\gamma = \beta^\sigma \in \Delta^\sigma = \Delta$, therefore $\Delta = \Omega$.

(2) Follows at once from the definition.

(3) The fact that S^Ω is n-transitive is obvious. Put $\Omega = \{1, \ldots, n\}$. Let $(i_1, \ldots, i_{n-2})$ be a sequence of $n - 2$ elements of Ω and put $\Omega - \{i_1, \ldots, i_{n-2}\} = \{k, l\}$. Defining $\sigma_1, \sigma_2 \in S_n$ by

$$\sigma_1 = \begin{pmatrix} 1 & 2 \ldots n-2 & n-1 & n \\ i_1 & i_2 \ldots i_{n-2} & k & l \end{pmatrix} \text{ and}$$

$$\sigma_2 = \begin{pmatrix} 1 & 2 \ldots n-2 & n-1 & n \\ i_1 & i_2 \ldots i_{n-2} & l & k \end{pmatrix},$$

then $\sigma_1 \sigma_2^{-1} = (n-1, n)$ and either σ_1 or σ_2 is even, i.e. is contained in A^Ω. Combining this with the primitivity of A_3, we conclude that A^Ω is $(n-2)$-primitive. The last statement is proved by induction on n. For $n = 2, 3$ this is obvious, so we may assume $n \geq 4$. Now suppose the statement is true for $1, 2, \ldots, n-1$. From $|S^{\Omega - \{1\}} : G_1| \leq 2$, and the induction hypothesis we conclude $G_1 \geq A^{\Omega - \{1\}}$. If $G \ngeq A^\Omega$, then $A^\Omega \gneq A^\Omega \cap G \geq A^{\Omega - \{1\}}$, hence $A^\Omega \cap G = A^{\Omega - \{1\}}$ since A^Ω is primitive. Further, since $A^\Omega G \gneq A^\Omega$ we have $S^\Omega = A^\Omega G$. Therefore $n = |A^\Omega : A^{\Omega - \{1\}}| = |A^\Omega : A^\Omega \cap G| = |A^\Omega G : G| - |S^\Omega : G| = 2$. Contradiction.

(4) Follows from:

$$\alpha^{\sigma_i G_\alpha} \cap \alpha^{\sigma_j G_\alpha} \neq \varnothing \Leftrightarrow \sigma_i \tau \sigma_j^{-1} \in G_\alpha \quad \text{for some } \tau \in G_\alpha$$
$$\Rightarrow G_\alpha \sigma_i G_\alpha \cap G_\alpha \sigma_j G_\alpha \neq \varnothing \Leftrightarrow G_\alpha \sigma_i G_\alpha = G_\alpha \sigma_j G_\alpha. \quad ■$$

In studying multiply transitive permutation groups the following theorem of Witt is often very useful. We first give a definition: Let (G, Ω) be t-transitive, and let $\Delta \subset \Omega$ be such that $|\Delta| = t$. A subset X of G_Δ is said to satisfy the Witt-condition (with regard to (G, G_Δ)) if it satisfies the following condition:

If the subset Y of G_Δ is G-conjugate to X, then Y is G_Δ-conjugate to X.

Theorem 3.4.3 (Witt) *Let (G, Ω) be t-transitive and let Δ be a* *subset of Ω consisting of t elements. If $H \subset G_\Delta$ satisfies the Witt-condition with regard to (G, G_Δ), then $\mathcal{N}_G(H)$ operates t-transitively on $F_\Omega(H)$.*

Proof Put $\Delta = \{\alpha_1, \ldots, \alpha_t\}$. For t arbitrary elements $\beta_1, \ldots, \beta_t \in F_\Omega(H)$ there is a $\sigma \in G$ such that $\beta_i^\sigma = \alpha_i$ $(i = 1, \ldots, t)$ since (G, Ω) is t-transitive. Since

$$\alpha_i^{H\sigma} = \beta_i^{\sigma\sigma^{-1}H\sigma} = \beta_i^{H\sigma} = \beta_i^\sigma = \alpha_i$$

we get $H^\sigma \subseteq G_\Delta$. Hence, there exists an element $\rho \in G_\Delta$ such that $H^\sigma = H^\rho$ by the Witt-condition. Therefore $\sigma\rho^{-1} \in \mathcal{N}_G(H)$ and $\beta_i^{\sigma\rho^{-1}} = \alpha_i^{\rho^{-1}} = \alpha_i$. Thus $(\mathcal{N}_G(H), F(H))$ is t-transitive. ■

Corollary 3.4.4 *Let (G, Ω) be t-transitive and let Δ be a subset of Ω consisting of t elements.*

(1) *$(\mathcal{N}_G(G_\Delta), F_\Omega(G_\Delta))$ is t-transitive.*

(2) *If P is a Sylow subgroup of G_Δ, then $(\mathcal{N}_G(P), F_\Omega(P))$ is t-transitive.*

Proof Both G_Δ and P satisfy the Witt-condition with regard to (G, G_Δ). ■

Theorem 3.4.5 *A $\frac{3}{2}$-transitive group is either primitive or a Frobenius group.*

Proof Let (G, Ω) be a $\frac{3}{2}$-transitive, imprimitive permutation group. We have to prove that (G, Ω) is a Frobenius group. Let $\{\psi_1, \ldots, \psi_s\}$ be a non-trivial system of sets of imprimitivity of (G, Ω). Putting $|\psi_1| = \ldots = |\psi_s| = t$, we have $t, s > 1$. Since (G, Ω) is $\frac{3}{2}$-transitive, the length of all orbits of $(G_\alpha, \Omega - \{\alpha\})$, for $\alpha \in \Omega$, is the same. Let us call this length m, then $m > 1$ and m is independent of the choice of α. Notice that it follows from this fact that $|G_{\alpha,\beta}|$ is independent of the choice of $\alpha, \beta \in \Omega$ $(\alpha \neq \beta)$. We divide the rest of the proof into five steps:

(*a*) $(m, t) = 1$. (Proof: Pick $\alpha \in \psi_1$ and $\sigma \in G_\alpha$. Since $\psi_1^\sigma \cap \psi_1 \neq \varnothing$, we have $\psi_1^\sigma = \psi_1$, i.e. ψ_1 is an invariant domain of G_α and $\psi_1 - \{\alpha\}$

is the direct sum of a certain number of orbits of $(G_\alpha, \Omega - \{\alpha\})$. Therefore, $t \equiv 1 \pmod m$, hence $(t, m) = 1$.)

(*b*) If $\beta \in \psi_i$ and $\alpha \notin \psi_i$, then $\beta^{G_\alpha} \cap \psi_i = \{\beta\}$. (Proof: Since $\psi_i^{G_\alpha}$ is invariant under the operation of G_α, $|\psi_i^{G_\alpha}| \equiv 0 \pmod m$. Since ψ_i is a set of imprimitivity, $\psi_i^{G_\alpha}$ can be written as the direct sum of certain sets from among $\{\psi_1, \ldots, \psi_s\}$. Therefore $|\psi_i^{G_\alpha}| \equiv 0 \pmod t$ and hence $|\psi_i^{G_\alpha}| \equiv 0 \pmod{mt}$ by (*a*). On the other hand, we have $|\psi_i^{G_\alpha}| \leq tm$ since $|\psi_i| = t$ and since for any $\beta' \in \psi_i$, β'^{G_α} is an orbit of $(G_\alpha, \Omega - \{\alpha\})$. Therefore, $|\psi_i^{G_\alpha}| = tm$. Hence $\psi_i^{G_\alpha} = \sum_{\gamma \in \psi_i} \gamma^{G_\alpha}$ and $\beta^{G_\alpha} \cap \psi_i = \{\beta\}$.)

(*c*) If $\beta \in \psi_i$ and $\alpha \notin \psi_i$, then $G_{\alpha,\beta} = G_{\psi_i}$. (Proof: Pick $\beta' \in \psi_i$, then $\beta' \in \beta'^{G_{\alpha,\beta}} \subseteq \psi_i^{G_\beta} \cap \beta'^{G_\alpha}$. Since $\beta \in \psi_i$, we have $\psi_i^{G_\beta} = \psi_i$. Hence $\psi_i^{G_\beta} \cap \beta'^{G_\alpha} = \psi_i \cap \beta'^{G_\alpha} = \{\beta'\}$ by (*b*). So $\beta' = \beta'^{G_{\alpha,\beta}}$. Therefore $G_{\alpha,\beta} \leq G_{\psi_i}$. On the other hand, $|\psi_i| \geq 2$, therefore $G_{\alpha,\beta} = G_{\psi_i}$.)

(*d*) If $\alpha, \beta \in \psi_i$, then $G_{\alpha,\beta} = G_{\psi_i}$. (Proof: $G_{\alpha,\beta} \geq G_{\psi_i}$ since $|\psi_i| \geq 2$. On the other hand, for an arbitrary $\gamma \in \Omega - \psi_i$ we have $G_{\alpha,\gamma} = G_{\psi_i}$ by (*c*). Therefore, $|G_{\alpha,\beta}| = |G_{\psi_i}|$, hence $G_{\alpha,\beta} = G_{\psi_i}$.)

(*e*) (G, Ω) is a Frobenius-group. (Proof: Pick $\alpha, \beta \in \Omega (\alpha \neq \beta)$. If $\alpha \in \psi_i$, then $G_{\alpha,\beta} = G_{\psi_i}$ by (*c*) and (*d*). Let γ denote an arbitrary element of Ω. If $\gamma \in \psi_i$, then $\gamma \in F(G_{\alpha,\beta})$ since $G_{\alpha,\beta} = G_{\psi_i}$. If $\gamma \in \psi_j \neq \psi_i$, then $G_{\alpha,\gamma} = G_{\psi_i} = G_{\alpha,\beta}$ by (*c*), hence $\gamma \in F(G_{\alpha,\beta})$. So in either case $G_{\alpha,\beta}$ leaves all points of Ω fixed.) ∎

A k-transitive group (G, Ω), such that the stabilizer of k points of Ω is $\{1\}$, is called 'sharply k-transitive'. To be sharply 1-transitive is nothing but to be regular. Since the regular permutation representation of a finite group is regular, all finite groups can be represented as sharply 1-transitive permutation groups. S_n and A_n are sharply n-transitive and $(n-2)$-transitive groups respectively. Sharply 2-transitive groups are nothing but complete Frobenius groups. For $k \geq 2$ all sharply k-transitive groups have been determined completely, but here we will discuss only a few examples.

Examples: Finite affine transformation groups and finite fractional transformation groups.

Let K be a finite near field (§1.3). For $a, b \in K$ $(a \neq 0)$ we have a mapping $f_{a,b}: K \to K$ defined by

$$f_{a,b}(x) = ax + b \quad (x \in K)$$

and $f_{a,b}$ is called an affine transformation of K. From

$$f_{a,b}(x) = f_{a,b}(y) \Leftrightarrow xa = ya \Leftrightarrow x = y$$

we conclude that $f_{a,b}$ defines a permutation of K. If $f_{a,b} = f_{a',b'}$, then $b = f_{a,b}(0) = f_{a',b'}(0) = b'$ and $a = f_{a,b}(1) - b = f_{a',b'}(1) - b' = a'$. The collection

$$L(K) = \{f_{a,b} \mid a, b \in K, a \neq 0\}$$

of all affine transformations of K is a subset of S^K consisting of $|K|(|K|-1)$ elements. For $f_{a,b}$ and $f_{a',b'} \in L(K)$ their product as elements of S^K is given by

$$(f_{a,b}f_{a',b'})(x) = x(aa') + ba' + b',$$

i.e.

$$f_{a,b}f_{a',b'} = f_{aa',ba'+b'}.$$

Therefore $L(K)$ is closed under the multiplication defined on S^K. Indeed, $L(K)$ is a group under this multiplication, $f_{1,0}$ is the identity element and $f_{a^{-1},-ba^{-1}}$ is the inverse element of $f_{a,b}$. $L(K)$ is called the affine transformations group on K. Since $f_{1,b}(0) = b$, $(L(K), K)$ is transitive and since also $f_{a,0}(1) = a$, $(L(K), K)$ is 2-transitive. Since $|L(K)| = |K|(|K|-1)$, we conclude that $(L(K), K)$ is a complete Frobenius group. We have proved (1) of the following theorem:

Theorem 3.4.6 (1) *Let K be a finite near field and let $L(K)$ denote the group of affine transformations on K. Then the permutation group $(L(K), K)$ is a perfect Frobenius group.*

(2) *Let (G, Ω) be a complete Frobenius group, then there exists a finite near field K such that $(G, \Omega) \simeq (L(K), K)$.*

Proof (2) Let N be the Frobenius kernel of (G, Ω). According to theorem 2.11.7, N is an elementary Abelian group and $|N| = p^r = |\Omega|$, where p is prime. Pick two elements from Ω and denote them by 0 and e. For any $\alpha \in \Omega$ there exists a unique element $\sigma_\alpha \in N$

such that $\sigma_\alpha(0) = \alpha$. The element σ_0 is the identity element of N. For an arbitrary $\beta \in \Omega - \{0\}$ there exists a unique element $\tau_\beta \in G_0 = H$ such that $\tau_\beta(e) = \beta$. The element τ_e is the identity element of H. For $\alpha, \beta \in \Omega$ we define the sum $\alpha + \beta$ by

$$\alpha + \beta = \sigma_\beta(\alpha),$$

and the product $\alpha\beta$ by

$$\alpha\beta = \begin{cases} \tau_\beta(\alpha) & \text{if } \alpha \neq 0 \text{ and } \beta \neq 0, \\ 0 & \text{if } \alpha = 0 \text{ or } \beta = 0. \end{cases}$$

We will show that Ω provided with these operations becomes a near field. Let $\alpha, \beta, \gamma \in \Omega$. Since N is Abelian, we have

$$\begin{aligned} \beta + \alpha &= \sigma_\alpha(\sigma_\beta(0)) = (\sigma_\beta \sigma_\alpha)(0) = (\sigma_\alpha \sigma_\beta)(0) \\ &= \sigma_\beta(\sigma_\alpha(0)) = \alpha + \beta \end{aligned}$$

Therefore: $\alpha + \beta = \beta + \alpha$ and $\sigma_\alpha \sigma_\beta = \sigma_{\alpha+\beta}$. Further,

$$\begin{aligned} (\alpha + \beta) + \gamma &= \sigma_\gamma(\alpha + \beta) = \sigma_\gamma(\sigma_\beta(\alpha)) = (\sigma_\beta \sigma_\gamma)(\alpha) \\ &= \sigma_{\beta+\gamma}(\alpha) = \alpha + (\beta + \gamma), \end{aligned}$$

and $0 + \beta = \sigma_0(\beta) = \beta$. Since σ_α is a permutation of Ω, there is a $\beta \in \Omega$ such that $0 = \sigma_\alpha(\beta) = \alpha + \beta$. Hence Ω is an Abelian group under addition with 0 as neutral element. Now, let $\alpha, \beta, \gamma \in \Omega - \{0\}$. Since:

$$(\tau_\beta \tau_\alpha)(e) = \tau_\alpha(\tau_\beta(e)) = \tau_\alpha(\beta) = \beta\alpha,$$

we have $\tau_\beta \tau_\alpha = \tau_{\beta\alpha}$. Therefore,

$$\begin{aligned} (\alpha\beta)\gamma &= \tau_\gamma(\alpha\beta) = \tau_\gamma(\tau_\beta(\alpha)) = (\tau_\beta \tau_\gamma)(\alpha) = \tau_{\beta\gamma}(\alpha) \\ &= \alpha(\beta\gamma). \end{aligned}$$

Further,

$$e\beta = \tau_\beta(e) = \beta \text{ and } \beta e = \tau_e(\beta) = \beta.$$

Since τ_α is a permutation of $\Omega - \{0\}$, there is a $\beta \in \Omega - \{0\}$ such that $e = \tau_\alpha(\beta) = \beta\alpha$. Hence $\Omega - \{0\}$ is a group under multiplication with e as identity element. The only thing that remains to be shown is the distributivity. Let $\alpha, \beta, \gamma \in \Omega$, we will show $(\beta + \gamma)\alpha = \beta\alpha + \gamma\alpha$.

This is obvious for $\alpha = 0$, so we may assume $\alpha \neq 0$. From

$$(\tau_\alpha^{-1}\sigma_\gamma\tau_\alpha)(0) = (\sigma_\gamma\tau_\alpha)(0) = \tau_\alpha(\gamma) = \gamma\alpha,$$

we conclude $\tau_\alpha^{-1}\sigma_\gamma\tau_\alpha = \sigma_{\gamma\alpha}$. Therefore,

$$\begin{aligned}(\beta + \gamma)\alpha &= \tau_\alpha(\beta + \gamma) = \tau_\alpha(\sigma_\gamma(\beta)) = (\sigma_\gamma\tau_\alpha)(\beta)\\ &= \{\tau_\alpha(\tau_\alpha^{-1}\sigma_\gamma\tau_\alpha)\}(\beta) = \sigma_{\gamma\alpha}(\beta\alpha) = \beta\alpha + \gamma\alpha.\end{aligned}$$

Thus Ω becomes a near field.

Since G is the semi-direct product of H and N, elements of G can be represented in a unique way as $\tau_\alpha\sigma_\beta$ ($\alpha, \beta \in \Omega, \alpha \neq 0$). Since $(\tau_\alpha\sigma_\beta)(\gamma) = \gamma\alpha + \beta$, the element $\tau_\alpha\sigma_\beta$ of G corresponds with $f_{\alpha,\beta} \in L(\Omega)$ and we obtain $(G, \Omega) \simeq (L(\Omega), \Omega)$. ∎

Next we will consider fractional transformation groups. Let K be a finite field, let ∞ denote a symbol, not contained in K and put $\tilde{K} = K \cup \{\infty\}$. For

$$\alpha = \begin{bmatrix} a & b \\ c & d \end{bmatrix} \in GL(2, K)$$

we define the mapping $f_\alpha : \tilde{K} \to \tilde{K}$ by

$$f_\alpha(x) = \begin{cases} \dfrac{ax + c}{bx + d} & \text{if } x \in K \text{ and } bx + d \neq 0, \\ \infty & \text{if } x \in K \text{ and } bx + d = 0, \\ \dfrac{a}{b} & \text{if } x = \infty \text{ and } b \neq 0, \\ \infty & \text{if } x = \infty \text{ and } b = 0. \end{cases}$$

The mapping f_α is called a linear fractional transformation on K. Since $ad - bc \neq 0$, it is easily verified that f_α defines a permutation of $\tilde{K}$. The collection of all linear fractional transformations on K is denoted by $LF(K)$. For $f_\alpha, f_\beta \in LF(K)$ their product as permutations on $\tilde{K}$ is

$$f_\alpha f_\beta = f_{\alpha\beta}$$

and so $LF(K)$ is closed under this multiplication, and it is easy to verify that $LF(K)$ together with this multiplication becomes a group, called the group of linear fractional transformations on K. The mapping from $GL(2, K)$ into $LF(K)$ given by $\alpha \to f_\alpha$ is a surjective homomorphism, and it is easy to check that the kernel of this homomorphism is

$$Z(2, K) = \left\{ \begin{bmatrix} a & 0 \\ 0 & a \end{bmatrix} \middle| a \in K, a \neq 0 \right\}.$$

Therefore,

$$LF(K) \simeq GL(2, K)/Z(2, K).$$

Theorem 3.4.7 *$(LF(K), \tilde{K})$ is sharply 3-transitive.*

Proof For

$$\alpha = \begin{bmatrix} a & 1 \\ 0 & 1 \end{bmatrix},$$

$f_\alpha(\infty) = a$, hence $LF(K)$ operates transitively on $\tilde{K}$. Let G denote the subgroup that leaves ∞ fixed, then

$$G = \left\{ f_\alpha \middle| \alpha = \begin{bmatrix} a & 0 \\ c & d \end{bmatrix} \in GL(2, K) \right\}.$$

The mapping from G into $L(K)$ defined by

$$f_\alpha \to f_{a/d, c/d}, \quad \text{where } \alpha = \begin{bmatrix} a & 0 \\ c & d \end{bmatrix},$$

is easily seen to set up an isomorphism between the permutation groups (G, K) and $(L(K), K)$. Therefore, $(LF(K), \tilde{K})$ is sharply 3-transitive by theorem 3.4.1. ∎

Finally we want to prove a lemma that we will need in chapter 5.

Lemma 3.4.8 *Let (G, Ω) be a 2-transitive group of degree $(n+1)$, where n is odd. If for some $a \in \Omega$ there exists a subgroup H_a of G_a satisfying:*

(1) $|H_a|$ is even;

(2) $(H_a, \Omega - \{a\})$ is a Frobenius group;

then (G, Ω) is 2-primitive.

Proof Since (G, Ω) is transitive, it is obvious that if there exists a group H_a satisfying (1) and (2) for one $a \in \Omega$, such groups exist for every $a \in \Omega$. Let K_a denote the Frobenius kernel of H_a. Since $|K_a| = n$ is odd, there is an element $\sigma \in H_a$ of order 2 and $\langle K_a, \sigma \rangle$ is a Frobenius group on $\Omega - \{a\}$ of order $2n$. Therefore, we may assume $|H_a| = 2n$ from the beginning. For $b, c \in \Omega - \{a\}$, there exists exactly one element of H_a of order 2 that maps b onto c. (For, H_a contains exactly n elements of order 2, say $\sigma_1, \dots, \sigma_n$. If σ_i and σ_j are such that $b^{\sigma_i} = b^{\sigma_j} = c$, then $b^{\sigma_i \sigma_j} = b$ and $c^{\sigma_i \sigma_j} = c$, hence $\sigma_i = \sigma_j$. Therefore, for $b \in \Omega - \{a\}$, the elements $b^{\sigma_1}, \dots, b^{\sigma_n}$ are all different elements of $\Omega - \{a\}$, proving the assertion.) Let us denote this element by $\sigma_{a;b,c}$. Now, we assume that $(G_a, \Omega - \{a\})$ is imprimitive and derive a contradiction from that assumption. Let $\Delta_1, \dots, \Delta_s$ denote a non-trivial system of sets of imprimitivity of $(G_a, \Omega - \{a\})$, i.e.

$$\Omega - \{a\} = \Delta_1 + \dots + \Delta_s$$

where $s > 1$ and $|\Delta_1| = \dots = |\Delta_s| = t > 1$. For all $\sigma \in G_a$ and for all Δ_i, Δ_i^σ coincides with some Δ_j. Putting $\Lambda = \{a\} \cup \Delta_1$, we have that $|\Lambda|$ is even since t is odd. For an arbitrary $c \in \Delta_s$, we put

$$\tilde{H} = \langle \sigma_{c;a,b} \mid b \in \Delta_1 \rangle.$$

Then $\Lambda^{\tilde{H}} = \Lambda$. (To prove this it suffices to prove:

$$\Lambda^{\sigma_{c;a,b}} = \Lambda \quad \forall b \in \Delta_1.$$

Suppose $\Lambda^{\sigma_{c;a,b}} \neq \Lambda$ and pick $d \in \Delta_1$ such that $d^{\sigma_{c;a,b}} = e \notin \Lambda$. Since $\sigma = \sigma_{c;a,b} \sigma_{e;a,b} \in G_{a,b}$, we have $b \in \Delta_1^\sigma \cap \Delta_1$ and so $\Delta_1^\sigma = \Delta_1$, hence $d^\sigma \in \Delta_1$. On the other hand, $d^\sigma = d^{\sigma_{c;a,b}\sigma_{e;a,b}} = e \notin \Delta_1$. Contradiction.) Since $\tilde{H}$ operates transitively on Λ and since $\tilde{H}$ is a subgroup of the Frobenius group $(H_c, \Omega - \{c\})$, $\tilde{H}$ operates faithfully on Λ and the stabilizer of two arbitrary points coincides with 1. Suppose $\tilde{H}_a \neq 1$. Then since $\tilde{H}_a \subseteq H_{a,c}$, $\tilde{H}_a$ is of order 2, and since $|\Lambda|$ is even, the elements of $\tilde{H}_a$ leave at least two points of Λ fixed, hence $\tilde{H}_a = 1$. Contradiction. Therefore $\tilde{H}$ is regular on Λ. Hence $\tilde{H}$ is a group of order $t + 1$. On the other hand since $\tilde{H}$ contains t elements of order 2, $\tilde{H}$ is a 2-subgroup of H_c. Since $2 \top |H_c|$, we get $t + 1 = 2$, contrary to $t > 1$. ■

3.5 Normal subgroups

In this section we study normal subgroups of transitive groups. For this section, let (G, Ω) denote a transitive permutation group and let N be a normal subgroup of G, $N \neq \{1\}$. Our first theorem is a consequence of theorem 3.3.6.

Theorem 3.5.1 (1) *(N, Ω) is $\frac{1}{2}$-transitive.*
(2) *If (G, Ω) is primitive, then (N, Ω) is transitive.*

Corollary 3.5.2 (1) *If (G, Ω) is primitive and (N, Ω) regular, then N is a minimal normal subgroup of G.*
(2) *If (G, Ω) is primitive and N Abelian, then (N, Ω) is regular.*

Theorem 3.5.3 *If (G, Ω) is $\frac{3}{2}$-transitive, then (N, Ω) is transitive or half-regular.*

Proof According to theorem 3.4.5, (G, Ω) is primitive or a Frobenius group. If (G, Ω) is primitive, (N, Ω) is transitive from theorem 3.5.1. So let (G, Ω) be a Frobenius group, and let K denote its Frobenius kernel. Since (K, Ω) is regular, (N, Ω) is half-regular if $K \geq N$. If $K \ngeq N$, then $N_\alpha \neq \{1\}$ for $\alpha \in \Omega$. Let Δ denote an orbit of (N, Ω) containing α, then $|N_\alpha| \,|\, |\Delta| - 1$ since $(N_\alpha, \Omega - \{\alpha\})$ is half-regular. If $\Delta \neq \Omega$, then $|N_\alpha| \,|\, |\Delta'|$ for an orbit $\Delta' \neq \Delta$ of (N, Ω). From $|\Delta| = |\Delta'|$, we conclude $|N_\alpha| = 1$. Contradiction. Hence $\Delta = \Omega$, so that (N, Ω) is transitive. ∎

Let (G, Ω) be multiply transitive. We distinguish two cases: where (N, Ω) is regular, and where (N, Ω) is not regular. If (N, Ω) is regular, i.e. if there exists a normal subgroup of G operating regularly on Ω, then the structure of (G, Ω) is determined to a large extent, as we will show. Let α denote an arbitrary point of Ω. Since (N, Ω) is regular, for $\beta \in \Omega$ there is a unique $\sigma \in N$ such that $\alpha^\sigma = \beta$ and conversely, the element β is uniquely determined by σ. Let $\varphi : N \to \Omega$ denote the one-to-one surjective mapping given by this association, i.e. for $\sigma \in N$ we have $\varphi(\sigma) = \alpha^\sigma$ and in particular $\varphi(1) = \alpha$. Since $N \trianglelefteq G$, we have $N = N^\tau$ for $\tau \in G$. Therefore, τ

defines a permutation $\begin{pmatrix} \sigma \\ \sigma^\tau \end{pmatrix}$ of $N - \{1\}$. If we take $\tau \in G_\alpha$, then

$$\varphi(\sigma^\tau) = \alpha^{\sigma^\tau} = \alpha^{\tau^{-1}\sigma\tau} = \alpha^{\sigma\tau} = \varphi(\sigma)^\tau.$$

Therefore, $(G_\alpha, N - \{1\}) \simeq (G_\alpha, \Omega - \{\alpha\})$, hence G_α is isomorphic to a subgroup of Aut N. We have proved:

Theorem 3.5.4 *If (N, Ω) is regular and $\alpha \in \Omega$, then:*

(1) *The map φ assigning $\alpha^\sigma \in \Omega$ to $\sigma \in N$ is a one-to-one correspondence between $N - \{1\}$ and $\Omega - \{\alpha\}$.*

(2) *The map assigning the automorphism of $N : \sigma \to \sigma^\tau$ to each $\tau \in G_\alpha$ is an isomorphism from G_α onto a subgroup of* Aut N. *By this map G_α can be regarded as a permutation group on $N - \{1\}$.*

(3) *Using the correspondences defined in* (1) *and* (2), *we have:*

$$(G_\alpha, N - \{1\}) \simeq (G_\alpha, \Omega - \{\alpha\}).$$

Theorem 3.5.5 *Let M be a finite group, H a subgroup of* Aut M, *regarded as a permutation group on $M - \{1\}$.*

(1) *If $(H, M - \{1\})$ is transitive, then M is an elementary Abelian group, and hence $|M| = p^n$ (p prime).*

(2) *If $(H, M - \{1\})$ is primitive, then $|M| = 2^n$ or $|M| = 3$.*

(3) *If $(H, M - \{1\})$ is $\frac{3}{2}$-transitive, then $|M| = 2^n$.*

(4) *If $(H, M - \{1\})$ is 2-primitive, then $|M| = 4$ or $|M| = 3$.*

(5) *If $(H, M - \{1\})$ is 3-transitive, then $|M| = 4$.*

(6) *It is impossible for $(H, M - \{1\})$ to be t-transitive with $t \geq \frac{7}{2}$.*

Proof (1) The orders of all elements of $M - \{1\}$ are the same. Hence M is a p-group and the order of each element of $M - \{1\}$ equals p. Since the centre of M is non-trivial and since H is transitive on $M - \{1\}$, we conclude that M is an elementary Abelian group.

(2) For $\sigma \in M$ ($\sigma \neq 1$), the set $\{\sigma, \sigma^{-1}\}$ is a set of imprimitivity of $(H, M - \{1\})$. Since $(H, M - \{1\})$ is primitive, we have $|\{\sigma, \sigma^{-1}\}| = 1$ or $\{\sigma, \sigma^{-1}\} = M - \{1\}$. In the former case, we have $p = 2$, in the latter case, $|M| = 3$.

(3) For $\sigma \in M$ ($\sigma \neq 1$) H_σ leaves σ^{-1} fixed. Since $(H, M - \{1\})$ is $\frac{3}{2}$-transitive, we have $\sigma = \sigma^{-1}$, i.e. $\sigma^2 = 1$.

(4) Put $|M| = 2^m$, then $m \geq 2$. If σ, τ are two different elements of $M - \{1\}$, then $\{\tau, \sigma\tau\}$ is a set of imprimitivity of $(H_\sigma, M - \{\sigma, 1\})$. Hence $\{\tau, \sigma\tau\} = M - \{\sigma, 1\}$. Therefore, $|M| = 4$.

(5) For $\sigma, \tau \in M - \{1\}$, $H_{\sigma,\tau}$ leaves $\sigma\tau$ fixed. Hence by our assumption: $M - \{1, \sigma, \tau\} = \{\sigma\tau\}$.

(6) Obvious. ■

The following theorem follows directly from theorems 3.5.4 and 3.5.5.

Theorem 3.5.6 *Let (N, Ω) be regular, then:*

(1) *If (G, Ω) is 2-transitive, then N is an elementary Abelian group. Hence $|N| = p^m$ (p prime).*

(2) *If (G, Ω) is 2-primitive, then $|N| = 2^n$ or $|N| = 3$.*

(3) *If (G, Ω) is $\frac{5}{2}$-transitive, then $|N| = 2^m$.*

(4) *If (G, Ω) is 3-primitive, then $|N| = 4$ or $|N| = 3$.*

(5) *If (G, Ω) is 4-transitive, then $|N| = 4$.*

(6) *It is impossible for (G, Ω) to be t-transitive with $t \geq \frac{9}{2}$.*

Theorem 3.5.7 *Let (G, Ω) be primitive and let, for $\alpha \in \Omega, \Omega - \{\alpha\}$ be divided into two orbits under G_α. If (N, Ω) is regular, then N is an elementary Abelian group.*

Proof According to theorem 3.5.4 the elements of N have, at most, two different orders. Therefore, the order of N is either a power of a prime p or the product of the powers of two primes p and q. According to theorem 2.3.4 and theorem 2.10.17, in both cases N is solvable, hence there exists a characteristic subgroup N_0 of N, that is elementary Abelian. Since $N_0 \lhd G, (N_0, \Omega)$ is transitive (theorem 3.5.1). Since (N, Ω) is regular, we conclude $N = N_0$. ■

Next, let us consider the case where (N, Ω) is not regular. Notice that N is not Abelian by corollary 3.2.12. If (G, Ω) is 2-transitive, then $G_\alpha \trianglerighteq N_\alpha \neq 1$ for $\alpha \in \Omega$. Therefore $(N_\alpha, \Omega - \{\alpha\})$ is $\frac{1}{2}$-transitive hence (N, Ω) is $\frac{3}{2}$-transitive. According to theorem 3.4.5 (N, Ω) is primitive or a Frobenius group. This is the content of the first half of the next theorem:

Theorem 3.5.8 *Let (G,Ω) be 2-transitive and let (N,Ω) be not regular. Then:*

(1) *(N,Ω) is primitive or a Frobenius group.*

(2) *If N is a minimal normal subgroup, then N is simple and (N,Ω) is primitive.*

Proof (2) If N is a minimal normal subgroup, then N cannot be a Frobenius group, hence has to be primitive. Suppose N is not simple, then N is the direct product of at least two isomorphic simple groups by theorem 2.2.2. Let N_1 denote one of those groups and write $N = N_1 \times N_2$ as a direct product. Since (N,Ω) is primitive, (N_1,Ω) and (N_2,Ω) are transitive. Since $[N_1,N_2]=1, (N_1,\Omega)$ and (N_2,Ω) are both regular by theorem 3.2.9, and we have that if $N = N_1 \times N_2 = N_1 \times N_3$ then $N_2 = N_3$. Hence putting $H = \mathcal{N}_G(N_1)$ we have $|G:H| = 2$. Therefore $|G_\alpha : H_\alpha| = 2$ for $\alpha \in \Omega$ and so $\Omega - \{\alpha\}$ is divided into, at most, two H_α orbits. According to theorems 3.5.6 and 3.5.7, N_1 is Abelian in both cases, hence $N = N_1 \times N_2$ is also Abelian which is a contradiction. ∎

Theorem 3.5.9 *Let (N,Ω) be not regular and let $k \geq 2$. Then:*

(1) *If (G,Ω) is k-transitive and $G \nsupseteq S^\Omega$, then (N,Ω) is $(k-\frac{1}{2})$-transitive or sharply $(k-1)$-transitive.*

(2) *If (G,Ω) is k-primitive and $G \nsupseteq S^\Omega$, then (N,Ω) is k-transitive.*

Proof Let α be an element of Ω.

(1) If $k=2, (G_\alpha, \Omega - \{\alpha\})$ is transitive and $(N_\alpha, \Omega - \{\alpha\})$ is $\frac{1}{2}$-transitive, hence (N,Ω) is $\frac{3}{2}$-transitive. If $k=3$, we distinguish two cases. If $(N_\alpha, \Omega - \{\alpha\})$ is not regular, we reduce this case to the case $k=2$ and conclude that (N,Ω) is $\frac{5}{2}$-transitive. If $(N_\alpha, \Omega - \{\alpha\})$ is regular, then (N,Ω) is sharply 2-transitive. If $k \geq 4$, we apply induction on k. If $(N_\alpha, \Omega - \{\alpha\})$ is not regular, we are reduced to the case $k-1$ and the assertion of the theorem is true. Suppose that $(N_\alpha, \Omega - \{\alpha\})$ is regular, then (N,Ω) is a Frobenius group. Hence there exists a characteristic subgroup N_0 of N such that

(N_0, Ω) is regular. Since $N_0 \lhd G$ we conclude from theorem 3.5.6 that $k=4$ and $|\Omega|=4$, therefore $G=S^\Omega$. Contradiction.

(2) If $k=2$, $(G_\alpha, \Omega-\{\alpha\})$ is primitive, hence $(N_\alpha, \Omega-\{\alpha\})$ is transitive. Therefore, (N, Ω) is 2-transitive. If $k \geq 3$, apply induction on k. If $(N_\alpha, \Omega-\{\alpha\})$ is not regular, we are reduced to the case $k-1$ and the theorem is true. If $(N_\alpha, \Omega-\{\alpha\})$ is regular, (N, Ω) is a Frobenius group and we arrive at the same contradiction as in (1). ■

Corollary 3.5.10 *Let (G, Ω) be k-transitive $(k \geq 2)$ and let $G \neq S^\Omega$. If (N, Ω) is not $(k-1)$-transitive, then $k=3$, $|\Omega|=2^m$, N is an elementary Abelian group of order 2^m and (N, Ω) is regular.*

Proof Follows from theorems 3.5.1, 3.5.6 and 3.5.9. ■

Corollary 3.5.11 *If $|\Omega|=n \geq 5$, the alternating group A^Ω is simple.*

Proof Let N be such that $\{1\} \neq N \trianglelefteq A^\Omega$. Since A^Ω is k-primitive $(k \geq 3)$, (N, Ω) is not regular by theorem 3.5.6. Since A^Ω is $(n-2)$-primitive, (N, Ω) is $(n-2)$-transitive by theorem 3.5.9. Therefore, $|N|=n!/2=|A^\Omega|$, i.e. $N=A^\Omega$. ■

3.6 Permutation groups of prime degree

In this section we want to state and prove two classic results on permutation groups of prime degree.

Theorem 3.6.1 (Galois) *Let (G, Ω) be a transitive permutation group of prime degree p, then the following conditions are equivalent.*

(1) *The Sylow p-subgroups of G are normal.*

(2) *G is solvable.*

(3) *G is isomorphic to a subgroup of the group $L(F_p)$ of affine transformations on F_p (the finite field consisting of p elements).*

(4) *(G, Ω) is a Frobenius group or $|G|=p$.*

Proof $(1) \Rightarrow (2)$. Since $p \,\mathrm{T}\, |G|$, the order of a Sylow p-subgroup P of G equals p and (P, Ω) is regular. Therefore, $\mathscr{C}_G(P)=P$, for example by theorem 3.2.10, and G/P is a subgroup of Aut P. Therefore, G/P is Abelian and G is solvable.

(2) ⇒ (3). If N is a minimal normal subgroup of G, then N is Abelian. Since (G, Ω) is primitive, (N, Ω) is transitive, hence (N, Ω) is regular and N is a Sylow p-subgroup of G of order p. The subgroup N is isomorphic to the additive group underlying F_p. Fix an isomorphism between N and F_p and denote the image of $\sigma \in N$ under that isomorphism by $\tilde{\sigma}$. Choose and fix a point from Ω and denote it 0. For $i \in \Omega$ there is a unique $\sigma_i \in N$ such that $0^{\sigma_i} = i$. So we have a one-to-one correspondence $i \leftrightarrow \tilde{\sigma}_i$ between Ω and F_p. Since G can be written as the semi-direct product of G_0 and N, every element $\rho \in G$ can be written in a unique way as:

$$\rho = \tau\sigma \quad \tau \in G_0, \sigma \in N.$$

Noticing that $\tau^{-1}\sigma_i\rho = (\tau^{-1}\sigma_i\tau)\sigma \in N$, we define the operation of $\rho \in G$ on F_p as

$$\tilde{\sigma}_i^\rho = \widetilde{\tau^{-1}\sigma_i\rho}.$$

Then since

$$i^\rho = 0^{\sigma_i\rho} = 0^{\tau^{-1}\sigma_i\rho}$$

we have $(G, F_p) \simeq (G, \Omega)$. Since, for $\tau \in G_0$,

$$\begin{aligned}(\tilde{\sigma}_i + \tilde{\sigma}_j)^\tau = \widetilde{\sigma_i\sigma_j}^\tau &= \widetilde{\tau^{-1}\sigma_i\sigma_j\tau} = \widetilde{\tau^{-1}\sigma_i\tau\tau^{-1}\sigma_j\tau} \\ &= \widetilde{\tau^{-1}\sigma_i\tau} + \widetilde{\tau^{-1}\sigma_j\tau} = \tilde{\sigma}_i^\tau + \tilde{\sigma}_j^\tau,\end{aligned}$$

we see that τ induces a linear transformation of F_p. Hence there exists an element $\alpha(\tau) \in F_p$ such that $\tilde{\sigma}_i^\tau = \alpha(\tau)\tilde{\sigma}_i$. Therefore, for $\rho = \tau\sigma \in G$

$$\tilde{\sigma}_i^\rho = \tau^{-1}\sigma_i\tau\sigma = \tau^{-1}\sigma_i\tau + \tilde{\sigma} = \alpha(\tau)\tilde{\sigma}_i + \tilde{\sigma}.$$

Therefore, ρ defines an affine transformation on F_p, from which the assertion of the theorem follows easily.

(3) ⇒ (4). Since $L(F_p)$ is a complete Frobenius group by theorem 3.4.6, (G, Ω) obviously satisfies (4).

(4) ⇒ (1). Obvious. ■

We need some preparations for our next theorem. Let (G, Ω)

be a transitive permutation group of prime degree p and let $F = F_p$ be the field consisting of p elements. Let V be a p-dimensional vector space over F and let $\{v_a | a \in \Omega\}$ be a basis for V the elements of which have indices in Ω. Letting $\sigma \in G$ operate on V in the following way:

$$(\sum \lambda_a v_a)^\sigma = \sum \lambda_a v_{a^\sigma},$$

it is easy to see that V becomes a G-group. The commutator ring of a G-group V, i.e.

$$\mathrm{Hom}_G(V, V) = \{\alpha \in \mathrm{Hom}_F(V, V) | \alpha(v^\sigma) = (\alpha(v))^\sigma, \forall \sigma \in G\}$$

is an F-submodule of the F-module $\mathrm{Hom}_F(V, V)$. We have the following lemma:

Lemma 3.6.2 $\dim_F(\mathrm{Hom}_G(V, V))$ *equals the number of orbits of* (G_a, Ω), *where* a *denotes an arbitrary point of* Ω.

Proof Let $\Delta_1, \ldots, \Delta_s$ represent the orbits of (G_a, Ω). If we define the F-submodule V_a of V by

$$V_a = \left\{ v \in V \,\middle|\, v = \sum_{i=1}^{s} \lambda_i \left(\sum_{x \in \Delta_i} v_x \right) \right\},$$

then $\dim_F V_a = s$. Hence, in order to prove our lemma, it suffices to prove the existence of an isomorphism between F_p-vector spaces $\mathrm{Hom}_G(V, V)$ and V_a. Let $\alpha \in \mathrm{Hom}_G(V, V)$ and put $\alpha(v_a) = \sum_{x \in \Omega} \lambda_x v_x$. Since, for $\sigma \in G_a$,

$$\alpha(v_a) = \alpha(v_{a^\sigma}) = (\alpha(v_a))^\sigma = \sum_{x \in \Omega} \lambda_x v_{x^\sigma},$$

we have $\lambda_x = \lambda_y$ for elements x and y of the same orbit Δ_i, therefore we can write: $\alpha(v_a) = \sum_{i=1}^{s} \lambda_i (\sum_{x \in \Delta_i} v_x)$. So we can define a homomorphism $f: \mathrm{Hom}_G(V, V) \to V_a$ for the F-modules by $f(\alpha) = \alpha(v_a)$. Suppose $f(\alpha) = \alpha(v_a) = 0$. Since (G, Ω) is transitive, there exists an element $\sigma \in G$ such that $a^\sigma = b$ for every $b \in \Omega$. Therefore,

$$\alpha(v_b) = \alpha(v_{a^\sigma}) = (\alpha(v_a))^\sigma = 0 \quad \forall b \in \Omega,$$

i.e. $\alpha = 0$. Hence f is one-to-one. To prove that f is surjective, let v_0 be an arbitrary element of V_a. We define $\alpha \in \mathrm{Hom}_F(V, V)$ by

$$\alpha(v_b) = v_0^\sigma,$$

where $\sigma \in G$ is such that $b = a^{\sigma}$. (It is almost clear that this definition is independent from the choice of σ.) Then $f(\alpha) = \alpha(v_a) = v_0$, i.e. f is surjective. Therefore $\mathrm{Hom}_G(V, V) \simeq V_a$ (as F-modules), from which the theorem follows. ∎

Let P be a Sylow p-subgroup of G and let σ_0 be a generator of P. Since $|\Omega| = p$, we can identify Ω with F by choosing a one-to-one map from Ω onto F, and we can regard G as a permutation group on F. Choosing a suitable map we may assume that $\sigma_0 = (p-1, p-2, \ldots, 1, 0) \in S^F$.

We denote the collection of all maps from F into F by $U(F, F)$. For $f, g \in U(F, F)$ and for $\lambda \in F$ we define:

$$\left.\begin{aligned} (f+g)(x) &= f(x) + g(x) \\ (fg)(x) &= f(x)g(x) \\ (\lambda f)(x) &= \lambda f(x) \end{aligned}\right\} \quad x \in F.$$

It is easy to verify that with these definitions $U(F, F)$ becomes an F-algebra. For $i \in F$ we define $f_i \in U(F, F)$ by

$$f_i(a) = \begin{cases} 1 & \text{if } a = i, \\ 0 & \text{if } a \neq i. \end{cases}$$

Then the elements $f_0, \ldots, f_{p-1}$ form a basis for $U(F, F)$, therefore $\dim_F U(F, F) = p$. Defining for $v = \sum_{a \in \Omega} \lambda_a v_a \in V$ the element $f_v \in U(F, F)$ by

$$f_v(a) = \lambda_a \quad a \in F,$$

we get a map $T_1 : V \to U(F, F)$, that is an isomorphism from V onto $U(F, F)$ as F-modules.

For $f(x) \in F[x]$, define $f_{|F} \in U(F, F)$ by

$$f_{|F}(\alpha) = f(\alpha) \quad \forall \alpha \in F.$$

The map $T_2 : F[x] \to U(F, F)$ defined by $T_2(f) = f_{|F}$ is obviously an F-algebra homomorphism. Moreover this map is surjective. (For, let K denote the kernel of T_2. From

$$f(x) \in K \Leftrightarrow f(a) = 0 \quad \forall a \in F \Leftrightarrow \prod_{a \in F} (x - a) \mid f(x)$$

and

$$\prod_{a\in F}(x-a)=x^p-x,$$

we conclude that K is the principal ideal generated by x^p-x. Since $F[x]/(x^p-x)$ as an F-module is p-dimensional, $F[x]/(x^p-x)\simeq U(F,F)$, hence T_2 is surjective.) Putting:

$$M_t=\{f(x)\in F[x]\,|\,\deg f(x)\le t\},$$

M_t is an F-submodule of $F[x]$ of dimension $t+1$. We see that, $F[x]/(x^p-x)$ is isomorphic to M_{p-1} (as F-modules).

We have proved the following lemma:

Lemma 3.6.3 *The following natural isomorphisms exist between the F-modules V, $U(F,F)$, $F[x]/(x^p-x)$ and M_{p-1}:*

$$\begin{array}{ccccccc}
V & \underset{T_1}{\rightleftarrows} & U(F,F) & \underset{T_2}{\leftrightarrows} & F[x]/(x^p-x) & \underset{T_3}{\rightleftarrows} & M_{p-1} \\
\Cup & & \Cup\ \ \Cup & & \Cup & & \Cup \\
v & \longrightarrow & f_v, f_{|F} & \longleftarrow & \overline{f(x)}\equiv f(x)\bmod(x^p-x) & \longrightarrow & \tilde{f}(x).
\end{array}$$

Here, for $v=\sum\lambda_a v_a$, f_v is defined by $f_v(a)=\lambda_a(\forall a\in F)$, and $\tilde{f}(x)$ is the remainder of the division of $f(x)$ by x^p-x. The F-module M_{p-1} inherits, via T_3, an F-algebra structure. T_2 is also an algebra isomorphism.

Since V is a G-group, the other F-modules can also be made into G-groups using these isomorphisms. For $\sigma\in G$, and $f\in U(F,F)$ f^σ is given by

$$f^\sigma(a)=f(a^{\sigma^{-1}})\quad \forall a\in F. \tag{3.6.1}$$

In particular,

$$f^{\sigma_0}(a)=f(a^{\sigma_0^{-1}})=f(a+1)\quad \forall a\in F,$$

where σ_0 is the generating element of P, fixed at the beginning of this discussion. For $f,g\in U(F,F)$ we have

$$(fg)^\sigma=f^\sigma g^\sigma, \tag{3.6.2}$$

i.e. σ is a ring-automorphism of $U(F,F)$.

Next, using T_2, $F[x]/(x^p - x)$ can also be made into a G-group. To write down concretely the way an arbitrary $\sigma \in G$ operates on $F[x]/(x^p - x)$ is quite troublesome, but for σ_0 it is simple. Using (3.6.1), we get for $\overline{f(x)} \in F[x]/(x^p - x)$:

$$\overline{f(x)}^{\sigma_0} = \overline{f(x+1)}.$$

For $f(x) \in M_{p-1}$ we have the similar result:

$$f(x)^{\sigma_0} = f(x+1).$$

Also, using (3.6.1), for $f(x) \in M_{p-1}$, $c \in F$ and $\sigma \in G$ we have

$$f^{\sigma}(c) = f(c^{\sigma^{-1}}).$$

It is easy to check that σ defines a ring-automorphism for M_{p-1}. Each element of M_0 is G-invariant, i.e. $\lambda^{\sigma} = \lambda$ for $\lambda \in M_0$, $\sigma \in G$ since λ corresponds to $\lambda \sum_{a \in \Omega} v_a \in V$.

Lemma 3.6.4 (1) *If the F-submodule $M \neq 0$ of M_{p-1} is G-invariant (i.e. if for all $\sigma \in G$, $M^{\sigma} = M$), then $M = M_i$ for some $i (0 \leq i \leq p-1)$.*

(2) *$0, M_0, M_{p-2}$ and M_{p-1} are G-invariant submodules of M_{p-1}.*

(3) *If (G, Ω) is not 2-transitive, then M_{r-1}, M_r are G-invariant for some $r \leq p-2$.*

Proof (1) Let $f(x) = \lambda_i x^i + \lambda_{i-1} x^{i-1} + \ldots + \lambda_0$ $(\lambda_i \neq 0)$ denote a polynomial of maximal degree in M. Since $f^{\sigma_0}(x) \in M$, we have $f^{\sigma_0}(x) - f(x) \in M$. Since $f^{\sigma_0}(x) - f(x) = f(x+1) - f(x) = \mu_{i-1} x^{i-1} + \ldots + \mu_0$ where $\mu_{i-1} = i\lambda_i \neq 0$, we see that M (if $M \neq M_0$) also contains a polynomial of degree $i-1$. Therefore M contains polynomials of degree $i, i-1, \ldots, 1, 0$, hence $M \supseteq M_i$ and since M does not contain any polynomial of degree $i+1$, we have $M = M_i$.

(2) We have already noted that M_0 is G-invariant. The G-invariance of 0 and M_{p-1} is obvious. If $f(x) \in M_{p-1}$ is such that $\sum_{c \in F} f(c) = 0$, then for $\sigma \in G$,

$$\sum_{c \in F} f^{\sigma}(c) = \sum_{c \in F} f(c^{\sigma^{-1}}) = \sum_{d \in F} f(d) = 0,$$

i.e.

$$\tilde{M} = \left\{ f(x) \in M_{p-1} \,\middle|\, \sum_{c \in F} f(c) = 0 \right\}$$

is a G-invariant F-submodule of M_{p-1} and its dimension equals $p-1$. (For, define a F-homomorphism $T: M_{p-1} \to F$ by

$$T(f(x)) = \sum_{c \in F} f(c).$$

Then Ker $T = \tilde{M}$, so it suffices to show that T is not the zero-homomorphism. Suppose T is the zero-homomorphism. Then $\sum_{c\in F} c^i = 0, \forall i, 1 \leq i \leq p-1$, hence, letting a represent a generating element of $F - \{0\}$ we have $\sum_{t=1}^{p-1} a^{it} = 0$. This means that for $g(x) = x^{p-1} + \ldots + x$, we have $g(c) = 0$, $\forall c \in F - \{0\}$. Therefore $\prod_{c \in F - \{0\}} (x - c) = x^{p-1} - 1$ divides $g(x)$. Contradiction.) So by (1) $\tilde{M}$ coincides with M_{p-2} and the result follows.

(3) For $h(x) \in M_{p-1}$, we define polynomials $h_i(x) (i \leq p-1)$ inductively as follows:

$$h_{p-1}(x) = h(x), \quad h_i(x) = h_{i+1}(x+1) - h_{i+1}(x) \quad (i \leq p-2).$$

If deg $h(x) = r$, then deg $h_{p-2}(x) = r - 1$, as we saw in (1) and deg $h_i(x) = i - (p - 1 - r)$ (however, if $i < p - 1 - r$, then $h_i(x) = 0$). Since (G, Ω) is not 2-transitive, we conclude from lemma 3.6.2 that $\dim_F \operatorname{Hom}_G(M_{p-1}, M_{p-1}) \geq 3$. Let $f(x)$ denote a polynomial of degree $p-1$ of M_{p-1}. Then $M = \{\alpha f(x) | \alpha \in \operatorname{Hom}_G(M_{p-1}, M_{p-1})\}$ is a F-submodule of M_{p-1} and if $\alpha f(x) = 0$ then $\alpha = 0$. (For, since $\deg f_i(x) = i$, $M_{p-1} = \langle f_{p-1}(x) (= f(x)), \ldots, f_0(x) \rangle$ (as an F-module). If $\alpha f_i(x) = 0$, then $\alpha f_{i-1}(x) = \alpha(f_i(x+1) - f_i(x)) = \alpha(f_i^{\sigma_0}(x)) = (\alpha f_i(x))^{\sigma_0} = 0$. Therefore $\alpha M_{p-1} = 0$.) Therefore, the assignment $\alpha \to \alpha f(x)$ defines an isomorphism from $\operatorname{Hom}_G(M_{p-1}, M_{p-1})$ into M_{p-1}, hence dim $M \geq 3$. Therefore, $M \nsubseteq Fx^{p-1} + F (\subseteq M_{p-1})$ and there is an element $\alpha \in \operatorname{Hom}_G(M_{p-1}, M_{p-1})$ such that $\alpha f(x) = g(x)$ has degree r with $1 \leq r \leq p-2$. We have $\alpha f_{p-1}(x) = g_{p-1}(x)$, and if $\alpha f_i(x) = g_i(x)$, then

$$\begin{aligned}\alpha f_{i-1}(x) &= \alpha(f_i(x+1) - f_i(x)) = \alpha(f_i^{\sigma_0}(x) - f_i(x)) \\ &= (\alpha f_i(x))^{\sigma_0} - \alpha f_i(x) = g_i^{\sigma_0}(x) - g_i(x) = g_{i-1}(x),\end{aligned}$$

therefore: $\alpha M_{p-1} = \alpha \langle f_{p-1}(x), \ldots, f_0(x) \rangle = \langle g_{p-1}(x), \ldots, g_0(x) \rangle = M_r$. Since $M_{p-2} = \langle f_{p-2}(x), \ldots, f_0(x) \rangle$ we have $\alpha M_{p-2} = M_{r-1}$. Since M_{p-1} and M_{p-2} are G-invariant and since $\alpha \in \operatorname{Hom}_G(M_{p-1}, M_{p-1})$, we conclude that M_r and M_{r-1} are also G-invariant. ■

Theorem 3.6.5 (Burnside) *Let (G, Ω) be a permutation group of prime degree p. If (G, Ω) is transitive, but not 2-transitive, then (G, Ω) is regular or a Frobenius group.*

Proof. We use the same notation as before. We shall show that G is isomorphic to a subgroup of the group of affine transformations of F. According to lemma 3.6.4 there is an $r \leq p-2$ such that M_{r-1} and M_r are G-invariant. Putting

$$M = \{f(x) \in M_{p-1} \mid f(x)M_{r-1} \subseteq M_r\}$$

we see that M is a F-submodule of M_{p-1} and since elements of G define ring-automorphisms for M_{p-1}, M is G-invariant. We have $x \in M$, but $x^2 \notin M$ since $r \leq p-2$. Therefore $M = M_1$, hence M_1 is G-invariant. Hence for $\sigma \in G$, x^σ can be written in a unique way as:

$$x^\sigma = \lambda(\sigma)x + \mu(\sigma) \quad \lambda(\sigma), \mu(\sigma) \in F.$$

Now, define a map from G into the group of affine transformations of F by $\sigma \to f_\sigma$, where

$$f_\sigma(x) = \lambda(\sigma^{-1})x + \mu(\sigma^{-1}).$$

It is easy to check that this is an injective isomorphism. Then the result follows from theorem 3.6.1. ■

3.7 Primitive permutation groups

We saw in §3.3 that loosely speaking, imprimitive permutation groups are composed of primitive permutation groups. Therefore, many problems concerning permutation groups have to do with the clarification of the properties of primitive permutation groups. Primitive permutation groups have very different properties depending on whether they are multiply transitive or not. Historically speaking, it seems that most attention has been devoted to problems concerning multiple transitivity, but interesting results have also been achieved concerning groups not multiply transitive. Especially recently, the research concerning these groups has entered a new phase with the discovery of new simple groups.

In this section we will state some of the fundamental properties of primitive permutation groups that are not multiply transitive.

3.7.1 Orbitals and graphs of transitive groups

Let (G,Ω) denote a transitive permutation group. For $(a_1,\dots,a_n)\in\Omega^n=\underbrace{\Omega\times\dots\times\Omega}_{n\text{ times}}$ and $\sigma\in G$, we define

$$(a_1,\dots,a_n)^\sigma=(a_1^\sigma,\dots,a_n^\sigma).$$

Therefore (G,Ω^n) becomes a permutation group. The number r of orbits of (G,Ω^2) is called the rank of (G,Ω). The subset $\{(a,a)\,|\,a\in\Omega\}$ is clearly an orbit of (G,Ω^2). Denoting this orbit by $\Delta^{(1)}$, the remaining orbits of (G,Ω^2) are denoted by $\Delta^{(2)},\dots,\Delta^{(r)}$. The $\Delta^{(i)}$ are called orbitals of (G,Ω); in particular, $\Delta^{(1)}$ is called the trivial orbital.

Let $a\in\Omega$ and let us put,

$$\Delta_0^{(i)}(a)=\{b\,|\,(a,b)\in\Delta^{(i)}\}\quad\text{for } i=1,\dots,r.$$

Since (G,Ω) is transitive, $\Delta^{(i)}(a)\neq\varnothing$ for $i=1,\dots,r$ and for $a\in\Omega$. We have $\Delta^{(1)}(a)=\{a\}$ and

$$\Omega=\{a\}+\Delta^2(a)+\dots+\Delta^{(r)}(a)$$

is the orbit decomposition of (G_a,Ω). Therefore, there exists a one-to-one correspondence $\Delta^{(i)}\leftrightarrow\Delta^{(i)}(a)$ between the orbitals of (G,Ω) and the orbits of (G_a,Ω). Choosing an element $\sigma_i\in G$ for each i such that $a^{\sigma_i}\in\Delta^i(a)$ then we have for $\sigma\in G$

$$a^\sigma\in\Delta^{(i)}(a)\Leftrightarrow(a,a^\sigma)\in\Delta^{(i)}\Leftrightarrow\exists\beta\in G_a,\text{ such that } a^{\sigma_i}=a^{\sigma\beta}\Leftrightarrow$$

$$\exists\beta\in G_a,\text{ such that }\sigma\beta\sigma_i^{-1}\in G_a\Leftrightarrow$$

$$\sigma\in G_a\sigma_iG_a\Leftrightarrow G_a\sigma G_a=G_a\sigma_iG_a.$$

Hence,

$$G_a\sigma_iG_a=\{\sigma\in G\,|\,a^\sigma\in\Delta^{(i)}(a)\},$$

and the element $G_a\sigma_iG_a$ of $G_a\backslash G/G_a$ is uniquely determined by $\Delta^{(i)}$. Therefore there exists a one-to-one correspondence $\Delta^{(i)}\leftrightarrow G_a\sigma_iG_a$

between the orbitals of (G, Ω) and the double cosets of G with regard to G_a. If Δ is an orbital of (G, Ω), then $\{(\alpha, \beta) | (\beta, \alpha) \in \Delta\}$ is also an orbital of (G, Ω), called the paired orbital of Δ and denoted by ${}^t\Delta$.

For $a \in \Omega$, let Δ be the orbital of (G, Ω) corresponding to $G_a \sigma G_a \in G_a \backslash G / G_a$. Since

$$a^\sigma \in \Delta(a) \Leftrightarrow (a, a^\sigma) \in \Delta \Leftrightarrow (a, a^{\sigma^{-1}}) = (a^\sigma, a)^{\sigma^{-1}} \in {}^t\Delta \Leftrightarrow a^{\sigma^{-1}} \in {}^t\Delta(a)$$

we have: if $G_a \sigma G_a$ corresponds to Δ, then

$$\left.\begin{array}{c} G_a \sigma^{-1} G_a \text{ corresponds to } {}^t\Delta \\ \text{and} \\ {}^t\Delta(a) = \{a^\tau | a^{\tau^{-1}} \in \Delta(a), \tau \in G\}. \end{array}\right\} \quad (3.7.1)$$

We have proved:

Theorem 3.7.1 *Let (G, Ω) be a transitive permutation group, and let $a \in \Omega$.*

(1) *There exist one-to-one correspondences between the collection of orbitals of (G, Ω), the collection of orbits of (G_a, Ω) and $G_a \backslash G / G_a$, given by*

$$\Delta \leftrightarrow \Delta(a) \leftrightarrow G_a \sigma G_a,$$

where σ is an element of G such that $(a, a^\sigma) \in \Delta$. Therefore:

(2) *Denoting the rank of (G, Ω) by r,*

$$r = \text{number of orbits of } (G_a, \Omega) = |G_a \backslash G / G_a|.$$

(3) *In particular: (G, Ω) is 2-transitive $\Leftrightarrow r = 2$.*

(4) *If Δ is an orbital of (G, Ω), then*

$${}^t\Delta = \{(x, y) | (y, x) \in \Delta\}$$

is also an orbital of (G, Ω) and

$${}^t\Delta(a) = \{a^\tau | a^{\tau^{-1}} \in \Delta(a), \tau \in G\}.$$

If $G_a \sigma G_a$ is the element of $G_a \backslash G / G_a$ corresponding to Δ, then $G_a \sigma^{-1} G_a = (G_a \sigma G_a)^{-1}$ corresponds to ${}^t\Delta$.

The orbit $\Delta^{(i)}(a)$ of (G_a, Ω) is called a suborbit of (G, Ω) (with regard to a) and $n_i = |\Delta^{(i)}(a)|$ is called its length. Since (G, Ω) is transitive and since

$$\Delta^{(i)}(a^\sigma) = \Delta^{(i)}(a)^\sigma \quad \sigma \in G,$$

the number n_i is independent of the choice of a, completely determined by $\Delta^{(i)}$ and $|\Delta^{(i)}| = n_i|\Omega|$.

An orbital Δ satisfying ${}^t\Delta = \Delta$ is called self-paired or symmetric. Obviously, $\Delta^{(1)}$ is self-dual.

Theorem 3.7.2 (G, Ω) *has a self-dual, non-trivial orbital if and only if* $|G|$ *is even.*

Proof Let $|G|$ be even and let $\sigma \in G$ be an element of order 2. Then there are $a, b \in \Omega$ $(a \neq b)$ such that $a^\sigma = b$, $b^\sigma = a$. The orbital Δ of (G, Δ) containing (b, a) is non-trivial and self-dual since $(a, b) \in \Delta \cap {}^t\Delta$. Conversely, let Δ be a non-trivial, self-dual orbital of (G, Ω). If $(a, b) \in \Delta$, then $(b, a) \in \Delta$, hence $a^\sigma = b$ and $b^\sigma = a$ for some $\sigma \in G$ and σ is an element of even order. ■

Suppose we represent the elements of Ω as (different) points (for example, in the plane). If Δ is a relation on Ω (i.e. $\Delta \subseteq \Omega \times \Omega$), then this relation can be represented by drawing an arrow from a to b if $(a, b) \in \Delta$. The resulting diagram is called the graph of Δ and represented by (Ω, Δ). The points of Ω are called the vertices of the graph, the points of Δ are called the edges. If

$$\dot{a} \longrightarrow \dot{b} \text{ or } \dot{a} \longleftarrow \dot{b}$$

occurs in the graph, we say that the vertices a and b are connected by an edge. A sequence of vertices $a = a_0, a_1, \ldots, a_n = b$, such that that a_0 and a_1, a_1 and $a_2, \ldots, a_{n-1}$ and a_n are connected by an edge, is called a path of length n between a and b. Moreover if for all pairs a_i, a_{i+1}, we have $\dot{a}_i \rightarrow \dot{a}_{i+1}$, the path is called a directed path from a to b. We define an equivalence relation $\sim$ on Ω by

$$a \sim b \Leftrightarrow \text{there is a path between } a \text{ and } b.$$

If Ω itself is an equivalence class under $\sim$, then the graph is called connected. If Ω_1 is an equivalence class under $\sim$, then $(\Omega_1, \Delta \cap \Omega_1^2)$

is easily seen to be a connected graph, called a connected component of the graph (Ω, Δ).

If the permutation σ of Ω, regarded as a permutation of $\Omega \times \Omega$, leaves Δ invariant, then σ is called an automorphism of the graph (Ω, Δ). The collection of all automorphisms of (Ω, Δ), i.e. $\{\sigma \in S^{\Omega} \mid \Delta^{\sigma} = \Delta\}$, is a group, called the automorphism group of (Ω, Δ) and denoted by $\mathrm{Aut}(\Omega, \Delta)$. If the subgroup G of $\mathrm{Aut}(\Omega, \Delta)$ operates transitively on Ω, we say that G operates transitively on the vertices and we call (Ω, Δ) a vertex-transitive graph. If G operates transitively on Δ, we say that G operates transitively on the edges and we call (Ω, Δ) an edge-transitive graph.

Theorem 3.7.3 *Let (Ω, Δ) be a vertex-transitive graph. If $a, b \in \Omega$ are such that there is a path between a and b, then there is a directed path from a to b.*

Proof It suffices to prove, that if $\dot{a} \to \dot{b}$, then there exists a directed path from b to a. Put:

$$\Gamma(a) = \{b \in \Omega \mid \text{there is a directed path from } a \text{ to } b\}.$$

Then $b \in \Gamma(a)$ and obviously $\Gamma(b) \subseteq \Gamma(a)$. Since $G = \mathrm{Aut}(\Omega, \Delta)$ operates transitively on the vertices, we have $a^{\sigma} = b$ for some $\sigma \in G$. Since $\Gamma(a^{\sigma}) = \Gamma(a)^{\sigma}$, we have $|\Gamma(b)| = |\Gamma(a)^{\sigma}| = |\Gamma(a)|$ and so $\Gamma(a) = \Gamma(b)$, hence $b \in \Gamma(b)$. Therefore:

$$a = b^{\sigma^{-1}} \in \Gamma(b)^{\sigma^{-1}} = \Gamma(b^{\sigma^{-1}}) = \Gamma(a) = \Gamma(b). \qquad \blacksquare$$

Corollary 3.7.4 *If (Ω, Δ) is a connected, vertex-transitive graph, then for every two points $a, b \in \Omega$ there exists a directed path from a to b.*

Let (G, Ω) be a transitive group and let Δ be a non-trivial orbital of (G, Ω). Obviously G operates transitively on the vertices and the edges of the graph (Ω, Δ). Conversely, if the graph (Ω, Δ) is vertex-transitive and edge transitive, then (G, Ω), where $G = \mathrm{Aut}(\Omega, \Delta)$, is a transitive permutation group and Δ is one of its orbitals. The graph determined in this way by the orbital Δ of the transitive permutation group (G, Ω), is called the orbital graph corresponding to an orbital Δ of (G, Ω).

Theorem 3.7.5 *Let (G, Ω) be a transitive permutation group, then (G, Ω) is primitive if and only if all orbital graphs of (G, Ω) that are determined by non-trivial orbitals are connected.*

Proof Still using the notation introduced in the beginning of this section, let $\Delta^{(2)}, \ldots, \Delta^{(r)}$ denote the non-trivial orbitals of (G, Ω). First, let us suppose that (G, Ω) is imprimitive. Let ψ be a non-trivial set of imprimitivity and let $a \in \psi$. Since $a \in \psi \cap \psi^\sigma$ for $\sigma \in G_a$, we have $\psi^\sigma = \psi$, i.e. ψ is an invariant domain under G_a. Therefore, $\psi - \{a\}$ can be written as the direct sum of a certain number of orbits of $(G_a, \Omega - \{a\})$. We will show that the graph $(\Omega, \Delta^{(i)})$ is not connected if $\Delta^{(i)}(a) \subseteq \psi$. Suppose that $(\Omega, \Delta^{(i)})$ is connected. Choose $b \in \Delta^{(i)}(a)$ and $c \in \Omega - \psi$ (this is possible because $\psi \subsetneq \Omega$) and let

$$b = b_0, b_1, \ldots, b_t = c$$

be a path between b and c of the graph $(\Omega, \Delta^{(i)})$. Since $b \in \psi$ and $c \notin \psi$, there is a number $j \geq 1$ such that $b_0, b_1, \ldots, b_{j-1} \in \psi$ and $b_j \notin \psi$. Since (b_{j-1}, b_j) or (b_j, b_{j-1}) is an element of $\Delta^{(i)}$, there is an element $\sigma \in G$ such that

$$(a, b)^\sigma = (b_{j-1}, b_j) \text{ or } (a, b)^\sigma = (b_j, b_{j-1}).$$

In both cases we have $b_{j-1} \in \psi \cap \psi^\sigma$ and so $\psi = \psi^\sigma$. Therefore, $b_j \in \psi$, contradiction. Hence, (G, Ω) is primitive if all non-trivial orbit-graphs are connected.

Conversely, suppose that there is a non-trivial orbital graph $(\Omega, \Delta^{(i)})$ that is not connected. Let (ψ, Δ) denote a connected component of $(\Omega, \Delta^{(i)})$. Since $(\Omega, \Delta^{(i)})$ is not connected, we have $\psi \subsetneq \Omega$. Further, since for $a \in \psi$ there is a path between a and each element of $\Delta^{(i)}(a)$, we have $\Delta^{(i)}(a) \subseteq \psi$ and $|\psi| \neq 1$. Since for $\sigma \in G$ $(\psi^\sigma, \Delta^\sigma)$ is also a connected component of $(\Omega, \Delta^{(i)})$, we conclude $\psi = \psi^\sigma$ or $\psi \cap \psi^\sigma = \varnothing$. Hence, ψ is a non-trivial set of imprimitivity. ■

3.7.2 Suborbits of primitive permutation groups

Let (G, Ω) be a primitive permutation group of degree n. If (G, Ω) is not 2-transitive, it will be called uniprimitive to distinguish it from groups with multiple transitivity. Let $\Delta^{(1)}, \Delta^{(2)}, \ldots, \Delta^{(r)}$ represent the orbitals of (G, Ω) just as in §3.7.1 ($\Delta^{(1)}$ represents the trivial orbital). Moreover we assume that the order of $\Delta^{(1)}, \ldots, \Delta^{(r)}$ is such that

$$1 = n_1 \leq n_2 \leq \ldots \leq n_r$$

where $n_i = |\Delta^{(i)}(a)|$.

Theorem 3.7.6 (1) *If* $\Delta^{(i)} = {}^t\Delta^{(i)}$ *for some* $i > 1$, *then* nn_i *is even.*

(2) *Let* n *be odd. If* n_i *is odd, then the number of* $\Delta^{(l)}$ *such that* $n_l = n_i$ *is even.*

Proof (1) If $(a, b) \in \Delta^{(i)}$, then $(b, a) \in \Delta^{(i)}$, hence $|\Delta^{(i)}| = nn_i$ is even.

(2) Suppose the number of $\Delta^{(l)}$ such that $n_l = n_i$ is odd, then at least one of these $\Delta^{(l)}$ is self-dual, contrary to (1). ■

Theorem 3.7.7 *If* (G, Ω) *is a primitive permutation group, then* (G, Ω) *is regular or* $n_2 > 1$.

Proof Suppose $n_2 = 1$. Putting $\psi = F_\Omega(G_a)$ for $a \in \Omega$, ψ is a set of imprimitivity by theorem 3.3.4 and $|\psi| > 1$. Hence $\psi = \Omega$. ■

Theorem 3.7.8 *Let* (G, Ω) *be a primitive permutation group. Then we have·*

(1) $n_i \leq n_2 n_{i-1}$ *for* $2 \leq i \leq r$.

(2) *If* $n_i > 1$, *then* $(n_i, n_r) \neq 1$.

Proof (1) For $i = 2$ this is obviously true. Suppose $n_i > n_2 n_{i-1}$ for some $i > 2$. For $a, b, c \in \Omega$ we have

$$|c^{G_b}| = |G_b : G_{b,c}| \geq |G_{a,b} : G_{a,b,c}| = \frac{|G_a : G_{a,c}||G_{a,c} : G_{a,b,c}|}{|G_a : G_{a,b}|}$$

$$\geq \frac{|G_a : G_{a,c}|}{|G_a : G_{a,b}|} = \frac{|c^{G_a}|}{|b^{G_a}|}.$$

Now choose $b \in \Delta^{(2)}(a)$ and $c \in \Delta^{(j)}(a)$ $(j \geq i)$, then $|b^{G_a}| = |\Delta^{(2)}(a)|$

and $|c^{G_a}| = |\Delta^{(j)}(a)|$, hence:

$$|c^{G_b}| \geq \frac{|c^{G_a}|}{|b^{G_a}|} = \frac{n_j}{n_2} > n_{i-1}.$$

Therefore, if $c^{G_b} = \Delta^{(l)}(b)$, then $l \geq i$.

Hence, putting $\Gamma_a = \bigcup_{j\geq i} \Delta^{(j)}(a)$, we have $(\Gamma_a \subseteq)\Gamma_a^{G_b} \subseteq \Gamma_b = \bigcup_{j\geq i} \Delta^{(j)}(b)$ and from the transitivity of (G, Ω) we conclude $\Gamma_a = \Gamma_a^{G_b} = \Gamma_b$. Since obviously $\Gamma_a^{G_a} = \Gamma_a$, we get $\Gamma_a^{\langle G_a, G_b\rangle} = \Gamma_a$. On the other hand $\langle G_a, G_b \rangle = G$ since (G, Ω) is primitive. Hence $\Gamma_a \subsetneq \Omega$ is an invariant domain of G, contrary to the transitivity of (G, Ω).

(2) Suppose $(n_i, n_r) = 1$. Put $\Gamma_a = \bigcup_{n_l = n_r} \Delta^{(l)}(a)$ for $a \in \Omega$, choose $b \in \Delta^{(i)}(a)$ and $c \in \Gamma_a$. Then $|G_a : G_{a,b,c}| = |G_a : G_{a,c}|\,|G_{a,c} : G_{a,b,c}|$, hence $|G_a : G_{a,b,c}| \equiv 0 \pmod{n_r}$. Since on the other hand: $|G_a : G_{a,b,c}| = |G_a : G_{a,b}|\,|G_{a,b} : G_{a,b,c}|$ and $|G_a : G_{a,b}| = n_i$, we get $|G_{a,b} : G_{a,b,c}| \equiv 0 \pmod{n_r}$. Hence n_r divides $|c^{G_{a,b}}|$. Since $|c^{G_{a,b}}| \leq c^{|G_a|} = n_r$, we conclude $|c^{G_{a,b}}| = |c^{G_b}| = n_r$, i.e. we have

$$|c^{G_b}| = n_r \quad \forall b \in \Delta^{(i)}(a) \text{ and } \forall c \in \Gamma_a.$$

Hence $\Gamma_b \supseteq \Gamma_a$. Since (G, Ω) is transitive, $\Gamma_b = \Gamma_a$ and $\Gamma_a^{G_a} = \Gamma_a^{G_b} = \Gamma_a$. Hence $\Gamma_a^G = \Gamma_a$, i.e. $\Omega = \Gamma_a$, contrary to $a \notin \Gamma_a$. ■

Using the notion of graphs, the first part of the previous theorem can be extended in the following way.

Theorem 3.7.9 *Let (G, Ω) be a transitive permutation group. If the graph $(\Omega, \Delta^{(2)})$ is connected, then:*

(1) *$n_i \leq n_2 n_{i-1}$ for $2 \leq i \leq r$.*

(2) *Moreover if $\Delta^{(2)} = {}^t\Delta^{(2)}$, then $n_i \leq (n_2 - 1)n_{i-1}$* for $3 \leq i \leq r$.

Proof (1) This is obviously true for $i = 2$, so we may assume $i \geq 3$. Put $\Gamma = \bigcup_{j \leq i-1} \Delta^{(j)}(a)$ for $a \in \Omega$. Since $(\Omega, \Delta^{(2)})$ is connected, there exists for an arbitrary $b \in \Omega - \Gamma$ a directed path from a to b:

$$a = a_0, a_1, a_2, \ldots, a_t = b$$

by theorem 3.7.3. Since $a \in \Gamma$, $b \notin \Gamma$, there is a number $s \leq t$ such that: $a, a_1, \ldots, a_{s-1} \in \Gamma$ but $a_s \notin \Gamma$. Since $a_1 \in \Delta^{(2)}(a)$, we have $s \geq 2$.

Let $\Delta^{(p)}(a)$ and $\Delta^{(q)}(a)$ denote orbits of $(G_a, \Omega - \{a\})$ containing a_{s-1} and a_s respectively. Letting the elements of G_a operate on $(a_{s-1}, a_s) \in \Delta^{(2)}$, we find that there exists for every element $c \in \Delta^{(q)}(a)$ an element $d \in \Delta^{(p)}(a)$ such that $(d, c) \in \Delta^{(2)}$. Since the number of elements $x \in \Delta^{(q)}(a)$ such that $(d, x) \in \Delta^{(2)}$ for all $d \in \Delta^{(p)}(a)$, is at most n_2, we have

$$n_i \leq |\Delta^{(q)}(a)| \leq n_2 |\Delta^{(p)}(a)| \leq n_2 n_{i-1}.$$

(2) Using the same notation as above, we see that for the element $a_{s-1} \in \Delta^{(p)}(a)$, $(a_{s-1}, a_{s-2}) \in \Delta^{(2)}$ and $a_{s-2} \in \Gamma$. Therefore, the number of elements $x \in \Delta^{(q)}(a)$ such that $(a_{s-1}, x) \in \Delta^{(2)}$ is at most $n_2 - 1$. In the same way as above, we conclude: $n_i \leq (n_2 - 1) n_{i-1}$. ■

Theorem 3.7.10 *Let (G, Ω) be a primitive permutation group, let $a \in \Omega$ and let H be a subgroup of G_a such that $H \neq 1$. For every orbit $\Delta^{(i)}(a)\,(i \geq 2)$ of $(G_a, \Omega - \{0\})$, there exists a subgroup K of G_a satisfying:*

(1) *H and K are conjugate in G.*

(2) *$K^{\Delta^{(i)}(a)} \neq 1$ (i.e. $F_\Omega(K) \not\supseteq \Delta^{(i)}(a)$).*

Proof If $F_\Omega(H) \not\supseteq \Delta^{(i)}(a)$, we can take $K = H$. So we may assume $F_\Omega(H) \supseteq \Delta^{(i)}(a)$. For $c \in \Delta^{(i)}(a)$, since $c \neq a$ and (G, Ω) is primitive, there exists an element $\sigma \in G$ such that $a \in F_\Omega(H)^\sigma$ and $c \notin F_\Omega(H)^\sigma$ by theorem 3.3.8. Put $K = H^\sigma$. Then $a \in \Gamma_\Omega(H)^\sigma = F_\Omega(K)$ and $c \notin F_\Omega(K)$, hence $K \leq G_a$ and $F_\Omega(K) \not\supseteq \Delta^{(i)}(a)$. ■

Corollary 3.7.11 *All prime factors of n_i are less than or equal to n_2.*

Proof Let p be a prime factor of n_i. If P is a Sylow p-subgroup of G_a, then $P \neq 1$. Therefore, there exists a Sylow p-subgroup Q of G_a such that $Q^{\Delta^{(2)}(a)} \neq 1$ by theorem 3.7.10. Hence $p \leq n_2$. ■

Corollary 3.7.12 (1) *Let K be a composition factor of G_a. Then for any $i\,(2 \leq i \leq r)$ there exists a subgroup H of G_a such that $H^{\Delta^{(i)}(a)}$ has a composition factor isomorphic to K.*

(2) *If $G_a^{\Delta^{(i)}(a)}$ is solvable for some $i\,(2 \leq i \leq r)$, then so is G_a.*

(3) *The prime factors of* $|G_a^{\Delta^{(i)}(a)}|$ *are the same as those of* $|G_a|$ *for* $2 \leq i \leq r$.

(4) *If for some* $i\,(2 \leq i \leq r)\,G_a^{\Delta^{(i)}(a)}$ *is a p-group, then* G_a *is also a p-group.*

Proof (1) Let H be a subgroup of G_a of minimal order such that K is a composition factor of H. According to theorem 3.7.10 we may choose H satisfying $H^{\Delta^{(i)}(a)} \neq 1$. Therefore $H \gneq H_{\Delta^{(i)}(a)}$ and K can be represented as a composition factor of $H/H_{\Delta^{(i)}(a)} \simeq H^{\Delta^{(i)}(a)}$ or $H_{\Delta^{(i)}(a)}$ (theorem 2.3.1). Since K cannot be a composition factor of $H_{\Delta^{(i)}(a)}$ by the choice of H, the statement of the theorem follows.

(2), (3) and (4) are easy consequences of (1). ■

Theorem 3.7.13 *Let* (G, Ω) *be a primitive permutation group. If for some* $i\,(2 \leq i \leq r)\, n_i = p$ *(p prime), then* $|G_a| \not\equiv 0 \pmod{p^2}$.

Proof Let $b \in \Delta^{(i)}(a)$ and put $H^{(0)} = G_a$, $K^{(0)} = G_b$, $S^{(0)} = H^{(0)} \cap K^{(0)} = G_{a,b}$. We define subgroups $H^{(i)}$, $K^{(i)}$ and $S^{(i)}$ of G inductively by

$$H^{(i)} = \bigcap_{\sigma \in G_a} S^{(i-1)\sigma}, \quad K^{(i)} = \bigcap_{\sigma \in G_b} S^{(i-1)\sigma},$$

$$S^{(i)} = H^{(i)} \cap K^{(i)}.$$

Then

$$G_a \trianglerighteq H^{(i)} \geq S^{(i)} \geq H^{(i+1)}$$

$$G_b \trianglerighteq K^{(i)} \geq S^{(i)} \geq K^{(i+1)}.$$

From $S^{(i)} \trianglerighteq H^{(i+1)}$ and $S^{(i)} \trianglerighteq K^{(i+1)}$ we get

$$S^{(i)} \trianglerighteq S^{(i+1)}, H^{(i+1)} \trianglerighteq S^{(i+1)} \text{ and } K^{(i+1)} \trianglerighteq S^{(i+1)}.$$

Since $|G| < \infty$, $S^{(l)} = S^{(l+1)}$ for some l. We can prove that $S^{(l)} = 1$. (For, since

$$S^{(l)} \geq H^{(l+1)}, S^{(l)} \geq K^{(l+1)}$$

and

$$S^{(l)} \geq S^{(l+1)} = H^{(l+1)} \cap K^{(l+1)},$$

we have $S^{(l+1)} = H^{(l+1)} = K^{(l+1)}$. In particular, $G_a \trianglerighteq S^{(l)}$ and $G_b \trianglerighteq S^{(l)}$, hence $G = \langle G_a, G_b \rangle \trianglerighteq S^{(l)}$. Now suppose $S^{(l)} \neq 1$. Then $(S^{(l)}, \Omega)$ is transitive, since (G, Ω) is primitive. On the other hand, we have $S^{(l)} \leq G_a$, i.e. S_l has a fixed point. Contradiction.) We have

$$|H^{(0)} : S^{(0)}| = |G_a : G_{a,b}| = |\Delta^{(i)}(a)| = p$$

and in the same way we have

$$|K^{(0)} : S^{(0)}| = |{}^t\Delta^{(i)}(a)| = p$$

using theorem 3.2.2 (2). Since $H^{(1)}$ is the kernel of the permutation representation $(H^{(0)}, H^{(0)}/S^{(0)})$ and since $|H^{(0)} : S^{(0)}| = p$, we see that $|S^{(0)} : H^{(1)}|$ is relatively prime with p. In the same way, we find that $|S^{(0)} : K^{(1)}|$ is relatively prime with p. Suppose that for some i both $|S^{(i)} : H^{(i+1)}|$ and $|S^{(i)} : K^{(i+1)}|$ are relatively prime with p. From:

$$H^{(i+1)}/S^{(i+1)} = H^{(i+1)}/H^{(i+1)} \cap K^{(i+1)} \simeq K^{(i+1)}H^{(i+1)}/K^{(i+1)},$$

we conclude that in that case $|H^{(i+1)} : S^{(i+1)}|$ is also relatively prime with p. Since for all $\sigma \in G_a$, $H^{(i+1)} = H^{(i+1)\sigma} \trianglerighteq S^{(i+1)\sigma}$ we conclude from the isomorphism theorem that

$$H^{(i+1)}/H^{(i+2)} \left(= H^{(i+1)} / \bigcap_{\sigma \in G_a} S^{(i+1)\sigma}\right)$$

is a p'-group. In particular $|S^{(i+1)} : H^{(i+2)}|$ is relatively prime with p. In the same way, we find that $|S^{(i+1)} : K^{(i+2)}|$ is also relatively prime with p. Therefore we conclude that, $\forall i \geq 1$, the numbers $|S^{(i-1)} : H^{(i)}|$ and $|H^{(i)} : S^{(i)}|$ are all relatively prime with p. Since $|G_a| = \prod_{i=0}^{l} |H^{(i)} : S^{(i)}| \, |S^{(i)} : H^{(i+1)}|$, we get $|G_a| = p \cdot s$ where $(s, p) = 1$. ∎

Corollary 3.7.14 *If for some* $i\,(2 \leq i \leq r)\, n_i = p$ *(p prime), then* $n_j \not\equiv 0 \pmod{p^2}$ *for all j.*

Theorem 3.7.15 *Let* (G, Ω) *be a primitive permutation group. If for some* $i\,(2 \leq i \leq r)$ $n_i = p$ *(p prime) and if* $|G_a^{\Delta^{(i)}(a)}| = p$, *then G is a solvable Frobenius group and its degree is a power of some prime q, where* $q \neq p$.

Proof Since $|\Delta^{(i)}(a)| = p$, we have $|G_a| = p \cdot s$ with $(s, p) = 1$ by theorem 3.7.13, and hence we have $|G_a| = p$ by corollary 3.7.12 (3). Therefore $|G_a| = p$ for all $a \in \Omega$. For $a, b \in \Omega$ $(a \neq b)$ we have $G_a \neq G_b$, i.e. $G_{a,b} = 1$ since (G, Ω) is primitive. Therefore (G, Ω) is a Frobenius group. Let M $(\neq 1)$ be a characteristic subgroup of the Frobenius kernel N of G, then $G \trianglerighteq M$ since $G \trianglerighteq N \trianglerighteq M$ and (M, Ω) is transitive since (G, Ω) is primitive. On the other hand, since N operates regularly on Ω, we have $N = M$. From this and from nilpotency of N (theorem 2.11.5), we conclude that N is an elementary Abelian group and its order $(= |\Omega|)$ is a power of some prime q with $p \neq q$. ■

Theorem 3.7.16 (Sims) *Let (G, Ω) be a primitive permutation group. If for some $i\,(2 \leq i \leq r)\, n_i = 3$, then for $a \in \Omega$, $|G_a| = 3 \times 2^m$, where $m \leq 4$ or $m = 6$.*

Proof Put $\Delta^{(i)} = \Delta$. Since $|\Delta(a)| = 3$, we have $(G_a^{\Delta(a)}, \Delta(a)) = A_3$ or S_3. If $(G_a^{\Delta(a)}, \Delta(a)) = A_3$ then $|G_a| = 3$ by corollary 3.7.12 (3) and theorem 3.7.13. So we may assume $(G_a^{\Delta(a)}, \Delta(a)) = S_3$ for all $a \in \Omega$. Since $|S_3| = 6$, we can write $|G_a| = 3 \times 2^m$ by corollary 3.7.12 (3) and theorem 3.7.13 and $|G_{a,b}| = 2^m$ for $(a, b) \in \Delta$.

Let $P^{(k)}$ denote the collection of all directed paths of length k of the orbital graph (Ω, Δ). For $l = (a_0, a_1, \ldots, a_k) \in P^{(k)}$ and $\sigma \in G$ we define

$$(a_0, a_1, \ldots, a_k)^\sigma = (a_0^\sigma, a_1^\sigma, \ldots, a_k^\sigma),$$

making G into a permutation group on $P^{(k)}$. The stabilizer G_l of $l = (a_0, a_1, \ldots, a_k) \in P^{(k)}$ coincides with $G_{a_0, a_1, \ldots, a_k}$. If $\hat{l} = (a_0, a_1, \ldots, a_k, a_{k+1}) \in P^{(k+1)}$ is such that $(a_0, a_1, \ldots, a_k) \in P^{(k)}$, then $\hat{l}$ is called an extension of $(a_0, a_1, \ldots, a_k)$. Since $|\Delta(a_k)| = 3$ there are three extensions of a path l and the collection of these three extensions will be denoted by $T(l)$. Since $G_l = G_{a_0, \ldots, a_k} \leq G_{a_{k-1}, a_k}$ $(k \geq 1)$, G_l is a 2-group that operates on $T(l)$. Hence there are two possibilities:

(i) G_l leaves all elements of $T(l)$ fixed, or

(ii) G_l leaves one element of $T(l)$ fixed and transposes the other two.

In other words: putting $\Delta(a_k)=\{b_1, b_2, b_3\}$ we have

(i)′ $$|G_{a_0,\ldots,a_k}:G_{a_0,\ldots,a_k,b_i}|=1 \quad \forall i\in\{1,2,3\}$$

(if (i) is true), or

(ii)′ $$|G_{a_0,\ldots a_k}:G_{a_0,\ldots,a_k,b_{i_0}}|=1 \quad \exists i_0\in\{1,2,3\}$$

and

$$|G_{a_0\ldots,a_k}:G_{a_0,\ldots,a_k,b_i}|=2 \quad \forall i\in\{1,2,3\}-\{i_0\}$$

(if (ii) is true).

To distinguish between these possibilities we need the following lemma.

Lemma 3.7.17 *If $k=1$ then case (i) cannot occur. In other words: Let $l=(a,b)$ be an element of $P^{(1)}$ and put $\Delta(b)=\{c_1, c_2, c_3\}$. Let $l_i=(a,b,c_i)$ $(i=1,2,3)$ be the three possible extensions of l. Then $T(l)=\{l_1, l_2, l_3\}$ is divided into two orbits by the operation of $G_{a,b}$, namely orbits of length 1 and length 2. Let l_1 be the fixed element (= the orbit of length 1) then:*

$$|G_{a,b}:G_{a,b,c_1}|=1 \text{ and } |G_{a,b}:G_{a,b,c_i}|=2 \quad (i=2,3).$$

Proof Suppose that case (i) occurs for $l=(a,b)\in P^{(1)}$. Since,

$$G_{a,b}=G_{a,b,\Delta(b)}\leq G_{b,\Delta(b)} \text{ and } G_b^{\Delta(b)}=G_b/G_{b,\Delta(b)}$$

we have

$$2^m=|G_{a,b}|\leq|G_{b,\Delta(b)}|=|G_b|/|G_b^{\Delta(b)}|=2^{m-1}.$$

Contradiction. (End of proof of lemma 3.7.17.) ■

We continue the proof of theorem 3.7.16 by defining $E^{(2)}\subseteq P^{(2)}$ by

$$E^{(2)}=\{(a_0, a_1, a_2)\in P^{(2)} \mid |G_{a_0,a_1}:G_{a_0,a_1,a_2}|=2\}.$$

$E^{(2)}\neq\varnothing$ by the previous lemma. We define $E^{(k)}\subseteq P^{(k)}$ for $k\geq 3$ by

$$E^{(k)}=\{(a_0,\ldots,a_k)\in P^{(k)} \mid |G_{a_0,\ldots,a_i}:G_{a_0,\ldots,a_i,a_{i+1}}|=2, \text{ for } 1\leq i\leq k-1\},$$

and $H^{(k)}\subseteq P^{(k)}$ for $k\geq 2$ by

$$H^{(k)}=\{(a_0,\ldots,a_k)\in P^{(k)} \mid (a_{i-2}, a_{i-1}, a_i)\in E^{(2)}, \quad \text{for } 2\leq i\leq k\}.$$

If $|G_{a_0,\ldots,a_i}:G_{a_0,\ldots,a_i,a_{i+1}}|=2$, then obviously $|G_{a_{i-1},a_i}:G_{a_{i-1},a_i,a_{i+1}}|=2$ (since the last number is always less than or equal to 2), hence $E^{(k)}\subseteq H^{(k)}$. Further, $E^{(2)}=H^{(2)}$. It is easy to check that $E^{(k)}$ and $H^{(k)}$ are both invariant domains of G and that G operates transitively on $E^{(k)}$. Since, for $(a_0,\ldots,a_k)\in E^{(k)}$,

$$2^m=|G_{a_0,a_1}|=\left(\prod_{i=1}^{k-1}|G_{a_0,\ldots,a_i}:G_{a_0,\ldots,a_i,a_{i+1}}|\right)|G_{a_0,\ldots,a_k}|$$
$$=2^{k-1}|G_{a_0,\ldots,a_k}|,$$

we conclude:

$$|G_{a_0,\ldots,a_k}|=2^{m-(k-1)}.$$

We have proved:

Lemma 3.7.18 *If $E^{(k)}\neq\varnothing$, then $|G_l|=2^{m-k+1}$ for $l\in E^{(k)}$. In particular, if $m+1<k$, then $E^{(k)}=\varnothing$.*

Before we can complete the proof of theorem 3.7.16 we must introduce and prove lemmas 3.7.19–23.

Lemma 3.7.19 *If $E^{(k)}\neq\varnothing$, then $H^{(k)}=E^{(k)}$.*

Proof The proof is by induction on k. The statement is true for $n=2$. Pick $(a_0,\ldots,a_k)\in H^{(k)}$, then $|G_{a_{k-2},a_{k-1}}:G_{a_{k-2},a_{k-1},a_k}|=2$. Hence putting $\Delta(a_{k-1})=\{a_k,c,d\}$, we may assume

$$|G_{a_{k-2},a_{k-1}}:G_{a_{k-2},a_{k-1},c}|=2$$

and

$$|G_{a_{k-2},a_{k-1}}:G_{a_{k-2},a_{k-1},d}|=1$$

by lemma 3.7.17. Since $(a_0,\ldots,a_{k-1})\in H^{(k-1)}$, we have $(a_0,\ldots,a_{k-1})\in E^{(k-1)}$ by the induction hypothesis. Pick $(b_0,\ldots,b_k)\in E^{(k)}$ $(E^{(k)}\neq\varnothing)$. Since $(b_0,\ldots,b_{k-1})\in E^{(k-1)}$ and G operates transitively on $E^{(k-1)}$, there exists an element $\sigma\in G$ such that $(b_0,\ldots,b_{k-1})^\sigma=(a_0,\ldots,a_{k-1})$. Then the element $b_k^\sigma\in\Delta(a_{k-1})$ and $b_k^\sigma\neq d$. (For

$$|G_{a_0,\ldots,a_{k-1}}:G_{a_0,\ldots,a_{k-1},b_k^\sigma}|=2$$

since

$$|G_{b_0,\ldots,b_{k-1}}:G_{b_0,\ldots,b_{k-1},b_k}|=2.$$

Therefore

$$|G_{a_{k-2},a_{k-1}}:G_{a_{k-2},a_{k-1},b_k^\sigma}|=2).$$

Therefore

$$|G_{a_0,\ldots,a_{k-1}}:G_{a_0,\ldots,a_{k-1},a_k}|=2.$$

Combining this with $(a_0,\ldots,a_{k-1})\in E^{(k-1)}$, we get $(a_0,\ldots,a_k)\in E^{(k)}$. (End of proof of lemma 3.7.19.) ■

For $(a,b)\in\Delta$, we define $V(a,b)\subseteq\Delta$ by

$$V(a,b)=\{(x,y)\in\Delta\,|\,\exists k,\ \exists(a_0,\ldots,a_k)\in H^{(k)},(a_0,a_1)$$
$$=(a,b),(a_{k-1},a_k)=(x,y)\}.$$

If $(c,d)\in V(a,b)$, then $V(c,d)\subseteq V(a,b)$, while also $|V(a,b)|=|V(c,d)|$ by the transitivity of (G,Δ), hence $V(a,b)=V(c,d)$. Further, $(c,d)\in V(c,d)$ and for any $(x,y)\in\Delta$ we have $(x,y)\in V(x,y)$ by the transitivity of (G,Δ). Therefore,

$$(c,d)\in V(a,b)\Leftrightarrow V(a,b)=V(c,d).$$

Hence the relation $\sim$, defined on Δ by

$$(a,b)\sim(c,d)\Leftrightarrow(c,d)\in V(a,b)$$

is an equivalence relation.

Lemma 3.7.20 *Δ itself is an equivalence class under $\sim$.*

Proof Since any two points of Ω can be connected by a directed path of the orbital graph (Ω,Δ) (theorems 3.7.3, 3.7.5), it suffices to prove that for any two elements (a,b) and $(b,c)\in\Delta$ $(b,c)\in V(a,b)$. If $(a,b,c)\in H^{(2)}$, then *obviously* $(b,c)\in V(a,b)$. So we may assume $(a,b,c)\notin H^{(2)}$. Put $\Delta(b)=\{c,d,e\}$, then (a,b,d), $(a,b,e)\in H^{(2)}$ by lemma 3.7.17. Since $(G_b,\Delta(b))$ is transitive there is an element $\sigma\in G$ such that $(b,d)^\sigma=(b,c)$. Since $(a,b,d)\in H^{(2)}$, we have $(a,b,d)^\sigma=(a^\sigma,b,c)\in H^{(2)}$. On the other hand since $(a,b,e)\in H^{(2)}$, we have $(a,b,e)^\sigma=(a^\sigma,b,e^\sigma)\in H^{(2)}$, hence $e^\sigma\in\{d,e\}$. Therefore:

$$(b,c)\sim(a^\sigma,b)\sim(b,e^\sigma)(=(b,d)\text{ or }(b,e))\sim(a,b),$$

hence $(b,c)\in V(a,b)$. (End of proof of lemma 3.7.20.) ■

Lemma 3.7.21 $E^{(m+1)} \neq \varnothing$.

Proof By lemma 3.7.18 $E^{(m')} \neq \varnothing$ and $E^{(m'+1)} = \varnothing$ for some $m' \leq m+1$. Pick $(a_0, \ldots, a_{m'}) \in E^{(m')}$. For an arbitrary a of Ω there exists an element b such that $(b, a) \in \Delta$. By lemma 3.7.20 there exist $l > m'$ and $a_{m'+1}, \ldots, a_{l-1}, a_l \in \Omega$ such that

$$(a_{l-1}, a_l) = (b, a) \text{ and } (a_0, \ldots, a_{m'}, a_{m'+1}, \ldots, a_{l-1}, a_l) \in H^{(l)}.$$

We will prove:

$$G_{a_0, \ldots, a_{m'}} = G_{a_1, \ldots, a_{m'+1}} = \ldots = G_{a_{l-m'}, \ldots, a_l}.$$

(For $G_{a_0, \ldots, a_{m'}} = G_{a_0, \ldots, a_{m'}, a_{m'+1}} \leq G_{a_1, \ldots, a_{m'+1}}$ because $E^{(m'+1)} = \varnothing$. Also $|G_{a_0, \ldots, a_{m'}}| = |G_{a_1, \ldots, a_{m'+1}}|$ because G operates transitively on $E^{(m')}$. Therefore $G_{a_0, \ldots, a_{m'}} = G_{a_1, \ldots, a_{m'+1}}$. The other equalities are proved in the same way.) Therefore, $G_{a_0, \ldots, a_{m'}}$ leaves a fixed and since a is an arbitrary element of Ω, we conclude $G_{a_0, \ldots, a_{m'}} = 1$. On the other hand, $|G_{a_0, \ldots, a_{m'}}| = 2^{m-(m'-1)}$ by lemma 3.7.18. Hence $m' = m+1$. (End of proof of lemma 3.7.21.) ■

Now, let $(a_0, \ldots, a_{m+1})$ be an element of $E^{(m+1)}$. Since $(a_0, \ldots, a_m)$ and $(a_1, \ldots, a_{m+1}) \in E^{(m)}$ and since G operates transitively on $E^{(m)}$, there exists an element $\sigma \in G$ such that $(a_0, \ldots, a_m)^\sigma = (a_1, \ldots, a_{m+1})$, i.e.

$$a_i^\sigma = a_{i+1} \quad (i = 0, \ldots, m).$$

Inductively, we put $a_i = a_{i-1}^\sigma$ for all integers $i > m+1$ and $a_i = a_{i+1}^\sigma$ for all integers $i < 0$. Then we have an infinite sequence $\{a_i | i \in \mathbb{Z}\}$ consisting of elements of Ω. Since $(a_l, \ldots, a_{l+m}) = (a_0, \ldots, a_m)^{\sigma^l} \in E^{(m)\sigma^l} = E^{(m)}, \forall l \in \mathbb{Z}$, we have $|G_{a_l, \ldots, a_{l+m}}| = 2$ by lemma 3.7.18 and so $G_{a_l, \ldots, a_{l+m}}$ is generated by a unique element x_l of order 2. Since

$$G_{a_l, \ldots, a_{l+m}} = G_{a_0^{\sigma^l}, \ldots, a_m^{\sigma^l}} = G^{\sigma^l}_{a_0, \ldots, a_m},$$

we have $x_l = x_0^{\sigma^l}$. Further, since $G_{a_0, \ldots, a_{m+1}} = 1$, for all l,

$$G_{a_l, a_{l+1}, \ldots, a_{l+m+1}} = 1.$$

Lemma 3.7.22 *For arbitrary i and l $(1 \le l \le m+1)$ we have*

$$x_{i+l+1} \notin \langle x_{i+1}, \ldots, x_{i+l} \rangle = G_{a_{i+l}, \ldots, a_{i+m+1}}.$$

In particular:

$$|\langle x_{i+1}, \ldots, x_{i+l} \rangle| = |G_{a_{i+l}, \ldots, a_{i+m+1}}| = 2^l \quad (1 \le l \le m),$$
$$\langle x_{i+1}, \ldots, x_{i+m+1} \rangle = G_{a_{i+m+1}},$$

and

$$\langle x_{i+1}, \ldots, x_{i+m+1}, x_{i+m+2} \rangle = G.$$

Proof It suffices to prove the assertion for $i = 0$.

$$\langle x_1, \ldots, x_l \rangle \subseteq \left\langle \bigcup_{j=1}^{l} G_{a_j, a_{j+1}, \ldots, a_{j+m}} \right\rangle \subseteq G_{a_l, \ldots, a_{m+1}},$$

since $\langle x_j \rangle = G_{a_j, a_{j+1}, \ldots, a_{j+m}}$. Suppose $x_{l+1} \in \langle x_1, \ldots, x_l \rangle$ then $x_{l+1} \in G_{a_l}$, hence,

$$x_{l+1} \in G_{a_l} \cap G_{a_{l+1}, \ldots, a_{l+m+1}} = G_{a_l, \ldots, a_{l+m+1}} = 1.$$

Contradiction, so we get $x_{l+1} \notin \langle x_1, \ldots, x_l \rangle$. Therefore, $|\langle x_1, \ldots, x_l \rangle| \lneq |\langle x_1, \ldots, x_l, x_{l+1} \rangle|$ and from this we have

$$2^{l+1} \le |\langle x_1, \ldots, x_{l+1} \rangle|.$$

For $l \le m-1$ we also have

$$|\langle x_1, \ldots, x_{l+1} \rangle| \le |G_{a_{l+1}, \ldots, a_{m+1}}| = 2^{m-(m-l-1)} = 2^{l+1}.$$

Hence $|\langle x_1, \ldots, x_{l+1} \rangle| = |G_{a_{l+1}, \ldots, a_{m+1}}|$, therefore $\langle x_1, \ldots, x_{l+1} \rangle = G_{a_{l+1}, \ldots, a_{m+1}}$. For $l = m$, we have

$$|\langle x_1, \ldots, x_{m+1} \rangle| \le |G_{a_{1+m}}| = 3 \cdot 2^m,$$

hence $\langle x_1, \ldots, x_{m+1} \rangle = G_{a_{m+1}}$. Since (G, Ω) is primitive, $\langle x_1, \ldots, x_{m+1} \rangle = G_{a_{m+1}}$ is a maximal subgroup of G and since $x_{m+2} \notin \langle x_1, \ldots, x_{m+1} \rangle$ we have $\langle x_1, \ldots, x_{m+1}, x_{m+2} \rangle = G$. (End of proof of lemma 3.7.22.) ■

Lemma 3.7.23 *We have the following relations for all $i \in \mathbb{Z}$:*

(1) $[x_i, x_{i+1}] = 1$.

(2) $[x_i, x_{i+l}] \in \langle x_{i+1}, \dots, x_{i+l-1} \rangle$ $(1 \le l \le m-1)$.
(3) *If* $m \ge 4$, *then* $[x_i, x_{i+2}] = 1$.
(4) *If* $m \ge 5$ *but* $m \neq 6$, *then* $[x_i, x_{i+l}] = 1$ *for* $1 \le l \le [m/2] + 1$. ($[m/2]$ *is the greatest integer that is less than or equal to* $m/2$).

Proof (1) According to lemma 3.7.22 we have $|\langle x_i, x_{i+1} \rangle| = 4$, therefore $\langle x_i, x_{i+1} \rangle$ is Abelian.

(2) Putting $T = \langle x_i, \dots, x_{i+l} \rangle$, $S = \langle x_i, \dots, x_{i+l-1} \rangle$ and $R = \langle x_{i+1}, \dots, x_{i+l} \rangle$, we have $|T| = 2^{l+1}$ and $|R| = |S| = 2^l$ by lemma 3.7.22, hence $T \rhd R$ and $T \rhd S$. Therefore:

$$[x_i, x_{i+l}] \in [S, R] \le S \cap R = \langle x_{i+1}, \dots, x_{i+l-1} \rangle.$$

(3) It suffices to prove $[x_1, x_3] = 1$. Noting that the order of x_i is 2 and using theorem 1.2.6, we have

$$[x_1, x_3, x_4]^{x_3}[x_3, x_4, x_1]^{x_4}[x_4, x_1, x_3]^{x_1} = 1.$$

Now $[x_3, x_4] = 1$ by (1), hence $[x_3, x_4, x_1] = 1$. Since $m \ge 4$ we have $[x_4, x_1] \in \langle x_2, x_3 \rangle$ by (2), hence $[x_4, x_1, x_3] \in [\langle x_2, x_3 \rangle, x_3] = 1$. Therefore $[x_1, x_3, x_4] = 1$. Suppose $[x_1, x_3] \neq 1$, then $[x_1, x_3] = x_2$ by (2), hence $[x_2, x_4] = 1$. However, $[x_2, x_4] = [x_1^\sigma, x_3^\sigma] = [x_1, x_3]^\sigma = x_2^\sigma \neq 1$. Contradiction.

(4) It suffices to prove the assertion for $i = 1$. We have $[x_1, x_2] = [x_1, x_3] = 1$ by (1) and (3).

First let us prove that $[x_1, x_4] = 1$ if $m \ge 5$. From

$$[x_1, x_4, x_5]^{x_4}[x_4, x_5, x_1]^{x_5}[x_5, x_1, x_4]^{x_1} = 1,$$

we conclude in the same way as above that $[x_1, x_4, x_5] = 1$ and from

$$[x_1, x_4, x_0]^{x_4}[x_4, x_0, x_1]^{x_0}[x_0, x_1, x_4]^{x_1} = 1,$$

we conclude that $[x_1, x_4, x_0] = 1$. Suppose $[x_1, x_4] \neq 1$. Since $[x_0, x_3]$ and $[x_2, x_5]$ are both conjugate to $[x_1, x_4]$, these two elements are not equal to 1. Since $[x_1, x_4] \in \langle x_2, x_3 \rangle$ by (2), we can write

$$[x_1, x_4] = x_2^{\varepsilon_2} x_3^{\varepsilon_3},$$

where ε_2 and ε_3 are 0 or 1, but not both 0. If $\varepsilon_2 = 1$, then

$$[x_2, x_5]^{x_3^{\varepsilon_3}}[x_3^{\varepsilon_3}, x_5] = [x_2 x_3^{\varepsilon_3}, x_5] = 1,$$

since $[x_1, x_4, x_5] = 1$, hence $[x_2, x_5] = 1$. Contradiction. If $\varepsilon_3 = 1$ we find in the same way using $[x_1, x_4, x_0] = 1$ that $[x_0, x_3] = 1$. Contradiction. Therefore $[x_1, x_4] = 1$. Since $[\frac{5}{2}] + 1 = 3$, this proves (4) for $m = 5$.

Next we will prove that $[x_1, x_5] = 1$ for $m \geq 7$. From

$$[x_1, x_5, x_6]^{x_5}[x_5, x_6, x_1]^{x_6}[x_6, x_1, x_5]^{x_1} = 1,$$

and $[x_1, x_2] = [x_1, x_3] = [x_1, x_4] = 1$ we conclude in the same way as in (3) that $[x_1, x_5, x_6] = 1$. If we use x_7 and x_0 instead of x_6 we find in the same way as above:

$$[x_1, x_5, x_7] = [x_1, x_5, x_0] = 1. \tag{3.7.2}$$

Suppose $[x_1, x_5] \neq 1$. Since $[x_2, x_6]$, $[x_3, x_7]$ and $[x_0, x_4]$ are conjugate to $[x_1, x_5]$, these four elements are not equal to 1. Since $[x_1, x_5] \in \langle x_2, x_3, x_4 \rangle$, we can write:

$$[x_1, x_5] = x_2^{\varepsilon_2} x_3^{\varepsilon_3} x_4^{\varepsilon_4},$$

where $\varepsilon_2, \varepsilon_3, \varepsilon_4$ are 0 or 1, but not all 0. Since $[x_1, x_5, x_6] = 1$, we find $[x_2^{\varepsilon_2}, x_6] = 1$ (the proof is similar to the proof of $[x_1, x_4] = 1$), hence we have $\varepsilon_2 = 0$ since $[x_2, x_6] \neq 1$. Using the equalities of (3.7.2) we find in the same way $\varepsilon_3 = \varepsilon_4 = 0$. Contradiction, hence $[x_1, x_5] = 1$. Since $[\frac{7}{2}] + 1 = 4$, this proves the assertion for $m = 7$.

For $m \geq 8$ we can give a unified proof in a similar way to above. We know already that $[x_1, x_2] = [x_1, x_3] = [x_1, x_4] = [x_1, x_5] = 1$. So it suffices to prove that $[x_1, x_{l+1}] = 1$ under the assumption that

$$[x_1, x_2] = [x_1, x_3] = \ldots = [x_1, x_l] = 1,$$

where $5 \leq l \leq [m/2] + 1$.

For an arbitrary j, we have

$$[x_1, x_{l+1}, x_{l+j}]^{x_{l+1}}[x_{l+1}, x_{l+j}, x_1]^{x_{l+j}}[x_{l+j}, x_1, x_{l+1}]^{x_1} = 1. \tag{3.7.3}$$

Let j satisfy $1 \leq j \leq l - 2$. Then since $[x_{l+1}, x_{l+j}] = [x_1, x_j]^{\sigma^l} = 1$, we have

$$[x_{l+1}, x_{l+j}, x_1] = 1.$$

Since $l+j-1 \leq 2l-3 \leq 2([m/2]+1)-3 = 2[m/2]-1 \leq m-1$ we have $[x_{l+j}, x_1, x_{l+1}] \in [\langle x_2, \ldots, x_{l+j-1} \rangle, x_{l+1}] = 1$ by (2), hence $[x_{l+j}, x_1, x_{l+1}] = 1$. Therefore $[x_1, x_{l+1}, x_k] = 1$ for those k satisfying $l+1 \leq k \leq 2l-2$ (by (3.7.3)). In the same way, we find $[x_1, x_{l+1}, x_k] = 1$ for those k with $-l+4 \leq k \leq 1$. Now suppose $[x_1, x_{l+1}] \neq 1$. Since $[x_j, x_{l+j}]$ is conjugate to $[x_1, x_{l+1}]$ for all j, $[x_j, x_{l+j}] \neq 1$ $(\forall j)$. Since $[x_1, x_{l+1}] \in \langle x_2, \ldots, x_l \rangle$ by (2), we can write:

$$[x_1, x_{l+1}] = x_2^{\varepsilon_2} \ldots x_l^{\varepsilon_l},$$

where ε_i is 0 or 1, but not all ε_i are 0. Using $[x_1, x_{l+1}, x_k] = 1$ for $l+1 \leq k \leq 2l-2$ we conclude $\varepsilon_2 = \ldots = \varepsilon_{l-2} = 0$ in the same way as we did in the proof that $[x_1, x_5] = 1$. From $[x_1, x_{l+1}, x_k]$ for $-l+4 \leq k \leq 1$ we find $\varepsilon_4 = \ldots = \varepsilon_l = 0$. Since $l \geq 5$, we conclude $\varepsilon_i = 0$ for all i. Contradiction. (End of proof of lemma 3.7.23.) ■

Now we can finish the proof of theorem 3.7.16. $G = \langle x_0, x_1, \ldots, x_m, x_{m+1} \rangle$ by lemma 3.7.22 and $x_{[m/2]+1}$ commutes with $x_0, \ldots, x_{m+1}$ for $m = 5$ and $m \geq 7$ by lemma 3.7.23, hence $Z(G) \neq 1$. Since (G, Ω) is primitive, $(Z(G), \Omega)$ is transitive, and this is a contradiction (see, for example, corollary 3.2.12). (End of proof of theorem 3.7.16.) ■

For primitive permutation groups (G, Ω) theorems 3.7.15 and 3.7.16 permit us in certain simple cases to draw conclusions concerning the structure of (G, Ω) from the structure of a suborbit $\Delta(a)$ or $(G_a, \Delta(a))$. For a general investigation of this relation it is necessary to study the relation between several suborbits, in particular the relation between $(G_a, \Delta(a))$ and $(G_a, {}^t\Delta(a))$ and several results in this connection have been obtained.

Theorem 3.7.24 (Sims) *Let (G, Ω) be a transitive permutation group, Δ an orbital of (G, Ω), $a \in \Omega$ and $|\Delta(a)| > 1$. Then $G_a^{\Delta(a)}$ and $G_a^{{}^t\Delta(a)}$ have a common homomorphic image, that is not equal to 1, i.e. there are $N_1 \ntrianglelefteq G_a^{\Delta(a)}$ and $N_2 \ntrianglelefteq G_a^{{}^t\Delta(a)}$ such that $G_a^{\Delta(a)}/N_1 \simeq G_a^{{}^t\Delta(a)}/N_2$.*

Proof Suppose the theorem is not true. Let $P^{(m)}$ denote the collection of all directed paths of length m of the graph (Ω, Δ). Then the following assertions A_m and B_m are true for $m \geq 0$:

(A_m) G operates transitively on $P^{(m)}$

(B_m) $(G^{\Delta(a_m)}_{a_0,\ldots,a_m}, \Delta(a_m)) \simeq (G^{\Delta(a)}_a, \Delta(a))$

and

$$(G^{{}^t\Delta(a_0)}_{a_0,\ldots,a_m}, {}^t\Delta(a_0)) \simeq (G^{{}^t\Delta(a)}_a, {}^t\Delta(a)) \quad \text{for } (a_0, \ldots, a_m) \in P^{(m)}.$$

(In fact we can prove this by induction on m. Since A_0 and B_0 are obviously true, let us prove A_{m+1} and B_{m+1} under the assumption that A_m and B_m are true. Since A_m and B_m are true, $(G, P^{(m+1)})$ is transitive, hence A_{m+1} is true. Now, let us consider the operation of $G_{a_1,\ldots,a_{m+1}}$ on $\Gamma = \Delta(a_{m+1}) \cup {}^t\Delta(a_1)$ for $(a_0, \ldots, a_{m+2}) \in P^{(m+2)}$. Since $G^{\Gamma}_{a_1,\ldots,a_{m+1}}$ is a subgroup of $G^{\Delta(a_{m+1})}_{a_1,\ldots,a_{m+1}} \times G^{{}^t\Delta(a_1)}_{a_1,\ldots,a_{m+1}}$ satisfying the conditions theorem 2.2.3, the subgroups $N_1 = \{\sigma \in G^{\Gamma}_{a_1,\ldots,a_{m+1}} \mid x^\sigma = x, \forall x \in {}^t\Delta(a_1)\}$ and $N_2 = \{\sigma \in G^{\Gamma}_{a_1,\ldots,a_{m+1}} \mid x^\sigma = x, \forall x \in \Delta(a_{m+1})\}$ satisfy $N_1 \trianglelefteq G^{\Delta(a_{m+1})}_{a_1,\ldots,a_{m+1}}$, $N_2 \trianglelefteq G^{{}^t\Delta(a_1)}_{a_1,\ldots,a_{m+1}}$ and $G^{\Delta(a_{m+1})}_{a_1,\ldots,a_{m+1}}/N_1 \simeq G^{{}^t\Delta(a_1)}_{a_1,\ldots,a_{m+1}}/N_2$. Therefore $N_1 = G^{\Delta(a_{m+1})}_{a_1,\ldots,a_m}$ and $N_2 = G^{{}^t\Delta(a_1)}_{a_1,\ldots,a_{m+1}}$, i.e. $G^{\Delta(a_{m+1})}_{a_0,a_1,\ldots,a_{m+1}} = G^{\Delta(a_{m+1})}_{a_1,\ldots,a_{m+1}} \simeq G^{\Delta(a)}_a$ and $G^{{}^t\Delta(a_1)}_{a_1,\ldots,a_{m+2}} = G^{{}^t\Delta(a_1)}_{a_1,\ldots,a_{m+1}} \simeq G^{{}^t\Delta(a)}_a$. Therefore B_{m+1} is true.) Since $|P^{(m)}| = |\Omega| \, |\Delta(a)|^m$, we have $|G| \geq |\Omega| \, |\Delta(a)|^m$, $\forall m \geq 1$ by A_m. Since G is a finite group, we conclude $|\Delta(a)| = 1$. Contradiction.

■

Let $\mathfrak{p}$ represent a property of permutation groups. If for some permutation representation (G, Ω) the permutation group (G^Ω, Ω), where $G^\Omega = G/G_\Omega$, satisfies $\mathfrak{p}$, we will write: $(G, \Omega) \in \mathfrak{p}$.

A property $\mathfrak{p}$ of permutation groups satisfying (1), (2) and (3) below is called a C_m-property.

(1) if $(G, \Omega) \in \mathfrak{p} \Rightarrow |\Omega| = m$ and (G, Ω) is transitive,

(2) if $(G, \Omega) \in \mathfrak{p}$ and (H, Ω) are permutation representations such that $H^\Omega \geq G^\Omega$, then $(H, \Omega) \in \mathfrak{p}$,

(3) if (G, Ω) and (G, Δ) are transitive permutation representations of degree m, $(G, \Omega) \in \mathfrak{p}$ and $(G, \Delta) \notin \mathfrak{p}$, then $(G_a, \Omega) \in \mathfrak{p}$, $\forall a \in \Delta$,

If $\mathfrak{p}$ satisfies (1), (2) and (3) and:

(4) if $(G, \Omega) \in \mathfrak{p}$, then (G, Ω) is primitive,

(5) if $(G, \Omega) \in \mathfrak{p}$, $G \trianglerighteq K$ and (K^Ω, Ω) is regular, then G^Ω/K^Ω does not contain a normal subgroup that is isomorphic to K^Ω,

then $\mathfrak{p}$ is called a K_m-property. Since every K_m-property is a C_m-property, propositions concerning C_m-properties are automatically also true for K_m-properties.

Theorem 3.7.25 (Cameron) *Let $\mathfrak{p}$ be a C_m-property or a K_m-property for some $m > 1$. Let (G, Ω) be a transitive permutation group, Δ an orbital of (G, Ω) and $|\Delta(a)| = m\,(a \in \Omega)$, then:*

$$(G_a, \Delta(a)) \in \mathfrak{p} \Leftrightarrow (G_a, {}^t\Delta(a)) \in \mathfrak{p}.$$

Proof Suppose $(G_a, \Delta(a)) \in \mathfrak{p}$, but $(G_a, {}^t\Delta(a)) \notin \mathfrak{p}$. Let $P^{(k)}$ represent the collection of directed paths of length k of the graph (Ω, Δ). Then the following assertions A_k and B_k are true for $k \geq 1$:

(A_k): G operates transitively on $P^{(k)}$.

(B_k): $(G_{a_0,\ldots,a_{k-1}}, \Delta(a_{k-1})) \in \mathfrak{p}$ for $(a_0, \ldots, a_{k-1}) \in P^{(k-1)}$.

(For, since A_1 and B_1 are clearly true, let us assume that A_k and B_k are true. If $(a_1, \ldots, a_k) \in P^{(k-1)}$, then $(G_{a_1,\ldots,a_k}, \Delta(a_k)) \in \mathfrak{p}$ by B_k. From A_k we conclude, that $(G_{a_1,\ldots,a_k}, {}^t\Delta(a_1))$ is transitive. Since $G_{a_1}^{{}^t\Delta(a_1)} \geq G_{a_1,\ldots,a_k}^{{}^t\Delta(a)}$ and $(G_{a_1}, {}^t\Delta(a_1)) \notin \mathfrak{p}$ by assumption, we have $(G_{a_1,\ldots,a_k}, {}^t\Delta(a_1)) \notin \mathfrak{p}$ because $\mathfrak{p}$ satisfies (2). Since $\mathfrak{p}$ also satisfies (3), we conclude:

$$(G_{a_0,a_1,\ldots,a_k}, \Delta(a_k)) \in \mathfrak{p}, \forall a_0 \in {}^t\Delta(a_1),$$

i.e. B_{k+1} is true. A_{k+1} obviously follows from A_k and B_{k+1}.) From A_k we get $|G| \geq |\Omega|\,|\Delta(a)|^k, \forall k \geq 1$, contrary to $|G| < \infty$. ■

Theorem 3.7.26 *2-transitivity is a C_m-property for $m \geq 2$.*

Proof Conditions (1) and (2) are clearly satisfied, so let us prove (3). Let (G, Γ) be 2-transitive, let (G, Δ) be transitive but not 2-transitive and let $|\Gamma| = |\Delta| = m \geq 2$. If π_1 and π_2 are the characters of the permutation matrix representations of (G, Γ) and (G, Δ) respectively, then the character of the permutation matrix of $(G, \Gamma \times \Delta)$ equals $\pi_1\pi_2$. According to theorems 3.2.4 and 3.2.5 π_1 and π_2 are decomposed as

$$\pi_1 = 1_G + \chi \text{ and } \pi_2 = 1_G + \chi_1 + \ldots + \chi_t,$$

where 1_G is the identity character of $G, t \geq 2$ and $\chi, \chi_1, \ldots, \chi_t$ are irreducible characters not equal to 1_G. Therefore:

$$\sum_{\sigma \in G} \pi_1(\sigma)\pi_2(\sigma) = \sum_{\sigma \in G} \pi_1(\sigma)\pi_2(\sigma^{-1}) = |G|,$$

by the orthogonality relations. Hence $(G, \Gamma \times \Delta)$ is transitive by theorem 3.2.4. Therefore, $(G_a, \Gamma - \{a\})$ and (G_a, Δ) are both transitive for $(a, b) \in \Gamma \times \Delta$. Since $(|\Gamma - \{a\}|, |\Delta|) = 1$, we see that $(G_{a,b}, \Gamma - \{a\})$ is transitive and that (G_b, Γ) is 2-transitive. ■

As a consequence of theorem 3.7.25 and 3.7.26 we have:

Corollary 3.7.27 *Let (G, Ω) be a transitive permutation group, let Δ be an orbital of (G, Ω) such that $|\Delta(a)| > 1$ for $a \in \Omega$. If $(G_a, \Delta(a))$ is 2-transitive so is $(G_a, {}^t\Delta(a))$.*

Theorem 3.7.28 (Knapp) *Let $\mathfrak{p}$ be a K_m-property $(m > 1)$. Let (G, Ω) be a transitive permutation representation, let Δ be an orbital of (G, Ω) and let $a \in \Omega$ such that $|\Delta(a)| = m$. Then we have:*

$$\textit{if } (G_a, \Delta(a)) \in \mathfrak{p}, \textit{ then } G_a^{\Delta(a)} \simeq G_a^{{}^t\Delta(a)}.$$

Proof Suppose that the theorem is not true. Let (G, Ω) be a transitive permutation representation for which the theorem does not hold. Since there is only a finite number of transitive permutation representations of G, from transitive permutation representations of G for which the theorem is false we can select (for a fixed G) a (G, Ω) such that $|\Omega|$ is maximal. Let Δ be an orbital of (G, Ω) for which the theorem is false, i.e. let Δ be such that: $|\Delta(a)| = m$, $(G_a, \Delta(a)) \in \mathfrak{p}$, but $G_a^{\Delta(a)} \not\simeq G_a^{{}^t\Delta(a)}$. Putting: $K(a) = G_{a \cup \Delta(a)}$ and $\tilde{K}(a) = G_{a \cup {}^t\Delta(a)}$ we have $K(a) \trianglelefteq G_a$, $\tilde{K}(a) \trianglelefteq G_a$, $G_a/K_a = G_a^{\Delta(a)}$ and $G_a/\tilde{K}_a = G_a^{{}^t\Delta(a)}$, hence $K(a) \neq \tilde{K}(a)$. Let (a, b) be an element of Δ. Then we will have that: $\tilde{K}(a) \not\leq G_{a,b}$ or $K(b) \not\leq G_{a,b}$. (For, suppose that $\tilde{K}(a) \leq G_{a,b}$ and $K(b) \leq G_{a,b}$. From $\tilde{K}(a) \leq G_{a,b}$, we have

$$\tilde{K}(a) \leq \bigcap_{\sigma \in G_a} G_{a,b}^{\sigma} = G_{a,\Delta(a)} = K(a).$$

From $K(b) \leq G_{a,b}$ we get in the same way $K(b) \leq \tilde{K}(b)$. Therefore $|K(a)| = |\tilde{K}(a)|$, hence $K(a) = \tilde{K}(a)$. Contradiction.) We can see from theorem 3.7.25 that the case $K(b) \nleq G_{a,b}$ can be treated in a similar way to the case $\tilde{K}(a) \nleq G_{a,b}$. Hence we assume $\tilde{K}(a) \nleq G_{a,b}$ from now on. Then $1 \neq \tilde{K}(a)^{\Delta(a)} \trianglelefteq G_a^{\Delta(a)}$ and $(G_a, \Delta(a))$ is primitive by (4) of the definition of K_m-property. Therefore $(\tilde{K}(a), \Delta(a))$ is transitive hence $G_a = \tilde{K}(a)G_{a,b}$. Therefore

$$G_a^{t\Delta(a)} = G_{a,b}^{t\Delta(a)}. \tag{3.7.4}$$

Since $(G_a, {}^t\Delta(a)) \in \mathfrak{p}$ by theorem 3.7.25, we have $(G_{a,b}, {}^t\Delta(a)) \in \mathfrak{p}$. In particular, $G_{a,b}$ operates transitively on ${}^t\Delta(a)$.

Let $(c, a) \in \Delta$ and let Φ denote the orbital $\{(a, b), (c, a)\}^G$ of the permutation representation (G, Δ^2). Since $G_{a,b}$ operates transitively on ${}^t\Delta(a)$, we have

$$\Phi((a, b)) = \{(x, a) \mid (x, a) \in \Delta\}.$$

Via the correspondence $(x, a) \leftrightarrow x$ between $\Phi((a, b))$ and ${}^t\Delta(a)$ we have

$$(G_{a,b}, \Phi((a, b))) \simeq (G_{a,b}, {}^t\Delta(a)) \text{ and } G_{a,b}^{\Phi(a,b)} = G_{a,b}^{t\Delta(a)}. \tag{3.7.5}$$

Now we have the following situation: (G, Δ) is transitive, Φ is an orbital of (G, Δ), $|\Phi((a, b))| = m$ for $(a, b) \in \Delta$ and $(G_{(a,b)}, \Phi((a, b))) \simeq (G_{a,b}, {}^t\Delta(a)) \in \mathfrak{p}$. Since $|\Delta| > |\Omega|$, the theorem holds for (G, Δ) by the choice of Ω. Therefore we have

$$G_{a,b}^{\Phi((a,b))} \simeq G_{a,b}^{t\Phi((a,b))}. \tag{3.7.6}$$

Since,

$${}^t\Phi = \{((z, x), (x, y)) \mid (x, y), (z, x) \in \Delta\},$$

using the correspondence $(b, x) \leftrightarrow x$ between ${}^t\Phi((a, b))$ and $\Delta(b)$, we get: $(G_{a,b}, {}^t\Phi((a, b))) \simeq (G_{a,b}, \Delta(b))$. Hence $(G_{a,b}, \Delta(b))$ is transitive (since $(G_{a,b}, {}^t\Phi((a, b)))$ is transitive) and

$$G_{a,b}^{t\Phi((a,b))} = G_{a,b}^{\Delta(b)}. \tag{3.7.7}$$

From (3.7.4), (3.7.5), (3.7.6) and (3.7.7) we have

$$G_a^{t\Delta(a)} = G_{a,b}^{t\Delta(a)} = G_{a,b}^{\Phi((a,\, b))} \simeq G_{a,b}^{t\Phi((a,b))} = G_{a,b}^{\Delta(b)}. \tag{3.7.8}$$

Now, if $K(b) \nleq G_{a,b}$ then in a similar way to above $G_b^{\Delta(b)} = G_{a,b}^{\Delta(b)}$. Combining this with (3.7.8) we get $G_a^{t\Delta(a)} \simeq G_b^{\Delta(b)} \simeq G_a^{\Delta(a)}$, which is a contradiction. Hence $K(b) \leq G_{a,b}$ and

$$K(b) \leq \bigcap_{\sigma \in G_b} G_{a,b}^{\sigma} = G_{b,{}^t\Delta(b)} = \tilde{K}(b) \leq G_{a,b} < G_a.$$

Since $K(b) \neq \tilde{K}(b)$, we have $K(b) \lneqq \tilde{K}(b)$, and hence $\tilde{K}(b)^{\Delta(b)} = \tilde{K}(b)/K(b) \neq 1$. Since $(G_b, \Delta(b))$ is primitive by (4) of the definition of K_m-property and since $\tilde{K}(b) \lhd G_b$, we see that $(\tilde{K}(b), \Delta(b))$ is transitive. From

$$G_b/\tilde{K}(b) = G_b^{{}^t\Delta(b)} \simeq G_a^{{}^t\Delta(a)} \underset{3.7.4}{\simeq} G_{a,b}^{\Delta(b)} = G_{a,b}/K(b),$$

we get by theorem 3.7.8:

$$|\tilde{K}(b):K(b)| = \frac{|G_b:K(b)|}{|G_b:\tilde{K}(b)|} = \frac{|G_b:K(b)|}{|G_{a,b}:K(b)|} = |G_b:G_{a,b}| = m.$$

Hence $(\tilde{K}(b)^{\Delta(b)}, \Delta(b))$ is regular. We also have

$$\begin{aligned} G_b^{\Delta(b)}/\tilde{K}(b)^{\Delta(b)} &\simeq G_b/K(b)/\tilde{K}(b)/K(b) \simeq G_b/\tilde{K}(b) \\ &= G_b^{{}^t\Delta(b)} \simeq G_a^{{}^t\Delta(a)} \simeq G_{a,b}^{\Delta(b)} \\ &= G_{a,b}/K(b) \geq \tilde{K}(b)/K(b) = \tilde{K}(b)^{\Delta(b)}. \end{aligned}$$

Therefore $G_b^{\Delta(b)}/\tilde{K}(b)^{\Delta(b)}$ contains a subgroup that is isomorphic with $\tilde{K}(b)^{\Delta(b)}$, contrary to the assumption $(G_a, \Delta(a)) \in \mathfrak{p}$. ■

4

Examples–symmetric groups and general linear groups

Symmetric groups and general linear groups are the most basic examples of groups, and in a certain sense also the most general examples of groups. For example, every finite group can be regarded as a subgroup of these groups, and we can make clear the properties of a finite group by taking it as a properly chosen subgroup of these groups. Symmetric groups and general linear groups have been, and are, studied from different view points. Here we want to state and prove some basic facts concerning these groups that are relevant to this book.

4.1 Conjugacy classes and composition series of the symmetric and alternating group

To the end of §4.4, Ω will denote the set $\{1, 2, \ldots, n\}$ unless stated otherwise, and the symmetric group S_n of degree n will operate on Ω.

Let $\sigma \in S_n$ be of $(t_1, \ldots, t_n)$-type, i.e. σ can be written as the product of mutually independent cycles such that the number $t_i(\sigma)$ of cycles of length i equals t_i. The numbers $t_1, \ldots, t_n$ satisfy:

(i) the t_i are non-negative integers,

(ii) $\sum_{i=1}^{n} it_i = n.$

A vector $(t_1, \ldots, t_n)$ satisfying (i) and (ii) is called a Young partition of n. For every Young partition $n = \sum_{i=1}^{n} it_i$ there exists an element $\sigma \in S_n$ of type $(t_1, \ldots, t_n)$ and the number of such elements σ is equal to $n!/\prod_{i=1}^{n} i^{t_i}t_i!$. (For, according to theorem 3.1.2, two elements of S_n are conjugate if and only if they are of the same type. Therefore, the number of elements of type $(t_1, \ldots, t_n)$ is equal to $|S_n : \mathscr{C}_{S_n}(\sigma)|$, where σ represents an arbitrary element of type $(t_1, \ldots, t_n)$ and it is easy to verify that $|\mathscr{C}_{S_n}(\sigma)| = \prod_{i=1}^{n} i^{t_i}t_i!$) The type of a conjugacy class is defined as the type of any of its elements. We have proved:

Theorem 4.1.1 (1) *The type* $(t_1, \ldots, t_n)$ *of a conjugacy class of* S_n *is a Young partition of n. There is a one-to-one correspondence between conjugacy classes of* S_n *and Young partitions of n.*

(2) *The number of elements of a conjugacy class of type* $(t_1, \ldots, t_n)$ *equals* $n!/\prod_{i=1}^n i^{t_i} t_i!$.

Conjugacy classes in S_n of even permutations are subsets of A_n, but not necessarily conjugacy classes in A_n. In this connection we have the following result:

Theorem 4.1.2 (1) *The conjugacy class in* S_n *of even permutations is a subset of* A_n *and is divided into at most 2 conjugacy classes in* A_n.

(2) *The conjugacy class K in* S_n *of even permutations is divided into two conjugacy classes in* A_n *if and only if the type* $(t_1, \ldots, t_n)$ *of K satisfies*

$$t_{2m} = 0, \quad t_{2m+1} = 0 \text{ or } 1 \quad (\forall m). \tag{4.1.1}$$

In this case, K is divided into two conjugacy classes in A_n *having the same number of elements.*

Proof Let σ be an even permutation and let K and $\tilde{K}$ denote the conjugacy class containing σ in S_n and A_n respectively. Then $K \supseteq \tilde{K}$. Since $|S_n : A_n| = 2$, we have $|C_{S_n}(\sigma) : C_{A_n}(\sigma)| \leq 2$, therefore $|K| = |\tilde{K}|$ or $|K| = 2|\tilde{K}|$, and further we have:

K is divided into two

$$\text{conjugacy classes in } A_n \Leftrightarrow \mathscr{C}_{S_n}(\sigma) = \mathscr{C}_{A_n}(\sigma) \Leftrightarrow \mathscr{C}_{S_n}(\sigma) \leq A_n.$$

Suppose $t_{2m}(\sigma) \neq 0$, then there is a cycle τ of length $2m$ in the cycle decomposition of σ, hence $\mathscr{C}_{S_n}(\sigma)$ contains the odd permutation τ. Next, suppose $t_{2m+1}(\sigma) \geq 2$. Let $(i_1, \ldots, i_{2m+1})$ and $(j_1, \ldots, j_{2m+1})$ denote two cycle components of length $2m+1$ in the cycle decomposition of σ, then $\mathscr{C}_{S_n}(\sigma)$ contains the odd permutations $(i_1, j_1) \ldots (i_{2m+1}, j_{2m+1})$. Therefore, if K is divided into two conjugacy classes in A_n then (4.1.1) holds. Conversely, assume that (4.1.1) holds, then all cycles occurring in the cycle decomposition of σ

have different length. Hence the cycle components of σ are invariant under the operation of $\tau \in \mathscr{C}_{S_n}(\sigma)$. Hence, letting $\sigma = \sigma_1 \ldots \sigma_s$ denote the cycle decomposition of σ, the permutation τ can be written as $\tau = \sigma_1^{r_1} \ldots \sigma_s^{r_s}, (r_1, \ldots, r_s \in \mathbb{Z})$ (e.g. from corollary 3.2.12). Since $\sigma_1, \ldots, \sigma_s$ are all even, we conclude $\tau \in A_n$, therefore K is divided into two conjugacy classes in A_n. ■

We have already proved the next theorem – as corollary 3.5.11 – but here we will prove this using a different approach.

Theorem 4.1.3 *For $n \geq 5$, A_n is simple and if N is a normal subgroup of S_n then $N = \{1\}$, A_n or S_n.*

Proof We first show that A_5 is simple. According to theorem 4.1.2, A_5 has 5 conjugacy classes containing 1, 12, 12, 15 and 20 elements respectively. If N is a normal subgroup of A_5, then N is the union of a certain number of conjugacy classes of A_5. Since the conjugacy class consisting only of the identity element is contained in N, the order of N is the sum of 1 and one or more of the numbers 12, 12, 15 and 20. Since $|N|$ is also a divisor of $|A_5| = 60$, we conclude $|N| = 1$ or 60. To prove that $G = A_n$ is simple for $n \geq 6$ we apply induction on n. Suppose that N is such that $1 \neq N \ntrianglelefteq G$. For $a \in \Omega$ we have $N \cap G_a \trianglelefteq G_a \simeq A_{n-1}$, hence $N \cap G_a = 1$ or G_a by the induction hypothesis. Since (G, Ω) is primitive, (N, Ω) is transitive, hence $NG_a = G$. Therefore $N \cap G_a = 1$, hence (N, Ω) is regular, contrary to theorem 3.5.6. To prove the last part of the theorem, let N be such that $1 \neq N \trianglelefteq S_n$ and $A_n \nleq N$. Since $N \cap A_n \ntrianglelefteq A_n$, we have $N \cap A_n = 1$, hence:

$$N \simeq N/N \cap A_n \simeq A_n N/A_n = S_n/A_n.$$

Therefore: $|N| = 2$. Since S_n is primitive, (N, Ω) is transitive, hence $|\Omega| \leq |N| = 2$. Contradiction. ■

We see from this theorem that for $n \geq 5$, $S_n \supsetneqq A_n \supsetneqq 1$ is the only composition series of S_n, and hence the only principal composition series. It is easy to check that the only non-trivial normal subgroups of S_4 are A_4 and K, where K is given by

$$K = \{1, (1,2)(3,4), (1,3)(2,4), (1,4)(2,3)\}.$$

Therefore $S_4 \trianglerighteq A_4 \triangleright K \triangleright 1$ is the only principal composition series of S_4. Since $|S_4 : A_4|$ and $|A_4 : K|$ are prime numbers and since K has 3 normal subgroups that are not equal to 1, S_4 has three composition series. For S_3 the series $S_3 \gneqq A_3 \gneqq 1$ is the only principal composition series, while for S_2 the series $S_2 \gneqq 1$ is the only principal series. Summing up:

Theorem 4.1.4 (1) *For $n = 3$ and $n \geq 5$, $S_n \triangleright A_n \triangleright 1$ is the only principal composition series of S_n. For S_4 and S_2, the series $S_4 \triangleright A_4 \triangleright K \triangleright 1$ and $S_2 \ntrianglelefteq 1$ respectively are the only principal composition series. These series are characteristic normal series.*

(2) *Except for $n = 4$, these series are also the only composition series of S_n. For $n = 4$, the subgroup K has 3 subgroups of order 2, hence S_4 has three composition series.*

Corollary 4.1.5 (1) $D(S_n) = A_n, (n \geq 2)$.

(2) $D(A_n) = A_n, (n \geq 5)$, $\quad D(A_4) = K$, $\quad D(A_3) = 1$.

(3) $Z(S_n) = 1, (n \geq 3)$, $\quad Z(S_2) = S_2$.

(4) $Z(A_n) = 1, (n \geq 4)$ $\quad Z(A_3) = A_3$.

4.2 Conditions for being a symmetric or alternating group

In this section we try to find conditions that are sufficient to ensure that a primitive permutation group becomes the symmetric or alternating group.

Theorem 4.2.1 *Let (G, Ω) be a primitive permutation group.*

(1) *If G contains at least one transposition, then $G = S^{\Omega}$.*

(2) *If G contains at least one cycle of length 3, then $G = A^{\Omega}$ or $G = S^{\Omega}$.*

This theorem is a special case of the following theorem:

Theorem 4.2.2 (Jordan) *Let (G, Ω) be a primitive permutation* *group, such that* $\Omega = \Delta + \Gamma$ *(i.e.* $\Omega = \Delta \cup \Gamma$ *and* $\Delta \cap \Gamma = \varnothing$*) with* $1 < |\Gamma| = m < n = |\Omega|$. *If* (G_Δ, Γ) *is primitive, then* (G, Ω) *is* $(n-m+1)$*-primitive.*

(If a primitive permutation group (G, Ω) contains a transposition or cycle of length 3, then (G, Ω) is $(n-1)$-primitive or $(n-2)$-primitive by theorem 4.2.2 and we get theorem 4.2.1 from this and theorem 3.4.2.)

In order to prove theorem 4.2.2 we need the following lemma:

Lemma 4.2.3 *Let* (G, Ω) *be a primitive permutation group such that* $\Omega = \Gamma + \Delta$ *with* $1 < |\Gamma| < n$.

(1) *If* (G_Δ, Γ) *is transitive, then* (G, Ω) *is 2-transitive.*

(2) *If* (G_Δ, Γ) *is primitive, then* (G, Ω) *is 2-primitive.*

Proof We prove this by induction on $|\Delta|$. For $|\Delta| = 1$ the assertion is obviously true. Let (G_Δ, Γ) be transitive and choose $\sigma \in G$ such that:

$$\Delta^\sigma \neq \Delta \text{ and } \Delta^\sigma \cap \Delta \neq \varnothing \quad \text{if } n/2 \geq |\Delta|$$

$$\Delta^\sigma \neq \Gamma \text{ and } \Gamma^\sigma \cap \Gamma \neq \varnothing \quad \text{if } n/2 < |\Delta|.$$

(This is possible because (G, Ω) is primitive.) In either case we have

$$\Delta \cap \Delta^\sigma \neq \varnothing \text{ and } \Gamma \cap \Gamma^\sigma \neq \varnothing.$$

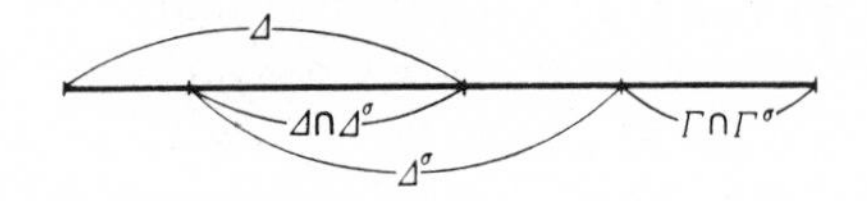

Therefore

$$\Lambda = \{\Delta \cap \Delta^\tau \mid \tau \in G, \Delta \neq \Delta^\tau, \Delta \cap \Delta^\tau \neq \varnothing, \Gamma \cap \Gamma^\tau \neq \varnothing\}$$

is not empty. If $\Delta \cap \Delta^\sigma \in \Lambda$, then $G_\Delta \leq G_{\Delta \cap \Delta^\sigma}$, $G_{\Delta^\sigma} \leq G_{\Delta \cap \Delta^\sigma}$, and since both (G_Δ, Γ) and $(G_{\Delta^\sigma}, \Gamma^\sigma)$ are transitive, $(G_{\Delta^\sigma \cap \Delta}, \Gamma^\sigma \cup \Gamma)$ is also transitive.

(1) Applying the induction hypothesis to $\Omega = (\Delta^\sigma \cap \Delta) + (\Gamma^\sigma \cup \Gamma)$, we get the 2-transitivity of (G, Ω).

(2) Let $\Delta \cap \Delta^\sigma$ denote a maximal element of Λ ordered by inclusion and put $\Phi = \Delta^\sigma - (\Delta \cap \Delta^\sigma)$. For $\tau \in G_{\Delta \cap \Delta^\sigma}$ we have $\Phi^\tau = \Phi$ or $\Phi^\tau \cap \Phi = \varnothing$ by the definition of $\Delta \cap \Delta^\sigma$. Since $G_\Delta \leq G_{\Delta \cap \Delta^\sigma}$, (G_Δ, Γ) is primitive and $\Phi \subsetneq \Gamma$, we have $|\Phi| = 1$. Therefore $|\Gamma \cup \Gamma^\sigma| = |\Gamma| + 1$ and $(G_{\Delta \cap \Delta^\sigma}, \Gamma \cup \Gamma^\sigma)$ is 2-transitive, hence primitive. So we get the lemma by the induction assumption. (End of proof of lemma 4.2.3.) ■

Proof of theorem 4.2.2 The proof is by induction on $|\Delta|$. If $|\Delta| = 1$, the truth of the theorem follows from lemma 4.2.3 (2). Now assume $|\Delta| > 1$. Since (G, Ω) is 2-primitive by lemma 4.2.3 (2), $(G_a, \Omega - \{a\})$ is primitive for $a \in \Delta$. Therefore $(G_a, \Omega - \{a\})$ is $((n-1) - m + 1)$-primitive by the induction hypothesis, from which the assertion of the theorem follows. (End of proof of theorem 4.2.2.) ■

Theorem 4.2.2 can be extended in several directions.

Theorem 4.2.4 (Marggraf) *Let (G, Ω) be a primitive permutation group, let $\Omega = \Gamma + \Delta$ and let (G_Δ, Γ) be transitive.*

(1) *If $1 < |\Gamma| \leq n/2$, then (G, Ω) is 3-transitive.*

(2) *If $1 < |\Gamma| \lneq n/2$, then $G = S^\Omega$ or $G = A^\Omega$.*

In order to prove theorem 4.2.4 we need the following lemma.

Lemma 4.2.5 *Let (G, Ω) be a primitive permutation group, let $\Omega = \Delta + \Gamma$ with $1 < |\Gamma| < n - 1$ and let (G_Δ, Γ) be transitive. If we put: Δ^* is a maximal element of the set*

$$\{\Delta \cap \Delta^\sigma \mid \sigma \in G, \Delta \neq \Delta^\sigma, \Delta \cap \Delta^\sigma \neq \varnothing, \Gamma \cap \Gamma^\sigma \neq \varnothing\}$$

ordered by inclusion (in the proof of lemma 4.2.3 we saw that this set is not empty)

$$\psi = \Delta - \Delta^*,$$

and

$$N = \mathcal{N}_G(G_\Delta),$$

then:

(1) $|\psi| < |\Gamma|$ *and* $|\psi| \mid |\Gamma|$.

(2) *$(G_{\Delta^*}, \Omega - \Delta^*)$, (N, Γ) and (N_{Δ^*}, ψ) are transitive.*
(3) *$N = N_a G_\Delta$ for $a \in \Gamma$, hence $N^\Delta = N_a^\Delta$.*
(4) *(N, Δ) is primitive.*

Proof (1) For $\tau \in G_{\Delta^*}$ we have $\psi^\tau = \psi$ or $\psi^\tau \cap \psi = \varnothing$ by the definition of Δ^*. Since $G_{\Delta^*} \geq G_{\Delta^\sigma}$ and $\Gamma^\sigma \geq \psi$, ψ is a set of imprimitivity of the transitive group $(G_{\Delta^\sigma}, \Gamma^\sigma)$, and so, we have $|\psi| \,\big|\, |\Gamma^\sigma| = |\Gamma|$. From $\Gamma \cap \Gamma^\sigma \neq \varnothing$ we conclude $|\psi| < |\Gamma|$.

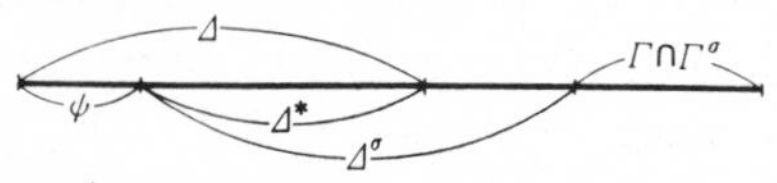

(2) Since $\Gamma \cap \Gamma^\sigma \neq \varnothing$, $(G_{\Delta^*}, \Omega - \Delta^*)$ is transitive. The transitivity of (N, Γ) is obvious. For $a, b \in \psi$, there exists an element $\tau \in G_{\Delta^*}$, such that $a^\tau = b$. Then $\Delta = \Delta^\tau$ since $\Delta \cap \Delta^\tau \supseteq \Delta^* \cup \{b\} \supsetneqq \Delta^*$. Therefore $\tau \in G_{\langle\Delta\rangle} \leq N$, hence $\tau \in N \cap G_{\Delta^*} = N_{\Delta^*}$ and (N_{Δ^*}, ψ) is transitive.

(3) For $\tau \in N$ we have $a^\tau = b \in \Gamma$. By the transitivity of (G_Δ, Γ) there exists $\rho \in G_\Delta$ such that $a^{\tau\rho} = a$. Thus $\tau\rho \in G_a N = N_a$ and $\tau \in N_a G_\Delta$, i.e. $N = N_a G_\Delta$.

(4) (G, Ω) is 2-transitive by lemma 4.2.3, therefore the assertion is true if $|\Delta| = 2$ by Witt's theorem (theorem 3.4.3). So let us proceed by induction on $|\Delta|$. We distinguish two cases:

First case: $|\Delta| \leq |\Gamma|$. If $a_1, a_2, b_1, b_2 \in \Delta$ such that $a_1 \neq a_2, b_1 \neq b_2$ then there exists an element $\tau \in G$ such that $b_1^\tau = a_1$ and $b_2^\tau = a_2$ by the 2-transitivity of (G, Ω). Since (G_Δ, Γ) and $(G_{\Delta^\tau}, \Gamma^\tau)$ are transitive and $\Gamma \cap \Gamma^\tau \neq \varnothing$, $H = G_{\Delta \cap \Delta^\tau}$ ($\geq G_\Delta$ and G_{Δ^τ}) operates transitively on $\Gamma \cup \Gamma^\tau = \Omega - (\Delta \cap \Delta^\tau)$. Since $H_{\Delta - (\Delta \cap \Delta^\tau)}$ $(= G_\Delta)$ and $H_{\Delta^\tau - (\Delta \cap \Delta^\tau)}$ $(= G_{\Delta^\tau})$ operate transitively on Γ and Γ^τ respectively, there exists an element $\rho \in H$ such that $\Gamma^{\tau\rho} = \Gamma$ by theorem 3.2.14. Therefore $\tau\rho \in G_{\langle\Delta\rangle} \leq N$ and $b_i^{\tau\rho} = a_i (i = 1, 2)$. Hence (N, Δ) is 2-transitive and therefore primitive.

Second case: $|\Delta| > |\Gamma|$. $|\Delta^*| \geq 2$ by (1). $(G_{\Delta^*}, \Omega - \Delta^*)$ is transitive by (2), hence $N^* = \mathcal{N}_G(G_{\Delta^*})$ operates primitively on Δ^* by the induction hypothesis. Now, $N_a^* \leq N, \forall a \in \psi$. (For, $\Delta = \Delta^\tau$ for $\tau \in N_a^*$ since $\Delta \cap \Delta^\tau \supseteq \Delta^* \cup \{a\} \supsetneqq \Delta^*$.) By (3) $N_a^{*\Delta^*} = N^{*\Delta^*}$ and N_a^* operates primitively on Δ^*. Since $|\psi| < \frac{1}{2}|\Gamma| < \frac{1}{2}|\Delta|$ we get $|\Delta^*| > \frac{1}{2}|\Delta|$. According to theorem 3.3.9 (N, Δ) is primitive. (End of proof of lemma 4.2.5.) ■

Proof of theorem 4.2.4 We use the same notation as in the proof of lemma 4.2.5.

(1) Take $a \in \Gamma$. $(G_a, \Omega - \{a\})$ is transitive by lemma 4.2.3 and the subgroup N_a of G_a operates primitively on its orbit Δ by lemma 4.2.5 (3) and (4). Since $|\Delta| > \frac{1}{2}|\Omega - \{a\}|$, $(G_a, \Omega - \{a\})$ is primitive according to theorem 3.3.9. Since (G, Ω) is transitive, $(G_b, \Omega - \{b\})$ is primitive for all $b \in \Omega$. Since $G_b \geq G_{b,\Delta - \{b\}} = G_\Delta$ for $b \in \Delta$ and since G_Δ operates transitively on Γ, $(G_b, \Omega - \{b\})$ is 2-transitive by lemma 4.2.3, hence (G, Ω) is 3-transitive.

(2) The proof is by induction on $|\Omega|$. We first prove that $N^\Delta = A^\Delta$ or S^Δ. We distinguish two cases: $|\psi| = 1$ and $|\psi| \geq 2$. If $|\psi| = 1$, then $(G_{\Delta*}, \Omega - \Delta^*)$ is 2-transitive by lemma 4.2.5(2), hence primitive. Therefore (G, Ω) is $|\Delta^*| + 1 = |\Delta|$-transitive by theorem 4.2.2. From Witt's theorem (theorem 3.4.3) we conclude $N^\Delta = S^\Delta$. If $|\psi| \geq 2$, then (N, Δ) is primitive by lemma 4.2.5. Since $\Delta = \Delta^* + \psi, |\psi| \leq \frac{1}{2}|\Gamma| < \frac{1}{2}|\Delta|$ and $(N_{\Delta*}, \psi)$ is transitive, hence (N^Δ, Δ) and $\Delta = \Delta^* + \psi$ satisfy the conditions of the theorem. Therefore, $N^\Delta = A^\Delta$ or S^Δ by the induction hypothesis.

For $a \in \Gamma$ we have $N_a^\Delta = N^\Delta$ by lemma 4.2.5 (3), hence $|N_a| \geq \frac{1}{2}|\Delta|!$. Since for $a \in \Gamma$ $|N_a^\Gamma| \leq (|\Gamma| - 1!)$, $N_a^\Gamma \simeq N_a/N_\Gamma$ and $|\Gamma| < |\Delta|$, we have $|N_\Gamma| \geq \frac{1}{2}|\Delta|(|\Delta| - 1)$. Now we assume $|\Delta| > 3$. Then $|N_\Gamma| > |\Delta|$ and N_Γ is not regular on Δ. On the other hand, since $N \geq N_\Gamma$ and $G_\Delta \cap N_\Gamma = 1$, we have $N_\Gamma \simeq N_\Gamma G_\Delta / G_\Delta \leq N/G_\Delta = A^\Delta$ or S^Δ. Hence $N_\Gamma^\Delta = A^\Delta$ or S^Δ by theorems 4.1.3 and 4.1.4. Therefore N_Γ (and thus also G) contains a transposition or a cycle of length 3 and $G = A^\Omega$ or S^Ω by theorem 4.2.1. If $|\Delta| \leq 3$, then $|\Omega| \leq 5$ since $|\Gamma| < |\Delta|$. By (1), (G, Ω) is 3-transitive, hence $G = A^\Omega$ or S^Ω. (End of proof of theorem 4.2.4.) ■

As another extension of theorem 4.2.1 we prove:

Theorem 4.2.6 (Jordan) *Let (G, Ω) be a primitive permutation group, let $|\Omega| = n = p + k$, where p is prime and $k \geq 3$. If G contains a cycle of length p, then $G = A^\Omega$ or S^Ω.*

Proof Let $\sigma = (1, \ldots, p)$ be the cycle of length p contained in G and put $\Gamma = \{1, \ldots, p\}$ and $\Delta = \Omega - \Gamma$. Now, (G_Δ, Γ) is primitive, because $|\Gamma|$ is prime, hence G is $(k+1)$-transitive by theorem 4.2.2. Since $\langle\sigma\rangle$ is a Sylow p-subgroup of G_Δ, $N = \mathcal{N}_G(\langle\sigma\rangle)$ operates k-transitively on Δ by Witt's theorem, hence $N^\Delta = S^\Delta$. Since $N = \langle\sigma\rangle N_a$ for $a \in \Gamma$, we have $N_a^\Delta = S^\Delta$. Since each element $\tau \in N_a^\Gamma$ induces an automorphism on $\langle\sigma\rangle$ and since $a^\tau = a$, we get that if τ and σ commute then $i^\tau = i$ $(i = 1, \ldots, p)$, i.e. $\tau = 1$. Therefore, N_a^Γ is isomorphic with a subgroup of $\mathrm{Aut}\langle\sigma\rangle$ and so N_a^Γ is an Abelian group. Hence, putting $[N_a, N_a] = K$, we have $K^\Gamma = 1$ and $K^\Delta = A^\Delta$ (corollary 4.1.5). Therefore K contains a cycle of length 3 and $G \geq A^\Omega$. ■

4.3 Subgroups and automorphism groups of S^Ω and A^Ω

The orders of primitive permutation groups on Ω that do not coincide with A^Ω or S^Ω cannot be too great, as shown by the following theorem.

Theorem 4.3.1 *Let (G, Ω) be a primitive permutation group that does not contain A^Ω, then:*

$$|S^\Omega : G| \geq \left[\frac{n+1}{2}\right]!.$$

In order to prove this theorem we need the following lemma.

Lemma 4.3.2 *If $|F_\Omega(\sigma) \cup F_\Omega(\tau)| = n - 1$ for permutations σ, τ of Ω, then $\sigma\tau\sigma^{-1}\tau^{-1}$ is a cycle of length* 3.

Proof Put $\Omega - (F_\Omega(\sigma) \cup F_\Omega(\tau)) = \{a\}$. Putting $a^{\sigma^{-1}} = b$ and $a^{\tau^{-1}} = c$ it is easy to check that $\sigma\tau\sigma^{-1}\tau^{-1} = (a, c, b)$. (End of proof of lemma 4.3.2.) ■

Proof of theorem 4.3.1 The set

$$\{k \mid k = |\Delta|, \Delta \subseteq \Omega \text{ and } (S^\Omega)_\Delta \cap G = 1\}$$

is not empty because $n = |\Omega|$ is certainly an element of this set. Let k_0 be the smallest number in that set, and choose $\Delta \subseteq \Omega$ such that $|\Delta| = k_0$ and $(S^\Omega)_\Delta \cap G = 1$. Then $k_0 \leq n/2$. (For, suppose $k_0 > n/2$. Putting $\Gamma = \Omega - \Delta$, we have $(S^\Omega)_\Gamma \cap G \neq 1$ since $|\Gamma| < n/2 < k_0$. Therefore there exists an element $\sigma \in G\ (\sigma \neq 1)$ such that $\Gamma \subseteq F_\Omega(\sigma)$. Choose $a \in \Delta$, such that $a^\sigma \neq a$. Then $(S^\Omega)_{\Delta - \{a\}} \cap G \neq 1$ by the definition of Δ, hence there exists an element $\tau \in G\ (\tau \neq 1)$ such that $F_\Omega(\tau) \supseteq \Delta - \{a\}$. Now, $F_\Omega(\tau) \not\supseteq \Delta$ since $(S^\Omega)_\Delta \cap G = 1$, hence $a^\tau \neq a$. Therefore $F_\Omega(\sigma) \cup F_\Omega(\tau) = \Omega - \{a\}$, hence G contains a cycle of length 3 by lemma 4.3.2 and $G \geq A^\Omega$. Contradiction.) Let $\sigma, \tau \in (S^\Omega)_\Delta$ belong to the same residue class of S^Ω/G, then $\sigma G = \tau G$, i.e. $\sigma^{-1}\tau \in G \cap (S^\Omega)_\Delta = 1$, hence $\sigma = \tau$. Therefore $|S^\Omega : G| \geq |(S^\Omega)_\Delta| = (n - k_0)!$. Since $n - k_0 \geq n/2$ and is an integer, we get the desired inequality. (End of proof of theorem 4.3.1.) ∎

Theorem 3.3.3 gives the following result concerning the order of imprimitive transitive permutation groups.

Theorem 4.3.3 *Let (G, Ω) be an imprimitive, transitive permutation group. Let Δ be a set of imprimitivity of (G, Ω) and put $|\Delta| = m_1$ and $n = m_1 m_2$. Then the order of G is a divisor of $(m_1)^{m_2} m_2\,!$.*

Concerning maximal subgroups of S_n we have the following result.

Theorem 4.3.4 *Let H be a proper subgroup of S_n such that $H \neq A_n$ and such that $|H|$ is maximal. Then:*

(1) If $n \neq 4$, then $|S_n : H| = n$.

(2) If $n = 4$, then H is a Sylow 2-subgroup of S_4 and $|S_4 : H| = 3$.

Proof From $|S^\Omega : (S^\Omega)_a| = n$ for $a \in \Omega$ we conclude $|S^\Omega : H| \leq n$. For $n \geq 5$, put $|S_n : H| = k$. Since $H \not\geq A_n$ we have $k \geq 3$ by theorem 3.4.2. Therefore the kernel N of the permutation representation $(S_n, S_n/H)$ is equal to $\{1\}$. Therefore S_n is isomorphic with a subgroup of S_k, hence $n \leq k$. Therefore $n = k$. The case $n \leq 4$ is almost trivial. ∎

Theorem 4.3.5 (1) *For $n \neq 6$, all subgroups of index n of S_n form an S_n-conjugate class of subgroups, the number of these subgroups* *equals n and these subgroups are just the stabilizers of the respective elements of* Ω.

(2) *For* $n = 6$, *there are just* 12 *subgroups of* S_6 *of index* 6. *These are all isomorphic to* S_5 *and they are divided into two sets of* 6 *subgroups each, such that the subgroups of each set make up one conjugacy class of subgroups.*

In order to prove this theorem we need the following lemma.

Lemma 4.3.6 (1) *If* $1 < t < n - 1$, *then:*

$$(n - t)!t! < (n - 1)!.$$

(2) *If* $n = m_1 m_2$, $m \neq 1$, $m_2 \neq 1$ *and* $n > 4$, *then:*

$$(m_1 !)^{m_2} m_2 ! < (n - 1)!.$$

(3) *If* $n > 4$ *and* $n \neq 6$, *then* $n < [(n + 1)/2]!$.

Proof (1) and (3) are obvious.

(2) Since $(im_1 + 1) \ldots (im_1 + m_1) > (i + 1)^{m_1} m_1 !$ we have:

$$\begin{aligned} n! &= 1 \times 2 \times \ldots \times m_1 \times (m_1 + 1) \ldots (m_1 + m_1) \\ &\quad \ldots \{(m_2 - 1)m_1 + 1\} \ldots (m_2 m_1) \\ &> m_1 ! 2^{m_1} m_1 ! 3^{m_1} \ldots m_2^{m_1} \times m_1 ! \\ &= (m_1 !)^{m_2} (m_2 !)^{m_1}. \end{aligned}$$

For $m_2 \not\geq 2$ we have $(m_2 !)^{m_1 - 1} \geq m_1 m_2 = n$, hence $(n - 1)! > (m_1 !)^{m_2} m_2 !$. If $m_2 = 2$, then:

$$\begin{aligned} n! &= 1 \times 2 \times \ldots \times m_1 \times (m_1 + 1) \times \ldots \times 2m_1 \\ &= (m_1 !)^2 (m_1 + 1)(m_1/2 + 1) \ldots (m_1/m_1 + 1) \\ &> (m_1 !)^2 \times 2 \times 2m_1 = (m_1 !)^2 \times 2n \end{aligned}$$

since $m_1 > 2$. (End of proof of lemma 4.3.6.) ■

Proof of theorem 4.3.5 It is clear that the set of all stabilizers of one element of Ω forms a conjugate class of subgroups of S_n.

(1) For $n \leq 4$ this is trivial, so assume $n \geq 5$ and $n \neq 6$. Let G be a subgroup of index n of S_n and suppose that G is not the stabilizer

of one point of Ω. First assume that G is not transitive, then we have $|G| \le t!(n-1)!$ where t represents the length of an orbit of G. Since G is not a stabilizer, we have $1 < t < n-1$. Therefore $|S_n : G| > n$ by lemma 4.3.6(1). Contradiction. Next assume that G is transitive. Suppose G is imprimitive, then $n = m_1 m_2$ for some integers $m_1 \neq 1$ and $m_2 \neq 1$ and $|G| \le (m_1!)^{m_2} m_2!$ by theorem 4.3.3. Therefore $|S_n : G| > n$ by lemma 4.3.6(2). Contradiction. Therefore G is primitive. (We remark that $n \neq 6$ has not been required so far.) From theorem 4.3.1 and lemma 4.3.6 (3) we conclude: $|S_n : G| \ge (n + 1/2)! > n$. Contradiction.

(2) Let H be the normalizer of a Sylow 5-subgroup of S_5. Then $|H| = 20$, $|S_5/H| = 6$ and $(S_5, S_5/H)$ is a transitive permutation representation of degree 6, and so S_5 is considered as a subgroup of $S^{S_5/H}$ ($\simeq$ the symmetric group of degree 6) which is not the stabilizer of one element of S_5/H. Conversely let G be a subgroup of S_6 of index 6 and assume that G is not the stabilizer of a point of Ω. It is clear from (1) that (G, Ω) is transitive. Let N denote the kernel of the permutation representation $(S_6, S_6/G)$. According to theorem 4.1.3 we have $N = 1$, hence $G \simeq S_5$ and (G, Ω) gives a transitive permutation representation of degree 6 of S_5. The order of the stabilizer of one element of Ω equals 20, namely, such a stabilizer is the normalizer of a Sylow 5-subgroup of S_5. Let G_1 and G_2 be subgroups of index 6 of S_6, neither of them a stabilizer of a point of Ω. Then $G_1 \simeq G_2 \simeq S_5$ and (G_1, Ω) and (G_2, Ω) are transitive permutation representations of degree 6 of S_5 and their stabilizers of one point are the normalizers of Sylow 5-subgroups of S_5. Therefore those two permutation representations of S_5 are equivalent by Sylow's theorem (theorem 2.1.1). Hence there exists an element $\sigma \in S_6$ such that $G_1^\sigma = G_2$ by theorem 1.2.12. Therefore all subgroups of index 6 of S_6 that are not stabilizers of a point of Ω are conjugate. It is obvious that the number of those subgroups is equal to 6. (End of proof of theorem 4.3.5.) ∎

Let us determine the group of automorphisms of S_n as an application of this theorem.

Theorem 4.3.7 *The order of* $\mathrm{Aut}(S_n)/\mathrm{Inn}(S_n) = \mathrm{Out}\ S_n$ *is equal* *to the number of conjugacy classes consisting of the subgroups of index n of S_n, i.e.*

$$|\mathrm{Aut}(S_n):\mathrm{Inn}(S_n)| = \begin{cases} 1 & (n \neq 6), \\ 2 & (n = 6). \end{cases}$$

Proof Put $S_n = G$ and $\Omega = \{1, 2, \dots, n\}$. Let f be an element of $\mathrm{Aut}(S_n)$ which satisfies $f(G_1) = G_j$ for some $j \in \Omega$. Then we have

$$\{f(G_1), \dots, f(G_n)\} = \{G_1, \dots, G_n\},$$

because $G_1 \underset{G}{\sim} G_i$ follows $f(G_1) \underset{G}{\sim} f(G_i)$. Hence there is an element $\sigma \in G$ such that $f(G_i) = G_{i\sigma}$, $(\forall i \in \Omega)$. Let I_σ denote the inner automorphism of G corresponding with σ, then $f(G_i) = I_\sigma(G_i)$, $(\forall i \in \Omega)$. Since $\langle (i,j) \rangle = \bigcap_{k \neq i,j} G_k$, $\forall i, j \in \Omega (i \neq j)$, we have

$$(I_\sigma^{-1} f)\langle (i,j) \rangle = \bigcap_{k \neq i,j} I_\sigma^{-1} f(G_k) = \bigcap_{k \neq i,j} G_k = \langle (i,j) \rangle.$$

Therefore $I_\sigma^{-1} f$ leaves all transpositions fixed. Since G is generated by the transpositions, we conclude $I_\sigma = f$. Hence, if $n \neq 6$ then $\mathrm{Aut}(S_n) = \mathrm{Inn}(S_n)$ by theorem 4.3.5(1). If $n = 6$, then besides $\mathscr{C}_1 = \{G_1, \dots, G_6\}$ there is another conjugate class $\mathscr{C}_2 = \{H_1, \dots, H_6\}$ of subgroups of S_6 of index 6 by theorem 4.3.5(2), and S_6 operates faithfully on $\mathscr{C}_2$ via conjugation. For $\sigma \in S_6$ let $\tilde{\sigma}$ denote the permutation of $\mathscr{C}_2$ defined by σ. Then the mapping $\sigma \to \tilde{\sigma}$ defines an isomorphism from S_6 onto S_6, i.e. an element of $\mathrm{Aut}(S_6)$, and so this induces a mapping from $\mathscr{C}_1$ onto $\mathscr{C}_2$. Generally, $f \in \mathrm{Aut}(S_6)$ induces a permutation $\tilde{f}$ of $\{\mathscr{C}_1, \mathscr{C}_2\}$ and the map: $f \to \tilde{f}$ is a homomorphism from $\mathrm{Aut}(S_6)$ into S_2. As we showed above, this map is surjective and its kernel is $\mathrm{Inn}(S_6)$. Therefore $|\mathrm{Aut}(S_6)/\mathrm{Inn}(S_6)| = 2$. ■

Next let us determine $\mathrm{Aut}(A_n)$. Let $\mathscr{C}$ denote the set of cycles of length 3 of A_n.

Lemma 4.3.8 *If $n \neq 6$, then $\mathscr{C}^\sigma = \mathscr{C}$ for all $\sigma \in \mathrm{Aut}\, A_n$.*

Proof For $n = 3, 4, 5$, the set $\mathscr{C}$ coincides with the set of elements of order 3, from which the theorem follows. So suppose $n \geq 7$. $\mathscr{C}$ is a conjugacy class of A_n by theorem 4.1.2. It is easy to verify that $\mathscr{C}$ has the following properties:

(i) if $x \in \mathscr{C}$, then $|x| = 3$,
(ii) if $x, y \in \mathscr{C}$, then $|xy| \leq 5$.

Conversely, let T be a conjugacy class of A_n satisfying (i) and (ii). Since all elements of T are of the same type and since the order of each element equals 3, all elements are of the type $(s, 0, r, \ldots, 0)$ for some s, r. Assume $\mathscr{C} \neq T$. Then $r \geq 2$. If $s > 0$, then T contains two elements x, y such that

$$x = (1, 2, 3)(4, 5, 6)(7)\ldots,$$
$$y = (1, 2, 4)(3, 5, 7)(6)\ldots,$$

but $xy = (1, 4, 7, 3, 2, 5, 6)\ldots$, which is contrary to (ii). Hence $s = 0$. Therefore $n \geq 9$ (since we assumed $n \geq 7$). Then T contains two elements x, y such that

$$x = (1, 2, 3)(4, 5, 6)(7, 8, 9)\ldots,$$
$$y = (1, 2, 4)(3, 5, 7)(6, 8, 9)\ldots,$$

but $xy = (1, 4, 7, 9, 3, 2, 5, 8, 6)\ldots$, which is contrary to (ii). Therefore $\mathscr{C} = T$, hence $\mathscr{C}^{\sigma} = \mathscr{C}$. ■

Lemma 4.3.9 *If $\sigma \in \mathrm{Aut}\ A_n$ is such that $\mathscr{C}^{\sigma} = \mathscr{C}$, then $\sigma = I_{\tau}$ for some $\tau \in S_n$. (I_{τ} denotes the inner automorphism of A_n defined by τ.)*

Proof This is clear for $n = 3$, so suppose $n \geq 4$. The product of two cycles (a, b, c) and (a', b', c') of length 3 of A_n can be an element of order 2 only if $|\{a, b, c\} \cap \{a', b', c'\}| = 2$. Putting $x = (1, 2, 3)$ and $y_i = (1, 2, i)\ (i \geq 4)$, we have $|xy_i| = |y_i y_j| = 2$ (if $i \neq j$), hence $|x^{\sigma} y_i^{\sigma}| = |y_i^{\sigma} y_j^{\sigma}| = 2$. Therefore $x^{\sigma} = (a_1, a_2, a_3)$ and $y_i^{\sigma} = (a_1, a_2, a_i)$, where a_1, a_2, a_3 and a_i are elements of Ω. Now, since

$$(i, j, k) = (1, 2, j)^2 (1, 2, i)^2 (1, 2, j)(1, 2, k)^2 (1, 3, i)(1, 2, k),$$

we get $(i, j, k)^{\sigma} = (a_i, a_j, a_k)$. Therefore $\sigma = I_{\tau}$ with $\tau = \begin{pmatrix} i \\ a_i \end{pmatrix} \in S_n$. ■

Theorem 4.3.10

$$\mathrm{Aut}\ A_n = \begin{cases} \text{a group of order } 2 & \text{if } n = 3 \\ \simeq \mathrm{Aut}\ S_n & \text{if } n > 3. \end{cases}$$

Proof (1) This is obvious for $n = 3$.

(2) If $n > 3$, but $n \neq 6$, then Aut $A_n \simeq$ Inn $S_n =$ Aut S_n by lemmas 4.3.8 and 4.3.9.

(3) If $n = 6$, then A_6 has 2 conjugacy classes of order 3. Therefore $|\text{Aut } A_6 : \text{Inn } S_6| \leq 2$ by lemma 4.3.9. Since on the other hand elements of Aut S_6 induce automorphisms of A_6, we have a homomorphism from Aut S_6 into Aut A_6 and the kernel of this homomorphism is 1. Therefore Aut $A_6 \simeq$ Aut S_6 by theorem 4.3.7. ■

4.4 Generators and fundamental relations for S_n and A_n

As we saw in §3.1 every element of S_n can be written as a product of transpositions. Putting $a_i = (i, i+1)$ $(1 \leq i \leq n-1)$, the transposition $(i, j)\,(i < j)$ can be written as

$$(i, j) = a_i a_{i+1} \cdots a_{j-2} a_{j-1} a_{j-2} \cdots a_{i+1} a_i.$$

Therefore $\{a_i \mid 1 \leq i \leq n-1\}$ is a system of generators of S_n. It is easy to verify that the a_i satisfy the following relations:

$$\left.\begin{array}{ll} a_i^2 = 1 & 1 \leq i \leq n-1, \\ (a_i a_{i+1})^3 = 1 & 1 \leq i \leq n-2, \\ (a_i a_j)^2 = 1 & |i-j| > 1. \end{array}\right\} \qquad (4.4.1)$$

We will show that these relations form a system of fundamental relations for S_n.

Theorem 4.4.1 *Let G be the group generated by $x_1, \ldots, x_{n-1}$, satisfying the following fundamental relations:*

$$\begin{array}{ll} x_i^2 = 1 & 1 \leq i \leq n-1, \\ (x_i x_{i+1})^3 = 1 & 1 \leq i \leq n-2, \\ (x_i, x_j)^2 = 1 & |i-j| > 1, \end{array}$$

then $G \simeq S_n$.

Proof The proof is by induction on n. For $n = 2$ the theorem is clearly true, so suppose $n \geq 3$. Let H be the subgroup of G generated by $x_1, \ldots, x_{n-2}$. There exists a homomorphism from S_{n-1} onto H by the induction hypothesis, hence $|H| \leq (n-1)!$. Let K

denote the union of the left cosets H, Hx_{n-1}, $Hx_{n-1}x_{n-2}, \ldots$, $Hx_{n-1}x_{n-2}\cdots x_2x_1$ in G with respect to H. Then K contains the generators $x_1, \ldots, x_{n-1}$ of G. Now we will prove that K is a subgroup of G. To prove this it suffices to prove that:

$$(Hx_{n-1}x_{n-2}\cdots x_{i+1}x_i)x_j \subseteq K$$

for all i, j such that $1 \leq i, j \leq n-1$. We distinguish several cases:

$j < i-1 \Rightarrow (Hx_{n-1}\cdots x_i)x_j = Hx_{n-1}\cdots x_i \subseteq K.$

$j = i-1 \Rightarrow (Hx_{n-1}\cdots x_i)x_{i-1} \subseteq K.$

$j = i \Rightarrow (Hx_{n-1}\cdots x_i)x_i = Hx_{n-1}\cdots x_{i+1} \subseteq K.$

$j > i \Rightarrow (Hx_{n-1}\cdots x_i)x_i = (Hx_{n-1}\cdots x_{j+1})(x_jx_{j-1}x_j)(x_{j-2}\cdots x_i)$
$= (Hx_{n-1}\cdots x_{j+1})(x_{j-1}x_jx_{j-1})(x_{j-2}\cdots x_i) = Hx_{n-1}\cdots x_i \subseteq K.$

We conclude that K is a subgroup of G, hence $K = G$. Therefore, $|G:H| \leq n$, hence $|G| \leq n!$. On the other hand, since there exists a homomorphism from G onto S_n, we conclude $G \simeq S_n$. ■

A_n is generated by the following $n-2$ elements:

$$a_1 = (1, 2, 3), \quad a_i = (1, 2)(i+1, +2)\ (2 \leq i \leq n-2).$$

(For, an arbitrary element of A_n can be written as the product of an even number of transpositions. Since $(i, j)(k, l) = (i, j)(1, 2)(1, 2)(k, l)$, it suffices to prove that elements of the form $(1, 2)(i, j)$ can be written as a product of the a_i. We distinguish the following cases:

$(1, 2)(1, 2) = 1,$

$(1, 2)(2, 3) = (1, 3, 2) = a_1^2,$

$(1, 2)(1, 3) = (1, 2, 3) = a_1,$

$$(1, 2)(1, k) = (1, 2, k) = \begin{cases} a_{k-2}a_{k-3}\cdots a_2a_1a_2\cdots a_{k-3}a_{k-2} & \text{(if } k \text{ is odd and } \geq 5) \\ (a_{k-2}a_{k-3}\cdots a_2a_1a_2\cdots a_{k-3}a_{k-2})^2 & \text{(if } k \text{ is even and } \geq 4), \end{cases}$$

$(1, 2)(2, k) = (1, k, 2) = (1, 2, k)^2 \quad \text{if } k \geq 3,$

$(1, 2)(i, j) = a_{i-1}\cdots a_{j-3}a_{j-2}a_{j-3}\cdots a_{i-1} \quad (3 \leq i < j).)$

It is easy to check that the a_i satisfy the following relations:

$$a_1^3 = a_i^2 = 1 \quad 2 \leq i \leq n-2,$$
$$(a_i a_{i+1})^3 = 1 \quad 1 \leq i \leq n-3,$$
$$(a_i a_j)^2 = 1 \quad |i-j| > 1.$$

We will prove that these relations form a system of fundamental relations for A_n.

Theorem 4.4.2 *Let G be the group generated by $x_1, \ldots, x_{n-2}$ satisfying the following fundamental relations:*

$$x_1^3 = x_i^2 = 1 \quad 2 \leq i \leq n-2,$$
$$(x_i x_{i+1})^3 = 1 \quad 1 \leq i \leq n-3,$$
$$(x_i x_j)^2 = 1 \quad |i-j| > 1,$$

then $G \simeq A_n$.

Proof The proof is by induction on n. Since the theorem is obviously true for $n = 3$, let us assume $n \geq 4$. Let H denote the subgroup of G generated by $x_1, \ldots, x_{n-3}$. There exists a homomorphism from A_{n-1} onto H by the induction hypothesis, hence $|H| \leq (n-1)!/2$. Let K denote the union of the left cosets $H, Hx_{n-2}, Hx_{n-2}x_{n-1}, \ldots, Hx_{n-2}x_{n-1} \cdots x_2 x_1, Hx_{n-2}x_{n-1} \cdots x_2 x_1^2$ of G with respect to H. Then K contains the generators $x_1, \ldots, x_{n-2}$. Now we shall prove that K is a subgroup of G. To prove that it suffices to show that:

$$(Hx_{n-1} \cdots x_i)x_j \subseteq K \quad 1 \leq i, j \leq n-2,$$
$$(Hx_{n-1} \cdots x_2 x_1^2)x_j \subseteq K \quad 1 \leq j \leq n-2.$$

For $(Hx_{n-1} \cdots x_i)x_j$ we have:

$$j < i-1 \Rightarrow (Hx_{n-2} \cdots x_i)x_j = Hx_{n-2} \cdots x_i \subseteq K,$$
$$j = i-1 \Rightarrow (Hx_{n-2} \cdots x_i)x_{i-1} \subseteq K,$$
$$j = i \Rightarrow (Hx_{n-2} \cdots x_i)x_i = Hx_{n-2} \cdots x_{i+1} \subseteq K,$$
$$j > i, j \geq 3 \Rightarrow (Hx_{n-2} \cdots x_i)x_j = Hx_{n-2} \cdots x_{j+1} x_j x_{j-1} x_j x_{j-2} \cdots x_i$$
$$= Hx_{n-2} \cdots x_{j+1} x_{j-1} x_j x_{j-1} x_{j-2} \cdots x_i = Hx_{n-2} \cdots x_i^{i} \subseteq K,$$

$$j=2, i=1 \Rightarrow (Hx_{n-2}\cdots x_2x_1)x_2 = Hx_{n-2}\cdots x_2x_1^2 \subseteq K.$$

For $(Hx_{n-1}\cdots x_2x_1^2)x_j$ we have:

$$j=1 \Rightarrow (Hx_{n-2}\cdots x_2x_1^2)x_1 = Hx_{n-2}\cdots x_2 \subseteq K,$$
$$j=2 \Rightarrow (Hx_{n-2}\cdots x_2x_1^2)x_2 = (Hx_{n-2}\cdots x_3)(x_1x_2x_1) = Hx_{n-2}\cdots x_2x_1 \subseteq K,$$
$$j>2 \Rightarrow (Hx_{n-2}\cdots x_2x_1^2)x_j = Hx_{n-2}\cdots x_2x_1^2 \subseteq K.$$

Therefore K is a subgroup of G, hence $K=G$. From $|G:H| \leq n$ we conclude $|G| \leq n!/2$. Since on the other hand there is a homomorphism from G onto A_n, we conclude $G \simeq A_n$. ■

For A_5 we have the following special results:

Theorem 4.4.3 *If G is the group generated by x_1, x_2 satisfying the fundamental relations*

$$x_1^5 = x_2^2 = (x_1x_2)^3 = 1,$$

then $G = A_5$.

Proof The elements $a_1 = (1,2,3,4,5)$, $a_2 = (1)(2,3)(4,5)$ are generators of A_5 satisfying $a_1^5 = a_2^2 = (a_1a_2)^3 = 1$. Therefore there exists a homomorphism from G onto A_5, hence $|G| \geq 60$. Putting $y_1 = x_1x_2$ and $y_2 = x_2x_1^3x_2x_1$ we have:

$$y_1^3 = y_2^3 = (y_1y_2)^2 = (y_2y_1)^2 = 1 = [y_1y_2, y_2y_1].$$

(For, since $(x_1x_2)^3 = 1$ we have:

$$y_2^3 = (x_2x_1^3x_2x_1)(x_2x_1^3x_2x_1)(x_2x_1^3x_2x_1) = (x_2x_1^2)(x_1x_2x_1x_2x_1x_2)x_2x_1$$
$$(x_1x_2x_1x_2x_1x_2)x_2x_1^2x_2x_1 = x_2x_1^2x_2x_1x_2x_1^2x_2x_1$$
$$= (x_2x_1)(x_1x_2x_1x_2x_1x_2)(x_2x_1x_2x_1) = 1.$$

The relations $(y_1y_2)^2 = (y_2y_1)^2 = 1$ are easy to check. Using these we have:

$$1 = y_2y_1(y_1y_2y_1y_2)y_2y_1 = y_2y_1^{-1}y_2y_1y_2^{-1}y_1 = y_2(y_2y_1y_2)y_2y_1y_2^{-1}y_1$$
$$= y_2^{-1}y_1y_2^{-1}y_1y_2^{-1}y_1,$$

hence

$$y_1y_2^{-1}y_1 = y_2y_1^{-1}y_2, \text{ i.e. } [y_1y_2, y_2y_1] = 1.)$$

Therefore, putting $K = \langle y_1y_2, y_2y_1 \rangle$, we have $|K| = 4$. Putting $H = \langle y_1, y_2 \rangle$, we have $H \rhd K$ and $|H:K| = 3$ (for, $y_1^{-1}(y_1y_2)y_1 = y_2y_1 \in K$ and $y_1^{-1}(y_2y_1)y_1 = y_1^{-1}y_2y_1^{-1} = (y_1y_2^{-1}y_1)^{-1} \in K$). Let L be the union of the left cosets $H, Hx_1, Hx_1^2, Hx_1^3, Hx_1^4$ of G with respect to H. The union L contains the generators of G and we will prove that L is a subgroup of G. To prove this it suffices to show that $Hx_1^ix_2 \subseteq L$ $(i = 1, \dots, 4)$. We have four cases to consider:

$$\begin{aligned}
x_1x_2 &= y_1 \in H,\\
x_1^2x_2 &= x_1x_1x_2 = x_1x_2x_2x_1x_2 = x_1x_2x_1^{-1}x_2x_1^{-1}\\
&= x_1x_2x_1x_2x_2x_1^3x_2x_1x_1^3 = y_1^2y_2x_1^3 \in L,\\
x_1^3x_2 &= x_1^2(x_1x_2) = x_1^2x_2x_1^{-1}x_2x_1^{-1} = y_1^2y_2x_1^3x_1^{-1}x_2x_1^{-1}\\
&= y_1^2y_2x_1^2x_2x_1^{-1} = y_1^2y_2y_1^2y_2x_1^3x_1^{-1} \in L,\\
x_1^4x_2 &= x_1x_2x_2x_1^3x_2x_1x_1^{-1} = y_1y_2x_1^4 \in L.
\end{aligned}$$

Therefore, $G = L$, hence $|G:H| \leq 5$ and $|G| \leq 60$. So we conclude that $G \simeq A_5$. ■

4.5 The structure of general semi-linear groups

Let $V = V(n+1, q)$ be an $(n+1)$-dimensional vector space over the field F_q containing q elements. Let $\mathbf{P} = \mathbf{P}(n, q)$ denote the n-dimensional projective space over F_q. We denote the points set underlying $\mathbf{P}$ by Ω and the collection of i-dimensional subspaces by $\mathfrak{B}^{(i)}$, i.e.

$$\begin{aligned}
\Omega &= \{\langle v \rangle \mid v \in V, v \neq 0\},\\
\mathfrak{B}^{(i)} &= \{[U] \mid U \text{ is an } (i+1)\text{-dimensional subspace of } V\}.
\end{aligned}$$

In §1.5 we proved:

(i) The elements of the general semi-linear group $\Gamma L(V)$ operate on Ω in the following natural way:

$$\langle v \rangle^\sigma = \langle v^\sigma \rangle \quad \text{for } \sigma \in \Gamma L(V) \text{ and } \langle v \rangle \in \Omega.$$

By this operation $(\Gamma L(V), \Omega)$ is a permutation representation with kernel $Z(V) = \{\lambda 1_V | \lambda \in F_q, \lambda \neq 0\}$, where 1_V is the identity transformation of V.

(ii) The elements of $\Gamma L(V)$ induce automorphisms for the designs $\mathbf{P}_i(n, q) = (\Omega, \mathfrak{B}^{(i)})$ $(i \geq 1)$. If dim $V = n + 1 \geq 3$, then $\Gamma L(V)/Z(V) \simeq$ Aut $\mathbf{P}_i(n, q) =$ Aut $\mathbf{P}$.

The group $\Gamma L(V)/Z(V)$ is denoted by $P\Gamma L(V)$ and called the projective general semi-linear group. The subgroups $PGL(V) = GL(V)/Z(V)$ and $PSL(V) = SL(V)Z(V)/Z(V) \simeq SL(V)/Z(V) \cap SL(V)$ of $P\Gamma L(V)$ are called the projective general linear group and the projective special linear group respectively.

By assigning to every $\sigma \in \Gamma L(V)$ its companion automorphism $\theta(\sigma)$ of Aut F_q we get a homomorphism $\theta : \Gamma L(V) \to \text{Aut } F_q$. For φ an arbitrary element of Aut F_q choosing a F_q-basis $u_0, \dots, u_n$ of V we define $\sigma : V \to V$ by

$$(\sum \lambda_i u_i)^\sigma = \sum \lambda_i^\varphi u_i.$$

Then $\sigma \in \Gamma L(V)$ and $\theta(\sigma) = \varphi$, i.e. θ is surjective. Since Ker $\theta = GL(V)$, we have $GL(V) \trianglelefteq \Gamma L(V)$ and $\Gamma L(V)/G(LV) \simeq$ Aut F_q. The mapping from $GL(V)$ onto $F_q^* = F_q - \{0\}$ defined by assigning to each element $\sigma \in GL(V)$ its determinant det σ is a homomorphism and its kernel is $SL(V)$. Therefore, $SL(V) \trianglelefteq GL(V)$ and $GL(V)/SL(V) \simeq F_q^*$. We sum up our results in the following lemma:

Lemma 4.5.1 (1) $GL(V) \trianglelefteq \Gamma L(V)$, $PGL(V) \trianglelefteq P\Gamma L(V)$ and $\Gamma L(V)/GL(V) \simeq P\Gamma L(V)/PGL(V) \simeq$ Aut F_q.

(2) $SL(V) \trianglelefteq GL(V)$, $PSL(V) \trianglelefteq PGL(V)$, $GL(V)/SL(V) \simeq F_q^*$ (= *a cyclic group) and* $PGL(V)/PSL(V) \simeq F_q^*/F_q^{*n+1}$. ■

(The last isomorphism follows from

$$PGL(V)/PSL(V) \simeq GL(V)/SL(V)Z(V).)$$

Lemma 4.5.2 *If $q = p^r$ (p prime), then:*

(1) $|\Gamma L(V)| = r|GL(V)| = r(q-1)|SL(V)|$

$$= rq^{n(n+1)/2} \prod_{i=1}^{n+1} (q^i - 1).$$

(2) $|P\Gamma L(V)| = r|PGL(V)| = r(q-1, n+1)|PSL(V)|$

$$= rq^{n(n+1)/2} \prod_{i=2}^{n+1} (q^i - 1).$$

Proof Let $u_0, \ldots, u_n$ be an F_q-basis for V. For $\sigma \in GL(V)$, $u_0^\sigma, \ldots, u_n^\sigma$ is also an F_q-basis and $(\sum \lambda_i u_i)^\sigma = \sum \lambda_i u_i^\sigma$. Conversely, if $v_0, \ldots, v_n$ is an F_q-basis for V, then the mapping from V into V defined by

$$\sum \lambda_i u_i \to \sum \lambda_i v_i$$

is a non-singular linear transformation of V. Therefore $|GL(V)|$ equals the number of F_q-bases for V, hence

$$|GL(V)| = \prod_{i=0}^{n} (q^{n+1} - q^i) = q^{n(n+1)/2} \prod_{i=1}^{n+1} (q^i - 1).$$

Using $|F_q^*| = q - 1 = |Z(V)|$, $|Z(V) \cap SL(V)| = (q-1, n+1)$ and $|\mathrm{Aut}\, F_q| = r$, we get the desired equalities by lemma 4.5.1. ■

Theorem 4.5.3 (1) *$SL(V)$ operates transitively on $\mathfrak{B}^{(i)}$ $(0 \le i \le n-1)$. Moreover,*

(2) *$SL(V)$ operates 2-transitively on $\Omega = \mathfrak{B}^{(0)}$ and $\mathfrak{B}^{(n-1)}$.*

(3) *If $n = 1$ (i.e. if $V = V(2, q)$), $GL(V)$ operates 3-transitively on Ω.*

Proof (1) Let $[U_1], [U_2]$ be two elements of $\mathfrak{B}^{(i)}$ (so U_1 and U_2 are $(i+1)$-dimensional subspaces of V). Let $\{u_0, \ldots, u_i\}$ and $\{v_0, \ldots, v_i\}$ denote F_q-bases for U_1 and U_2 respectively. Extend $\{u_0, \ldots, u_i\}$ to a basis $\{u_0, \ldots, u_i, u_{i+1}, \ldots, u_n\}$ for V and extend $\{v_0, \ldots, v_i\}$ to a basis $\{v_0, \ldots, v_i, v_{i+1}, \ldots, v_n\}$ for V. Let $\sigma: V \to V$ denote the mapping defined by $\sum \lambda_i u_i \to \sum \lambda_i v_i$. The mapping $\tau: V \to V$ defined by

$$\sum_{i=0}^{n} \lambda_i u_i \to \lambda_0 (\det \sigma)^{-1} v_0 + \sum_{i=1}^{n} \lambda_i v_i$$

is an element of $SL(V)$ such that $[U_1]^\tau = [U_2]$.

(2) Let $\langle u_0 \rangle, \langle u_1 \rangle$ be two different elements of Ω. So u_0 and u_1 are linearly independent vectors of V. If $\langle v_0 \rangle, \langle v_1 \rangle$ is another

pair of different elements of Ω, and if we define $\tau \in SL(V)$ with regard to $\{u_0, u_1\}$ and $\{v_0, v_1\}$ as in (1), then $\langle u_0 \rangle^\tau = \langle v_0 \rangle$ and $\langle u_1 \rangle^\tau = \langle v_1 \rangle$, hence $SL(V)$ operates 2-transitively on Ω. If $n = 1$, then $\mathfrak{B}^{(n-1)} = \Omega$ and so the proof is completed. If $n > 1$, then $\mathbf{P}_{n-1}(n, q) = (\Omega, \mathfrak{B}^{(n-1)})$ is a symmetric design and $PSL(V) \leq$ Aut $\mathbf{P}_{n-1}(n, q)$, therefore $SL(V)$ operates 2-transitively on $\mathfrak{B}^{(n-1)}$ by theorem 1.5.9.

(3) $GL(V)(\geq SL(V))$ operates 2-transitively on Ω by (2). So it suffices to prove, that for arbitrary points $\langle u_0 \rangle, \langle u_1 \rangle \in \Omega$ ($\langle u_0 \rangle \neq \langle u_1 \rangle$) the subgroup H of $GL(V)$ that leaves those two points fixed, operates transitively on $\Omega_0 = \Omega - \{\langle u_0 \rangle, \langle u_1 \rangle\}$. Since u_0 and u_1 are linearly independent vectors in V, all elements of Ω_0 can be represented as: $\langle u_0 + \lambda u_1 \rangle$ for some $\lambda \in F_q (\lambda \neq 0)$. For $\lambda (\neq 0) \in F_q$ if we define a mapping $\sigma_\lambda : V \to V$ by

$$(\lambda_0 u_0 + \lambda_1 u_1)^{\sigma_\lambda} = \lambda_0 u_0 + \lambda_1 \lambda u_1,$$

we have $\sigma_\lambda \in H$ and $\langle u_0 + u_1 \rangle^\sigma = \langle u_0 + \lambda u_1 \rangle$, hence H operates transitively on Ω_0. ∎

If $V = V(2, q)$, then $|PGL(V)| = (q+1)q(q-1)$ by lemma 4.5.2. Since $|\Omega| = q + 1$, we conclude from the previous theorem that $(PGL(V), \Omega)$ is sharply 3-transitive. Also $P\Gamma L(V)(\geq PGL(V))$ operates 3-transitively on Ω. If K denotes the subgroup of $P\Gamma L(V)$ that leaves three given, different points of Ω fixed, then $P\Gamma L(V) = K(PGL(V))$ and $K \cap PGL(V) = 1$, and so we have $K \simeq P\Gamma L(V)/PGL(V) \simeq \text{Aut}\, F_q$ (Aut F_q is the cyclic group of order r, where $q = p^r$ (p prime)). We have proved the following theorem:

Theorem 4.5.4 *If $V = V(2, q)$ where $q = p^r$(p prime) then:*

(1) *$(PGL(V), \Omega)$ is sharply 3-transitive.*

(2) *$(P\Gamma L(V), \Omega)$ is 3-transitive and the subgroup of $P\Gamma L(V)$ leaving three given, different points of Ω fixed, is the cyclic group of order r.*

(Note that (1) has been already proved in theorem 3.4.7.)

Since we know the structure of $P\Gamma L(V)/PGL(V)$ and $PGL(V)/PSL(V)$ by lemma 4.5.1, we have to determine the structure of $PSL(V)$ in order to know the structure of $P\Gamma L(V)$.

Let U be a hyperplane (i.e. n-dimensional subspace) of $V(n+1, q)$. If $\tau(\neq 1)\in GL(V)$ satisfies:

$$u^\tau = u \quad \forall u\in U, \quad v^\tau - v\in U \quad \forall v\in V$$

then τ is called a transvection of V with respect to U and U is called the axis of the transvection τ. Pick $v_0\in V - U$, then $v_0^\tau \neq v_0$ since $\tau \neq 1$. For $v\in V$ we can write: $v = \lambda_0 v_0 + u\,(\lambda_0\in F_q, u\in U)$, hence $v^\tau - v = \lambda_0(v_0^\tau - v_0)$. Therefore $\{v^\tau - v \mid v\in V\} = \langle v_0^\tau - v_0\rangle$, hence τ determines a one-dimensional subspace of U. This one-dimensional subspace (or its generating element) is called the direction of a transvection τ. If u_0 is the direction of τ, v^τ can be written as:

$$v^\tau = v + \lambda_v u_0 \quad \text{for some } \lambda_v\in F_q, \forall v\in V.$$

The mapping from V into F_q defined by $v \to \lambda_v$ is linear and its kernel is U. Therefore, if τ is a transvection, then we have

$$v^\tau = v + f(v)u_0 \quad \forall v\in V \tag{4.5.1}$$

with a non-zero linear mapping $f: V \to F_q$ and a non-zero element $u_0\in$ the axis of $\tau(=\operatorname{Ker} f)$. Conversely, if $f: V\to F_q$ is a linear map $(f\neq 0)$ and if $u_0(\neq 0)$ is an element of Ker f, then Ker f is a hyperplane of V and the map $\tau: V\to V$ defined by (4.5.1) is easily seen to be a transvection with axis Ker f and direction u_0. The transvection defined by (4.5.1) is denoted by τ_{f,u_0}. The following lemma is a straightforward consequence of the definition.

Lemma 4.5.5 (1) $\tau_{f,u}^{-1} = \tau_{f,-u}$, $\tau_{f,u} = \tau_{\lambda f,\lambda^{-1}u}$ $\quad \forall \lambda\in F_q^*$.

(2) $\tau_{f,u_1}\tau_{f,u_2} = \tau_{f,u_1+u_2}$ *if* $u_1 + u_2 \neq 0$

(3) $\tau_{f_1,u}\tau_{f_2,u} = \tau_{f_1+f_2,u}$ *if* $f_1 + f_2 \neq 0$

(4) $\tau_{f_1,u_1} = \tau_{f_2,u_2} \Leftrightarrow u_1 = \lambda u_2$ *and* $f_2 = \lambda f_1$ *for some* $\lambda\in F_q^*$

(5) $\tau_{f,u}{}^{\sigma} = \tau_{\sigma^{-1}f,u^\sigma}$ *for* $\sigma\in\Gamma L(V)$,

provided that $\sigma^{-1}f$ *is the linear map defined by* $(\sigma^{-1}f)(v) = f(v^{\sigma^{-1}})^\theta$, *where* θ *is the companion automorphism of* σ.

Lemma 4.5.6 (1) *The collection of all transvections of* V *forms a conjugacy class in* $GL(V)$.

(2) *Transvections of* V *are elements of* $SL(V)$. *Conversely, every* $\tau \in SL(V)(\tau \neq 1)$ *that leaves all the points of a hyperplane fixed, is a transvection of* V.

(3) *If* $\dim V = n+1 \geq 3$, *then the collection of all transvections forms a conjugacy class in* $SL(V)$.

Proof (1) Put $\tau_1 = \tau_{f,u_0}$ and $\tau_2 = \tau_{g,v_0}$. Starting with u_0 and v_0 it is possible to choose two bases $\{u_0, u_1, \ldots, u_n\}$ and $\{v_0, v_1, \ldots, v_n\}$ of V such that

(i) $\operatorname{Ker} f = \langle u_0, \ldots, u_{n-1} \rangle$, $\operatorname{Ker} g = \langle v_0, \ldots, v_{n-1} \rangle$.

(ii) $f(u_n) = g(v_n) = 1$.

If we define $\sigma \in GL(V)$ by $u_i^\sigma = v_i (i = 0, \ldots, n)$, then: $\sigma^{-1}\tau_1\sigma = \tau_2$. (For, $\sigma^{-1}\tau_1\sigma = \tau_{\sigma^{-1}f, u_0^\sigma}$ by lemma 4.5.5(5) and $u_0^\sigma = v_0$. For $u = \sum_{i=0}^n \lambda_i v_i \in V$ we have:

$$(\sigma^{-1}f)(u) = f(\textstyle\sum \lambda_i v_i^{\sigma^{-1}}) = \lambda_n = g(u),$$

hence $\sigma^{-1}f = g$.)

(2) Since all transvections are conjugate by (1), they all have the same determinant, say λ. According to lemma 4.5.5(2) we have $\lambda^2 = \lambda$, therefore $\lambda = 1$ since $\lambda \neq 0$. Hence all transvections are elements of $SL(V)$. Conversely, let $\tau \in SL(V)\,(\tau \neq 1)$ be such that τ leaves all the points of some hyperplane U fixed and let $u_0, \ldots, u_{n-1}$ be a basis of U. Then for any $u \in V - U$, $\{u_0, \ldots, u_{n-1}, u_n = u\}$ is a basis of V. By putting $u^\tau = \sum_{i=0}^n \lambda_i u_i$, we have $1 = \det \tau = \lambda_n$ since $u_i^\tau = u_i$ $(i = 0, \ldots, n-1)$. Therefore, $u^\tau - u \in U$, hence τ is a transvection.

(3) If the element σ of $GL(V)$ defined in (1) is not an element of $SL(V)$, then we choose σ' such that $u_i^{\sigma'} = v_i (\forall i \neq 1)$ and $u_1^{\sigma'} = \lambda^{-1} v_1$, where $\lambda = \det \sigma$. Then $\det \sigma' = 1$, and noting $n+1 \geq 3$ we have $\sigma'^{-1}\tau_1\sigma' = \tau_2$. ■

Theorem 4.5.7 *SL(V) is generated by the transvections.*

Proof For $\sigma \in SL(V)\,(\sigma \neq 1)$ put $I(\sigma) = \{u \in V \mid u^\sigma = u\}$. $I(\sigma)$ is a subspace of V. We will prove by induction on $\dim(V/I(\sigma))$ that σ can be written as a product of transvections. If $\dim(V/I(\sigma)) = 1$, then $I(\sigma)$ is a hyperplane of V and σ is a transvection by lemma 4.5.6(2). So assume $\dim(V/I(\sigma)) = t > 1$. We distinguish two cases:

(*a*) The case that there exists an element $u \in V(u \neq 0)$ such that u^σ and u are linearly independent mod $I(\sigma)$: Then $u^\sigma - u$ and u^σ are also linearly independent mod $I(\sigma)$. Therefore it is possible to choose an $(n-1)$-dimensional subspace W of V such that W contains $I(\sigma)$ and $\langle u - u^\sigma \rangle$, but does not contain $\langle u^\sigma \rangle$. Choosing a linear mapping f of V such that $f(W) = 0$ and $f(u^\sigma) = 1$, we have that $\sigma\tau_{f,u-u^\sigma}$ leaves all elements of $I(\sigma)$ fixed since $I(\sigma) \subseteq W$ and $u^{\sigma\tau_{f,u-u^\sigma}} = u^\sigma + f(u^\sigma)(u - u^\sigma) = u$, hence $I(\sigma) \subsetneq I(\sigma\tau_{f,u-u_\sigma})$. Therefore $\sigma\tau_{f,u-u_\sigma}$ (and hence also σ) is a product of transvections by the induction hypothesis.

(*b*) The case that for all $u \in V(u \neq 0)$ u^σ and u are linearly dependent modulo $I(\sigma)$: Choose $u \in V$ such that $u^\sigma \neq u$ and choose a hyperplane W of V such that $I(\sigma) \subseteq W$ and $u \notin W$. Let f be a linear map of V such that $f(W) = 0$ and $f(u^\sigma) = 1$ and let u_0 be an arbitrary element of $W - I(\sigma)$. Putting $\tau = \tau_{f,-u_0}$, we have $u^{\sigma\tau} = u^\sigma - f(u^\sigma)u_0 = u^\sigma - u_0$, therefore $u^{\sigma\tau}$ and u are linearly independent mod $I(\sigma)$. Since $f(I(\sigma)) = 0$, $I(\sigma) \subseteq I(\sigma\tau)$. If $I(\sigma) \subsetneq I(\sigma\tau)$, then σ is a product of transvections by the induction hypothesis. If $I(\sigma) = I(\sigma\tau)$, then $u^{\sigma\tau}$ and u are linearly independent mod $I(\sigma\tau)$, hence we are reduced to case (*a*). ∎

We have proved this theorem already in a stronger form in §1.4.2: the $A_{i,j:\lambda}$ are just the matrix representations of transvections, and therefore $SL(V)$ is generated by some (proper) subset of the collection of all transvections. However, the proof of theorem 4.5.7 presented above shows that each element of $SL(V)$ can be written as the product of $2n + 1$ or less transvections, a fact that is not included in the results of §1.4.2.

Theorem 4.5.8 *Let G be a subgroup of $GL(V)$ that is invariant under conjugation by all elements of $SL(V)$, then $G \leq Z(V)$ or $SL(V) \leq G$, except in the case that* $\dim V = 2$ *and* $F_q = F_2$ *or* $F_q = F_3$.

Proof We distinguish two cases. (*a*) For $\dim V \geq 3$. Since all transvections are conjugate in $SL(V)$, it suffices to prove that if $G \nleq Z(V)$, G contains at least one transvection.

We first show that there exists a non-identity element $\rho \in G \cap SL(V)$ such that $\langle v^\rho - v | v \in V \rangle \neq V$. (For, since $G \nsubseteq Z(V)$, there exist elements $\sigma \in G$ and $u \in V$ such that u^σ and u are linearly independent. Let $\tau = \tau_{f,u}$ be a transvection with direction u and $\rho = \tau^{-1}\tau^\sigma$ $(\in SL(V))$. Since $\tau^\sigma = \tau_{\sigma^{-1}f, u^\sigma}$ is a transvection with direction u^σ, we have $\rho \neq 1$. Since G is invariant under conjugation by transvections, $\rho = (\tau^{-1}\sigma^{-1}\tau)$, $\sigma \in G$. For $v \in V$:

$$v^\rho = v^{\tau^{-1}\tau^\sigma} = (v - f(v)u)^{\tau^\sigma} = v - f(v)u + f(v^{\sigma^{-1}} - f(v)u^{\sigma^{-1}})u^\sigma,$$

hence $\langle v^\rho - v | v \in V \rangle \subseteq \langle u, u^\sigma \rangle$, therefore $\langle v^\rho - v | v \in V \rangle \neq V$ since $\dim V \geq 3$.)

For such an element $\rho(\neq 1) \in G \cap SL(V)$ we take a hyperplane U of V containing $\langle v^\rho - v | v \in V \rangle$, then it is easily seen that $U^\rho = U$ and $v^\rho - v \in U$ for all $v \in V$. If $\tau = \tau_{f,u}$ denotes a transvection with axis U, then $\rho^{-1}f$ and f are both linear mappings of V with kernel U, hence $\rho^{-1}f = \lambda f$ for some $\lambda \in F_q^*$. Therefore we have

$$\tau^\rho = \tau_{\rho^{-1}f, u^\rho} = \tau_{\lambda f, u^\rho} = \tau_{f, \lambda u^\rho}.$$

Let us suppose that ρ and τ commute. Then $\tau_{f,u} = \tau_{f,\lambda u^\rho}$ from $\tau^\rho = \tau$, and so we have $u = \lambda u^\rho$. From this and $\tau\rho = \rho\tau$ we have $f(v - \lambda v^\rho) = 0$ for all $v \in V$, i.e. $v - \lambda v^\rho \in U$. Since $v - v^\rho \in U$ and $U \neq V$, we have $\lambda = 1$, i.e. $u = u^\rho$. Therefore, if ρ commutes with all transvections with axis U, then ρ leaves all points of U fixed and ρ is a transvection by lemma 4.5.6(2). If there exists a transvection $\tau = \tau_{f,u}$ with axis U, that does not commute with ρ, then $1 \neq [\tau, \rho] = \tau^{-1}\tau^\rho = \tau_{f,-u}\tau_{f,\lambda u^\rho} = \tau_{f,-u+\lambda u^\rho}$ is a transvection contained in G.

(*b*) For $\dim V = 2$. Suppose $G \nsubseteq Z(V)$. Let σ be an element of G that is not an element of $Z(V)$. Then there exists an element $u \in V$, such that u and u^σ are linearly independent, and since $\dim V = 2$, $v_1 = u$, $v_2 = u^\sigma$ form a basis for V. We denote the action of σ on v_1 and v_2 as follows,

$$v_1^\sigma = v_2 \; v_2^\sigma = \lambda_1 v_1 + \lambda_2 v_2 \quad \lambda_1, \lambda_2 \in F_q, \lambda_1 \neq 0.$$

Since $F_q \neq F_2, F_3$, we can choose $\lambda_0 \in F_q (\lambda_0 \neq 0)$ such that $\lambda_0^2 \neq -\lambda_1^{-1}$. Choose $\tau \in GL(V)$ such that

$$v_1^\tau = \lambda_0 \lambda_2 v_1 - \lambda_0 v_2, \quad u_2^\tau = \lambda_0^{-1} v_1,$$

then $\det \tau = 1$, i.e. $\tau \in SL(V)$. Putting $\rho = \tau^{-1}\sigma^{-1}\tau\sigma$ we have $\rho \in G \cap SL(V)$ and

$$v_1^\rho = \mu_1 v_1, \quad v_2^\rho = \mu_2 v_1 + \mu_1^{-1} v_2,$$

where $\mu_1 = -\lambda_0^2\lambda_1 \neq 0, 1$ and $\mu_2 = -\lambda_2(1 + \lambda_0^2\lambda_1) \neq 0$. Now we can prove that G contains a transvection with direction v_1. (For, if $\mu_1^2 = 1$, then $v_1^{\rho^2} = v_1$ and $v_2^{\rho^2} = 2\mu_1\mu_2 v_1 + v_2$ and $\rho^2 \in G$ is a transvection with direction v_1 since the characteristic of $F_q \neq 2$. If $\mu_1^2 \neq 1$, we choose $\tau_1 \in GL(V)$ such that $v_1^{\tau_1} = v_1$ and $v_2^{\tau_1} = v_1 + v_2$. Then $\tau_1 \in SL(V)$, hence $\rho_1 = \tau_1^{-1}\rho^{-1}\tau_1\rho \in G$ and we have $v_1^{\rho_1} = v_1$ and $v_2^{\rho_1} = (\mu_1^2 - 1)v_1 + v_2$, i.e. ρ_1 is a transvection with direction v_1.) So, let τ denote a transvection in G with direction v_1, then: $v_1^\tau = v_1$ and $v_2^\tau = \lambda v_1 + v_2$ for some $\lambda \in F_q$. For $\nu \in F_q (\nu \neq 0)$, define $\rho_\nu \in SL(V)$ by $v_1^{\rho_\nu} = \nu v_1$, $v_2^{\rho_\nu} = \nu^{-1} v_2$, then $\rho_\nu^{-1}\tau\rho_\nu \in G$ and $v_1^{\rho_\nu^{-1}\tau\rho_\nu} = v_1$ and $v_2^{\rho_\nu^{-1}\tau\rho_\nu} = \lambda\nu^2 v_1 + v_2$. According to theorem 1.3.12, for an arbitrary $\mu \in F_q^*$, there exist $\nu_1, \nu_2 \in F_q$ such that $\mu = \lambda(\nu_1^2 + \nu_2^2)$. Therefore, putting $\tilde{\tau} = \rho_{\nu_1}^{-1}\tau\rho_{\nu_1}\rho_{\nu_2}^{-1}\tau\rho_{\nu_2}$ we have $\tilde{\tau} \in G$ and

$$v_1^{\tilde{\tau}} = v_1, \quad v_2^{\tilde{\tau}} = \mu v_1 + v_2,$$

i.e. $\tilde{\tau}$ is a transvection. Therefore G contains all transvections with direction v_1. According to theorem 4.5.3 and lemma 4.5.5 (5) G contains all transvections, hence $G \geq SL(V)$. ■

Theorem 4.5.9 (1) *$PSL(V)$ is a non-Abelian simple group except when* $\dim V = 2$ *and* $F_q = F_2$ *or* F_3.

(2) *If $V = V(2, F_3)$, then $PSL(V) \simeq A_4$.*

(3) *If $V = V(2, F_2)$, then $PSL(V) \simeq S_3$.*

Proof (1) $PSL(V)$ is simple by theorem 4.5.8. Proof of the non-commutativity is easy.

(2) $PSL(2, F_3)$ operates 2-transitively on the point set underlying $\mathbf{P}(1, F_3)$ (theorem 4.5.3). Since $|\Omega| = 4$, and $|PSL(V)| = 4 \times 3$ we see that $PSL(V)$ is isomorphic with a subgroup of order 12 of S_4. Therefore $PSL(V) \simeq A_4$ by theorem 4.1.4.

(3) $PSL(2, F_2)$ operates on the point set underlying $\mathbf{P}(1, F_2)$, therefore we conclude $PSL(V) \simeq S_3$ by comparing the orders. ■

4.6 Properties of $PSL(V)$ as a permutation group (dim $V \geq 3$)

We use the same notation as in §4.5. We assume dim $V \geq 3$. According to theorem 1.5.16 $P\Gamma L(V) \simeq \text{Aut}\,\mathbf{P}$, i.e. each automorphism of $\mathbf{P}$ is induced by some element of $\Gamma L(V)$. We will denote the automorphism of $\mathbf{P}$ corresponding with $\tau \in \Gamma L(V)$ by $\bar{\tau}$. Now we will characterize the elements $\bar{\tau}$ of Aut $\mathbf{P}$ corresponding with a transvection τ of V. Put $\tau = \tau_{f,u_0}$. Since τ leaves all elements of $U = \text{Ker}\, f$ fixed, $\bar{\tau}$ leaves all points of the hyperplane $[U]$ of $\mathbf{P}$ fixed. Further, if $v \in V (v \neq 0)$, then $v^\tau \in \langle v, u_0 \rangle$, therefore $\langle v, u_0 \rangle^\tau = \langle v, u_0 \rangle$, i.e. $\bar{\tau}$ leaves all one-dimensional subspaces of $\mathbf{P}$, that contain the point $\langle u_0 \rangle (\in [U])$, fixed. So, if τ is a transvection, then $\rho = \bar{\tau} \in \text{Aut}\,\mathbf{P}$ satisfies:

(i) ρ leaves the points of some hyperplane $[U]$ of $\mathbf{P}$ fixed

(ii) there is a point $\langle u_0 \rangle \in [U]$, such that ρ leaves all one-dimensional subspaces (i.e. lines) of $\mathbf{P}$ containing $\langle u_0 \rangle$ fixed.

Conversely, let $\rho \in \text{Aut}\,\mathbf{P}$ satisfy conditions (i) and (ii). We will show that $\rho = \bar{\tau}$ for some transvection τ of V. Since $P\Gamma L(V) \simeq \text{Aut}\,\mathbf{P}$, there is a $\tau \in \Gamma L(V)$ such that $\rho = \bar{\tau}$, i.e. such that $\langle v^\rho \rangle = \langle v^\tau \rangle$ for $v \in V$. Since $\langle u \rangle^\rho = \langle u \rangle$ for $u \in U$ $(u \neq 0)$, we can write $u^\tau = \lambda_u u$ for some $\lambda_u \in F_q$. Let $u_1, \ldots, u_n$ be a basis for U. Putting $u = \sum_{i=1}^n u_i$ we have $u^\tau = \lambda_u u = \sum_{i=1}^n \lambda_u u_i$ and also $u^\tau = \sum_{i=1}^n u_i^\tau = \sum_{i=1}^n \lambda_{u_i} u_i$, hence $\lambda_{u_1} = \ldots = \lambda_{u_n} = \lambda_u$. Therefore, there exists a $\lambda_0 \in F_q$ such that $u^\tau = \lambda_0 u$ for all $u \in U$. If $v \in V - U$, then $[\langle v, u_0 \rangle]^\rho = [\langle v, u_0 \rangle]$, i.e. $v^\tau \in \langle v, u_0 \rangle$, hence we can write $v^\tau = \lambda v + \mu u_0$. Since dim $U \geq 2$, we can choose an element $u \in U$ such that u and u_0 are linearly independent. From $(v+u)^\tau = v^\tau + u^\tau$ we conclude $\lambda = \lambda_0$. Therefore picking $\tau_0 \in Z(V)$ such that $v^{\tau_0} = \lambda_0^{-1} v$ for $v \in V$, we have that $\tau\tau_0^{-1}$ is a transvection of V and $\overline{\tau\tau_0^{-1}} = \bar{\tau} = \rho$.

A non-identity element $\rho \in \text{Aut}\,\mathbf{P}$ satisfying (i) and (ii) is called a transvection (or elation) of $\mathbf{P}$ with centre $\langle u_0 \rangle$ and axis $[U]$. We have the following theorem:

Theorem 4.6.1· *If* dim $V = n + 1 \geq 3$, *then:*

(1) *$PSL(V)$ is generated by the transvections of* $\mathbf{P}$.

(2) *If τ is a transvection of V, then $\bar{\tau}$ is a transvection of* $\mathbf{P}$ *and every transvection of* $\mathbf{P}$ *can be obtained in this way. Hence, the*

assignment $\tau \to \bar{\tau}$ *sets up a one-to-one correspondence between the transvections of* V *and those of* $\mathbf{P}$.

(3) *For* $x \in \mathbf{P}$ *the set* $A(x) = \{\sigma \in \mathrm{Aut}\,\mathbf{P} \mid \sigma = 1$ *or* σ *is a transvection with centre* $x\}$ *is a subgroup of* $\mathrm{Aut}\,\mathbf{P}$. *The set* $A(x)$ *is an elementary Abelian group of order* $q^n = p^{rn}$ *and a normal subgroup of* $(\mathrm{Aut}\,\mathbf{P})_x$ ($=$ *the stabilizer of* x).

(4) *Let* $x \in \mathbf{P}$. *An element of order* p *of* $\mathrm{Aut}\,\mathbf{P}$, *that leaves* x *fixed and that leaves all lines through* x *fixed, is a transvection of* $\mathbf{P}$ *with centre* x. *Hence:*

$A(x) = \{\sigma \in \mathrm{Aut}\,\mathbf{P} \mid \sigma = 1$, *or* σ *is an element of order* p, *that leaves* x *fixed and that leaves all lines through* x *fixed*$\}$.

Proof (1) Follows from the fact that the transvections of V generate $SL(V)$.

(2) If $\bar{\tau} = \bar{\tau}'$ where τ is a transvection of V and $\tau' \in \Gamma L(V)$, then $\tau' = \tau\sigma$ for some $\sigma \in Z(V)$, therefore if τ' is also a transvection, then we have $\sigma = 1$. We have proved the other assertions of (2) already.

(3) $A(x)$ is a subgroup of Aut $\mathbf{P}$ by lemma 4.5.5(1) and (3), and $A(x) \trianglelefteq (\mathrm{Aut}\,\mathbf{P})_x$ by lemma 4.5.5(5). Let $x = \langle u_0 \rangle$ and define a mapping t from $A(x)$ into $\{f \in \mathrm{Hom}_{F_q}(V, F_q) \mid f(u_0) = 0\}$ such that:

$$t(\bar{\tau}) = \begin{cases} f & \text{if } \bar{\tau} = \bar{\tau}_{f,u_0} \\ 0 & \text{if } \bar{\tau} = 1. \end{cases}$$

Then this mapping is an isomorphism from $A(x)$ onto $\{f \in \mathrm{Hom}_{F_q}(V, F_q) \mid f(u_0) = 0\}$ by lemma 4.5.5(3). It is easily seen that $\{f \in \mathrm{Hom}_{F_q}(V, F_q) \mid f(u_0) = 0\}$ is an elementary Abelian group of order q^n.

(4) Let $\bar{\tau} \in \mathrm{Aut}\,\mathbf{P}$ be an element of order p satisfying the condition of the assertion. Put $x = \langle u_0 \rangle$ and choose a basis $u_0, u_1, \ldots, u_n$ of V. Then $u_0^\tau = \lambda_0 u_0$ since $\langle u_0 \rangle^{\bar{\tau}} = \langle u_0 \rangle$ and $u_i^\tau = \lambda_i u_i + \mu_i u_0$ since $\langle u_i \rangle^{\bar{\tau}} \in \langle u_i, u_0 \rangle$. From $(u_i + u_j)^\tau = u_i^\tau + u_j^\tau$ we conclude $\lambda_1 = \ldots = \lambda_n$. Picking $\sigma \in Z(V)$ such that $v^\sigma = \lambda_1 v$ for $v \in V$, we have $\bar{\tau} = \overline{\tau\sigma^{-1}}$ and $u_i^{\tau\sigma^{-1}} - u_i \in \langle u_0 \rangle$ $(i = 1, \ldots, n)$. Therefore we may assume that τ is such that

$$u_0^\tau = \lambda_0 u_0, \quad u_i^\tau = u_i + \mu_i u_0 \quad (i = 1, \ldots, n).$$

Then we have that $\lambda_0 = 1$. (For, suppose $\lambda_0 \neq 1$. Since $\bar{\tau}^p = 1$ we have

$$u_1^{\tau^p} = u_1 + \mu_1(1 + \lambda_0 + \ldots + \lambda_1^{p-1})u_0 \in \langle u_1 \rangle,$$

hence $\mu_1(1 + \lambda_0 + \ldots + \lambda_0^{p-1}) = 0$. Since $\lambda_0 \neq 1$, we have $1 + \lambda_0 + \ldots + \lambda_0^{p-1} = (\lambda_0^p - 1)/(\lambda_0 - 1) \neq 0$, hence $\mu_1 = 0$. Therefore $(u_1 + u_0)^{\tau^p} = u_1 + \lambda_0^p u_0 \in \langle u_0 + u_1 \rangle$, hence $\lambda_0^p = 1$, i.e. $\lambda_0 = 1$. Contradiction.) Now, defining the linear mapping $f: V \to F_q$ by

$$f(v) = \sum_{i=1}^{n} \xi_i \mu_i \quad \text{for} \quad v = \sum_{i=0}^{n} \xi_i u_i,$$

we have:

$$\tau = \tau_{f,u_0},$$

i.e. τ is a transvection with centre u_0. ■

$G = P\Gamma L(V)$ operates 2-transitively on the point set Ω underlying **P**, and any element $\sigma \in A(x)$ $(x \in \Omega)$ leaves the points of some hyperplane $[U]$ fixed. Putting $H = A(x)$, we have:

(i) H is a non-identity, Abelian, normal subgroup of G_x and $(H, \Omega - \{x\})$ is not half-regular.

Let $[U]$ be a hyperplane of **P** and put $K = G_{\langle [U] \rangle}$. Then we have $|G:K| = |\Omega|$ because $(G, \mathfrak{B}^{(n-1)})$ is (2-) transitive and $|\Omega| = |\mathfrak{B}^{(n-1)}|$. Further, since (G, Ω) is 2-transitive, K has no fixed points in Ω and since $K_{[U]}(= G_{[U]})$ contains a transvection, $K_{[U]} \neq 1$. Therefore:

(ii) K is a subgroup of G such that $|G:K| = |\Omega|$ and $F_\Omega(K) = \varnothing$, and K operates not faithfully on U.

In chapter 6 we will see that the existence of subgroups H and K satisfying properties (i) and (ii) is a fundamental property of $PSL(V)$ regarded as a permutation group on Ω(= point set underlying **P**).

4.7 Symmetric groups and general linear groups of low order

Some symmetric groups and general linear groups of low order have the same structure.

Theorem 4.7.1 (1) $PSL(2,2) = SL(2,2) = GL(2,2) \simeq S_3$.
(2) $PSL(2,3) \simeq A_4 \quad PGL(2,3) \simeq S_4$.
(3) $PGL(2,4) \simeq PSL(2,4) = SL(2,4) \simeq PSL(2,5) \simeq A_5, PGL(2,5) \simeq S_5$.

Proof (1) and (2) These are proved in theorem 4.5.9 except for $PGL(2,3)$. The group $PGL(2,3)$ operates faithfully on the point set Ω underlying $\mathbf{P}(1,3)$. Noticing $|\Omega| = 4$ and comparing orders yields $PGL(2,3) \simeq S_4$.

(3) Since $PGL(2,4)$ operates faithfully on the point set underlying $\mathbf{P}(1,4)$, it can be regarded as a subgroup of S_5. Comparing orders yields $PGL(2,4) \simeq A_5$. Since $PGL(2,5)$ operates faithfully on the point set underlying $\mathbf{P}(1,5)$, it is isomorphic to some subgroup H of S_6, such that $|S_6 : H| = 6$, hence $PGL(2,5) \simeq S_5$ by theorem 4.3.5. Since $PSL(2,5)$ is a subgroup of $PGL(2,5)$ of index 2, we get $PSL(2,5) \simeq A_5$. ■

Theorem 4.7.2 *A_5 is the only simple group of order* 60 *and among all non-Abelian simple groups A_5 has the lowest order.*

Proof Suppose G is a simple group satisfying $|G| = 60$ and $G \not\simeq A_5$. We first prove that G does not contain a subgroup of index 5. (For, suppose G contains a subgroup H of index 5, then $(G, G/H)$ is a permutation representation of degree 5 of G. Since G is simple, G is a subgroup of index 2 of S_5, hence $G \simeq A_5$ by theorem 3.4.2(3)). If P is a Sylow 2-subgroup of G, then $|G : \mathscr{N}_G(P)| =$ 3, 5 or 15, hence $|G : \mathscr{N}_G(P)| = 15$. Therefore, $\mathscr{N}_G(P)(= P) = \mathscr{C}_G(P)$ and G contains a normal 2-complement by corollary 2.8.5 (Burnside), contrary to the fact that G is simple.

Now, let G be a non-Abelian simple group such that $|G| < 60$, then by Burnside's theorem (theorem 2.10.17) there exist at least 3 different primes dividing $|G|$. Therefore $|G| = 7 \times 3 \times 2$ or $|G| = 5 \times 3 \times 2$. If P denotes a Sylow 2-subgroup of G, then $\mathscr{N}_G(P) = \mathscr{C}_G(P)$ in both cases, therefore G contains a normal 2-complement by corollary 2.8.5, contrary to the assumption that G is simple. ■

Theorem 4.7.3 *$PSL(2, 7)$ is the only simple group of order 168. The only non-Abelian simple group with order less then 168 is A_5.*

Proof Let G be a simple group of order $168(=7 \times 3 \times 8)$. Let $P = \langle \sigma \rangle$ be a Sylow 7 subgroup of G, then $|\mathcal{N}_G(P)| = 21 = 7 \times 3$ by Sylow's theorem. If $\rho \in \mathcal{N}_G(P)$ denotes an element of order 3, then we can write: $\mathcal{N}_G(P) = \langle \sigma, \rho \rangle$. Since G is simple, we have: $\mathcal{N}_G(P) \neq \mathcal{C}_G(P) = P$ (by, for example, corollary 2.8.5), hence $\mathcal{N}_G(P)$ is a Frobenius group by theorem 2.11.3. $Q = \langle \rho \rangle$ is a Sylow 3-subgroup of G and $|\mathcal{N}_G(Q)| = 3 \times 2$ or $|\mathcal{N}_G(Q)| = 3 \times 8$ by Sylow's theorem (theorem 2.1.1). If $|\mathcal{N}_G(Q)| = 3 \times 8$, then $|G : \mathcal{N}_G(Q)| = 7$ and G contains exactly 7 Sylow 3-subgroups. Since $\mathcal{N}_G(P)$ contains already 7 Sylow 3-subgroups and since $\mathcal{N}_G(P)$ is generated by the elements of those subgroups, we have $\mathcal{N}_G(P) \trianglelefteq G$. Contradiction. Therefore $|\mathcal{N}_G(Q)| = 6$. If $\tau \in \mathcal{N}_G(Q)$ is an element of order 2, then $\mathcal{N}_G(Q) = \langle \rho, \tau \rangle$ and $\mathcal{N}_G(Q) \neq \mathcal{C}_G(Q) = Q$ by the same reasoning as in the case of $\mathcal{N}_G(P)$ discussed above. Therefore $\tau\rho\tau = \rho^{-1}$. Since $G \geq \langle \sigma, \rho, \tau \rangle$ and $|G : \langle \sigma, \rho, \tau \rangle| \leq 4$, we conclude $G = \langle \sigma, \rho, \tau \rangle$ because G is simple.

Now let us regard G as operating on $\Omega = G/\langle \sigma, \rho \rangle$. Since G is simple, (G, Ω) is a permutation group. We denote the element $\langle \sigma, \rho \rangle$ of Ω by ∞. Since $G_\infty = \langle \sigma, \rho \rangle (= \mathcal{N}_G(P))$ and since $|\Omega - \{\infty\}| = 7$, $(G_\infty, \Omega - \{\infty\})$ is a Frobenius group of degree 7. According to theorem 3.6.1 it is possible to regard the group $\langle \sigma, \rho \rangle$ as an affine transformation group of F_7. Namely, by suitably identifying $\Omega - \{\infty\}$ with F_7, we can consider σ and ρ operating on $F_7 = \{0, 1, \ldots, 6\}$ as follows:

$$\sigma : x \to x + 1$$

$$\rho : x \to 2x.$$

Since $|\langle \sigma, \rho \rangle| \not\equiv 0 \pmod 2$, τ has no fixed points in Ω. Moreover since $\langle \rho \rangle^\tau = \langle \rho \rangle$ and since the orbits of $(\langle \rho \rangle, \Omega)$ are $\{\infty\}$, $\{0\}$, $\{1, 2, 4\}$ and $\{3, 5, 6\}$, we have

$$\tau : \infty \leftrightarrow 0,\ \{1, 2, 4\} \leftrightarrow \{3, 5, 6\}.$$

From $\rho\tau = \tau\rho^{-1}$ we get

$$2^\tau = (1^\rho)^\tau = (1^\tau)^{\rho^{-1}} = 2^{-1}1^\tau,$$
$$4^\tau = (2\rho)^\tau = (2^\tau)^{\rho^{-1}} = 2^{-1}2^\tau = 4^{-1} \times 1^\tau.$$

Therefore, putting $1^\tau = a$, we can write:

$$\tau : x \to \frac{a}{x} \quad \text{for} \quad \forall x \in F_7 \cup \{\infty\}.$$

Hence $G = \langle \sigma, \rho, \tau \rangle$ can be regarded as a subgroup of the group of linear fractional transformations on F_7, i.e. $PGL(2, F_7) = GL(2, F_7)/Z(2, F_7)$. In this identification σ, ρ and τ are the images of matrices $\begin{pmatrix} 1 & 0 \\ 1 & 1 \end{pmatrix}, \begin{pmatrix} 2 & 0 \\ 0 & 1 \end{pmatrix}$ and $\begin{pmatrix} 0 & 1 \\ a & 0 \end{pmatrix}$ respectively according to the natural homomorphism from $GL(2, F_7)$ onto $PGL(2, F_7)$. The determinant values of those matrices are 1, 2 and $-a$ and they belong to $\{1, 2, 4\}$. (Here we note that $a = 1^\tau \in \{3, 5, 6\}$, so $-a \in \{1, 2, 4\}$.) But since $\{1, 2, 4\} = \{1, 3^2, 2^2\}$, σ, ρ and τ can be considered as the images of some elements of $SL(2, F_7)$, so we get $G \leq PSL(2, F_7)$. Since $|PSL(2, F_7)| = 168$, hence $G = PSL(2, F_7)$.

Now let G be a non-commutative simple group such that $|G| < 168$. Let p be the greatest prime dividing $|G|$ and let P be a Sylow p-subgroup of G. According to Burnside's theorem (theorem 2.10.17) there exist at least three different primes dividing $|G|$. By Sylow's theorem we have $|G : \mathcal{N}_G(P)| \geq p + 1$ and by corollary 2.8.5, we have $\mathcal{N}_G(P) \supsetneqq \mathcal{C}_G(P)$. Therefore, $2p(p+1) \leq 167$, i.e. $p \leq 7$. If p^2 divides $|G|$, then $2p^2(p+1) \leq 167$, i.e. $p \leq 3$, hence $|G| = 2^a \times 3^b$, contradiction. Hence $p \top |G|$. If $p = 7$, then according to corollary 2.8.5, theorem 2.10.17 and Sylow's theorem, $|G| \geq 7 \times 3 \times 8 = 168$. Contradiction. Therefore $p = 5$ and $|G| = 5 \times 3^a \times 2^b$, where $(a, b) = (2, 1), (1, 1), (1, 2)$ or $(1, 3)$. If $b = 1$, then G is not simple by corollary 2.8.5. Therefore $|G| = 5 \times 3 \times 2^3$ or $5 \times 3 \times 2^2$. Suppose $|G| = 5 \times 3 \times 2^3$, and let Q be a Sylow 3-subgroup of Q, then $|\mathcal{N}_G(Q)| = 3$ or 12 by Sylow's theorem. We conclude from corollary 2.8.5, that $|\mathcal{N}_G(Q)| = 12$ and $|\mathcal{C}_G(Q)| = 6$. Therefore, G contains an element of order 6. If P denotes a Sylow 5-subgroup of G, then $|\mathcal{N}_G(P)| = 20$ by Sylow's theorem and $(G, G/\mathcal{N}_G(P))$ is a (faithful) permutation representation of degree 6 of G. An element of order 6 of a permutation group of degree 6 is an odd permutation. Therefore G contains a subgroup of index 2, contrary to the assumption that G is simple. Hence $|G| = 5 \times 3 \times 2^2 = 60$, and $G \simeq A_5$ by theorem 4.7.2. ■

Corollary 4.7.4 $PSL(2, 7) \simeq PSL(3, 2)$.

Proof Follows from the previous theorem and $|PSL(2, 7)| = |PSL(3, 2)|$. ■

Theorem 4.7.5 $PSL(4, 2) = GL(4, 2) \simeq A_8$.

Proof The first equality is obvious and it is easy to verify that the orders of these groups are the same. According to theorem 4.4.2, A_8 is generated by 6 elements $a_1, \ldots, a_6$ satisfying the fundamental relations:

$$a_1^3 = a_i^2 = 1 \quad (2 \leq i \leq 6),$$
$$(a_i a_{i+1})^3 = 1 \quad (1 \leq i \leq 5),$$
$$(a_i a_j)^2 = 1 \quad (|i - j| > 1).$$

If there are elements $b_1, \ldots, b_6 \in GL(4, 2)$ ($b_i \neq 1$, $i = 1, \ldots, 6$) satisfying the above relations, then $A_8 \simeq \langle b_1, \ldots, b_6 \rangle \leq GL(4, 2)$ because A_8 is simple and therefore we get $A_8 \simeq GL(4, 2)$. It is easy to verify that the following elements of $GL(4, 2)$ satisfy the desired relations:

$$b_1 = \begin{bmatrix} 1 & 1 & 1 & 1 \\ 0 & 0 & 0 & 1 \\ 1 & 1 & 0 & 0 \\ 0 & 1 & 0 & 1 \end{bmatrix}, \quad b_2 = \begin{bmatrix} 0 & 1 & 0 & 1 \\ 0 & 0 & 1 & 0 \\ 0 & 1 & 0 & 0 \\ 1 & 0 & 1 & 0 \end{bmatrix}, \quad b_3 = \begin{bmatrix} 0 & 1 & 1 & 1 \\ 0 & 1 & 0 & 1 \\ 1 & 1 & 0 & 0 \\ 0 & 0 & 0 & 1 \end{bmatrix},$$

$$b_4 = \begin{bmatrix} 1 & 0 & 1 & 0 \\ 0 & 1 & 0 & 0 \\ 0 & 0 & 1 & 0 \\ 0 & 1 & 0 & 1 \end{bmatrix}, \quad b_5 = \begin{bmatrix} 0 & 0 & 1 & 0 \\ 0 & 1 & 0 & 1 \\ 1 & 0 & 0 & 0 \\ 0 & 0 & 0 & 1 \end{bmatrix}, \quad b_6 = \begin{bmatrix} 0 & 1 & 1 & 1 \\ 0 & 0 & 1 & 0 \\ 0 & 1 & 0 & 0 \\ 1 & 1 & 1 & 0 \end{bmatrix}.$$

■

Consider $V = V(4, 2)$, i.e. a 4-dimensional vector space over F_2. According to theorem 4.5.3, $GL(4, 2)(= PGL(4, 2))$ operates 2-

transitively on $V - \{0\}$ ($=$ the point set underlying $\mathbf{P}(3, F_2)$). Since $GL(4, 2) \simeq A_8$, $GL(4, 2)$ contains subgroups of index 8 that are isomorphic to the alternating group A_7. Let us denote one of these subgroups by G. Then we have the following result:

Theorem 4.7.6 *$(G, V - \{0\})$ is 2-transitive (of degree 15).*

Proof Since $|G| = \frac{1}{2}7!$, we see that G contains an element σ of order 7 and an element τ of order 5. Since $|V - \{0\}| = 15$, the number of fixed points of σ in $V - \{0\}$ equals 1 or 8. Since $F_V(\sigma)$ is a subspace of V, $|F_V(\sigma)|$ must be a power of 2. Hence $|F_{V-\{0\}}(\sigma)| = 1$, namely σ has only one fixed point and two cycle-components of length 7. Similarly the number of fixed points of τ in $V - \{0\}$ equals 0, 5 or 10, and we can conclude $|F_{V-\{0\}}(\tau)| = 0$. Therefore σ is the product of 3 cycles of length 5. We have:

(*a*) $(G, V - \{0\})$ is transitive. (For, since $\sigma \in G$, the length of the orbits of $(G, V - \{0\})$ is 7, 8, 14 or 15, and since $\tau \in G$, the length of the orbits $(G, V - \{0\})$ is divisible by 5, from which the assertion follows.)

(*b*) $(G, V - \{0\})$ is 2-transitive. (For, let $1 \in V - \{0\}$, $\Lambda = V - \{0, 1\}$ and let H be the subgroup leaving 1 fixed. Since $|G:H| = 15$, H contains an element σ of order 7 and σ divides Λ into two orbits of length 7. Now, if $(G, V - \{0\})$ is not 2-transitive, then (H, Λ) has two orbits of length 7. Since $G \simeq A_7$ is simple, G contains an element ρ of order 2 that is not contained in H. Putting $1^\rho = i\ (\neq 1)$, we see that $H \cap \rho^{-1}H\rho = K$ is the subgroup leaving 1 and i fixed. Since the length of the orbits of (H, Λ) equals 7, we have: $|K| = |H|/7 = 24$. Since $K^\rho = K$, $\langle K, \rho \rangle$ is a subgroup of order $2 \times 24 = 48$ of G. Contradiction, since $48 \nmid |G| = \frac{1}{2}7!$.) ∎

Theorem 4.7.7 *$PSL(2, p^f)$ contains a subgroup that is isomorphic to A_5 if and only if*

(i) $p = 5$ *or*

(ii) $p^{2f} - 1 \equiv 0 \pmod 5$.

Proof Since

$$|PSL(2, p^f)| = \frac{1}{(2, p^f - 1)} p^f(p^{2f} - 1),$$

the validity of (i) or (ii) is a necessary condition. So let us suppose that condition (i) or (ii) holds. Since $PSL(2, 5) \simeq A_5$ we have $PSL(2, 5^f) \geq PSL(2, 5) \simeq A_5$, so we may assume $p^{2f} - 1 \equiv 0 \pmod 5$. We distinguish two cases:

(*a*) For $p^f - 1 \equiv 0 \pmod 5$. Let a be an element of order 5 of the multiplicative group $F_{p^f} - \{1\}$ (such an element exists, since $5 | p^f - 1$), pick $b \in F_{p^f}$ such that $b(a - a^4) = 1$ and choose $c, d \in F_{p^f}$ such that $b^2 + cd = -1$. Define $A, B \in SL(2, p^f)$ by

$$A = \begin{bmatrix} a & 0 \\ 0 & a^{-1} \end{bmatrix}, \quad B = \begin{bmatrix} b & c \\ d & -b \end{bmatrix}$$

and denote by $\bar{A}, \bar{B}$ the images of A, B respectively in $PSL(2, p^f) = SL(2, p^f)/SL(2, p^f) \cap Z(2, p^f)$. It is easy to verify that $\bar{A}$ and $\bar{B}$ satisfy:

$$\bar{A} \neq 1, \bar{B} \neq 1, \bar{A}^5 = \bar{B}^2 = (\bar{A}\bar{B})^3 = 1.$$

Therefore there exists a homomorphism from A_5 into $PSL(2, p^f)$, and this homomorphism is injective since A_5 is simple.

(*b*) For $p^f + 1 \equiv 0 \pmod 5$.

In order to consider this, we must first introduce a lemma.

Lemma 4.7.8 *Putting $\bar{a} = a^{p^f}$ for $a \in F_{p^{2f}}$, we have that the subset G of $GL(2, p^{2f})$, defined by:*

$$G = \left\{ \begin{bmatrix} a & b \\ -\bar{b} & \bar{a} \end{bmatrix} \middle| a, b \in F_{p^{2f}}, a\bar{a} + b\bar{b} = 1 \right\}$$

is a subgroup of $GL(2, p^{2f})$ and $G \simeq SL(2, p^f)$.

Proof It is easy to check that G is a subgroup of $GL(2, p^{2f})$. There are exactly $1 + p^f$ elements a in $F_{p^{2f}}$ such that $a\bar{a} = a^{1+p^f} = 1$. For such an a, there is exactly one b such that $a\bar{a} + b\bar{b} = 1$, namely $b = 0$. The number of elements a of $F_{p^{2f}}$ such that $a\bar{a} \neq 1$ equals $p^{2f} - p^f - 1$ and for such an a there are $p^f + 1$ different elements b such that $a\bar{a} + b\bar{b} = 1$. (For, since $1 - a\bar{a} \in F_{p^f} - \{0\}$ and since $F_{p^{2f}} - \{0\}$ is a cyclic group, the equation $x^{p^f+1} = 1 - a\bar{a}$ has at least one root in $F_{p^{2f}}$. Since $x^{p^f+1} = 1$ has $p^f + 1$ roots in $F_{p^{2f}}$, we conclude that $x^{p^f+1} = 1 - a\bar{a}$ also has $p^f + 1$ roots in $F_{p^{2f}}$.) Therefore

$$|G| = (p^f + 1) + (p^{2f} - p^f - 1)(p^f + 1) = (p^f + 1)p^f(p^f - 1)$$
$$= |SL(2, p^f)|.$$

In order to prove that $G \simeq SL(2, p^f)$ it suffices to prove that there exists an isomorphism from G into $SL(2, p^f)$. If $p \neq 2$, we choose $a_0 \in F_{p^{2f}}$ such that $a_0 \neq \bar{a}_0$, $a_0\bar{a}_0 = -1$ and we put

$$\alpha = \begin{bmatrix} 1 & a_0 \\ \bar{a}_0 & -1 \end{bmatrix}.$$

If $p = 2$, we choose $a_0 \in F_{p^{2f}}$ such that $a_0 \neq 1$ and $a_0\bar{a}_0 = 1$ and we put:

$$\alpha = \begin{bmatrix} a_0 & 1 \\ 1 & \bar{a}_0 \end{bmatrix}.$$

In both cases we can easily check up that $\alpha G \alpha^{-1} \leq SL(2, p^f)$. (End of proof of lemma 4.7.8.) ∎

We may now finish the proof of theorem 4.7.7. From lemma 4.7.8 it suffices to prove that $G/Z(G) = \bar{G}$ contains a subgroup isomorphic with A_5; here G denotes the group defined in lemma 4.7.8. Let $a \in F_{p^{2f}} - \{0\}$ be of order 5, put $b = (a - \bar{a})^{-1}$ and choose $c \in F_{p^{2f}}$ such that $c\bar{c} = 1 - b\bar{b}$. (Since $5 \nmid p^f - 1$ such an element b exists, and since $1 - b\bar{b} \in F_{p^f}$ such an element c clearly exists.) Defining $\alpha, \beta \in G$ by:

$$\alpha = \begin{bmatrix} a & 0 \\ 0 & \bar{a} \end{bmatrix}, \quad \beta = \begin{bmatrix} b & c \\ -\bar{c} & \bar{b} \end{bmatrix},$$

we can easily verify that the elements $\bar{\alpha}, \bar{\beta} \in \bar{G}$, that correspond with α and β, satisfy:

$$\bar{\alpha}^5 = \bar{\beta}^2 = (\bar{\alpha}\bar{\beta})^3 = 1.$$

(Note that $a\bar{a} = a^{p^f+1} = 1$ since $p^f + 1 \equiv 0 \pmod 5$)). Therefore $\bar{G}$ contains a subgroup isomorphic to A_5. (End of proof of theorem 4.7.7.) ∎

Theorem 4.7.9 $PSL(2,9) \simeq A_6$.

Proof By theorem 4.7.7 $PSL(2,9)$ contains a subgroup H isomorphic to A_5. Since $|PSL(2,9):H| = 6$, $PSL(2,9)$ has a permutation representation of degree 6. Since $PSL(2,9)$ is simple, $PSL(2,9)$ is isomorphic to a subgroup of S_6. Since $|PSL(2,9)| = |A_6|$, we have $PSL(2,9) \simeq A_6$. ■

Theorem 4.7.10 (1) *Both S_6 and $PGL(2,9)$ contain subgroups of index 2 which are isomorphic to the simple group A_6, but S_6 and $PGL(2,9)$ are not isomorphic.*

(2) *$PSL(4,2)$ and $PSL(3,4)$ are simple groups of the same order, but they are not isomorphic.*

Proof (1) The first part of the assertion is obvious from theorem 4.7.9. If a is a generating element of $F_9 - \{0\}$, then the image of $\begin{bmatrix} 1 & 0 \\ 0 & a \end{bmatrix}$ in $PGL(2,9)$ has order 8, while S_6 has no elements of order 8, hence $PGL(2,9) \not\simeq S_6$.

(2) $|PSL(4,2)| = |PSL(3,4)|$ follows from lemma 4.5.2.

$$P = \left\{ \begin{bmatrix} 1 & 0 & 0 & 0 \\ a_1 & 1 & 0 & 0 \\ a_2 & a_3 & 1 & 0 \\ a_4 & a_5 & a_6 & 1 \end{bmatrix} \middle| \, a_i \in F_2 \right\}$$

is a Sylow 2-subgroup of $PSL(4,2)$ and the centre of P has order 2. (For, the element of P for which $a_4 = 1$, $a_i = 0$ $(i \neq 4)$ is a generating element of the centre of P.) On the other hand

$$Q = \left\{ \begin{bmatrix} 1 & 0 & 0 \\ a_1 & 1 & 0 \\ a_2 & a_3 & 1 \end{bmatrix} \middle| \, a_i \in F_4 \right\}$$

is a Sylow 2-subgroup of $PSL(3,4)$. The centre of Q has order 4. (For, the elements of Q with $a_1 = a_3 = 0$ and a_2 arbitrary make up the centre of Q.) Therefore $PSL(4,2) \not\simeq PSL(3,4)$. ■

According to theorem 4.7.7, $PSL(2,11)$ contains a subgroup H

that is isomorphic to A_5, since $|PSL(2,11):H| = 11$ from lemma 4.5.2, the permutation representation $(PSL(2,11), PSL(2,11)/H)$ is of degree 11. We want to determine the number of subgroups of $PSL(2,11)$ that are isomorphic to A_5. The image of $\alpha \in SL(2,11)$ in $PSL(2,11)$ will be denoted by $\bar{\alpha}$. We first need the following lemma, the proof of which can be done from easy matrix calculations and is left to the reader:

Lemma 4.7.11 *For* $\alpha = \begin{bmatrix} a & b \\ c & d \end{bmatrix} \in SL(2,11)$ *we have:*

(1) *If* $\alpha^2 = 1$ *then* $\alpha = \pm \begin{bmatrix} 1 & 0 \\ 0 & 1 \end{bmatrix}$.

(2) $\alpha^4 = 1, \alpha^2 \neq 1$ *if and only if* $\alpha = \begin{bmatrix} a & b \\ c & -a \end{bmatrix}$ *with* $a^2 + bc = -1$ *and* $bc \neq 0$.

Let a_0 denote an element of $F_{11} - \{0\}$ of order 5, then $\bar{\alpha}_0 = \overline{\begin{bmatrix} a_0 & 0 \\ 0 & a_0^{-1} \end{bmatrix}}$ is an element of $PSL(2,11)$ of order 5. Further, elements of order 2 of $PSL(2,11)$ can be written as: $\bar{\beta}_0 = \overline{\begin{bmatrix} a & b \\ c & -a \end{bmatrix}}$ (with $a^2 + bc = -1$ and $bc \neq 0$) by lemma 4.7.11. For those elements $\bar{\alpha}_0$ and $\bar{\beta}_0$ we have the following lemma:

Lemma 4.7.12 $(\bar{\alpha}_0 \bar{\beta}_0)^3 = 1 \Leftrightarrow a(a_0^{-1} - a_0) = \pm 1$.

Proof Putting $(\alpha_0 \beta_0)^3 = \begin{bmatrix} u_1 & u_2 \\ u_3 & u_4 \end{bmatrix}$ we have:

$$u_1 = a_0 a(a_0^2 a^2 + bc) + a_0 abc(1 - a_0^{-2})$$
$$u_2 = a_0 b\{a^2(a_0^2 - 1) + a_0^{-2} a^2 + bc\}$$
$$u_3 = a_0^{-1} c\{a^2(a_0^{-2} - 1) + a_0^2 a^2 + bc\}$$
$$u_4 = a_0^{-1} abc(a_0^2 - 1) - a_0^{-1} a(a_0^{-2} a^2 + bc).$$

Hence:

$$u_2 = 0 \Leftrightarrow a^2(a_0^2 - 1) + a_0^{-2} a^2 - a^2 = 1 \Leftrightarrow a^2(a_0 - a_0^{-1})^2 = 1 \Leftrightarrow u_3 = 0$$

and if $a^2(a_0 - a_0^{-1})^2 = 1$, then: $u_1 = u_4 = a(a_0^{-1} - a_0)$, from which the lemma follows. ■

$P = \langle \bar{\alpha}_0 \rangle$ is a Sylow 5-subgroup of $PSL(2, 11)$ and we have for $\bar{\beta} \in PSL(2, 11)$:

$$P^{\bar{\beta}} = P \Leftrightarrow \beta = \begin{bmatrix} x & 0 \\ 0 & x^{-1} \end{bmatrix} \text{ or } \begin{bmatrix} 0 & y \\ -y^{-1} & 0 \end{bmatrix}.$$

Hence $|\mathcal{N}_{PSL(2,11)}(P):P| = 2$. Therefore $PSL(2, 11)$ contains 66 Sylow 5-subgroups, hence 66×4 elements of order 5. According to lemma 4.7.12:

$$|\{\bar{\beta} \in PSL(2, 11) \mid \bar{\beta}^2 = (\bar{\alpha}_0\bar{\beta})^3 = 1\}| = 10.$$

Therefore:

$$|\{(\bar{\alpha}, \bar{\beta}) \mid \bar{\alpha}, \bar{\beta} \in PSL(2, 11), \bar{\alpha}^5 = \bar{\beta}^2 = (\bar{\alpha}\bar{\beta})^3 = 1\}| = 10 \times 66 \times 4.$$

On the other hand A_5 contains 6 Sylow 5-subgroups, hence 6×4 elements of order 5. For a fixed element α of order 5, we have

$$|\{\beta \in A_5 \mid \beta^2 = (\alpha\beta)^3 = 1\}| = 5.$$

(For, assuming $\alpha = (1, 2, 3, 4, 5)$, there is exactly one β leaving i fixed and satisfying the conditions for $i = 1, \ldots, 5$). Hence:

$$|\{(\alpha, \beta) \mid \alpha, \beta \in A_5, \alpha^5 = \beta^2 = (\alpha\beta)^3 = 1\}| = 6 \times 4 \times 5.$$

Therefore $PSL(2, 11)$ contains 2×11 subgroups that are isomorphic to A_5. If H is a subgroup of $PSL(2, 11)$ that is isomorphic to A_5, then since H is a maximal subgroup of $PSL(2, 11)$ and since $PSL(2, 11)$ is simple, we have $\mathcal{N}_{PSL(2,11)}(H) = H$. Therefore the set of subgroups of $PSL(2, 11)$ that are isomorphic to A_5, is divided into two $PSL(2, F_{11})$-conjugate classes. Since $(PSL(2, 11), PSL(2, 11)/H)$ is a permutation group of prime degree and since $PSL(2, 11)$ is non-Abelian simple, this permutation group is 2-transitive by theorems 3.6.1 and 3.6.5. According to theorem 6.1.1 described later we can construct a symmetric design D (with 11 points) such that: $PSL(2, 11) \leq \operatorname{Aut} D$. The parameters of D are either $(11, 6, 3)$ or $(11, 5, 2)$ by theorem 1.5.1. Therefore D is a Hadamard design. Summarizing the above results:

Theorem 4.7.13 (1) *$PSL(2, 11)$ can be regarded as a 2-transitive permutation group of degree 11. The subgroups leaving one point fixed are isomorphic to A_5.*

(2) *$PSL(2, 11)$ has 22 subgroups that are isomorphic to A_5 and these subgroups are divided into two $PSL(2, 11)$-conjugacy classes.*

(3) *It is possible to construct a Hadamard design D, such that $PSL(2, 11) \leq \mathrm{Aut}\, D$.*

We showed in theorem 4.5.3 that $SL(2, q)$ always has a 2-transitive permutation representation of degree $q+1$. We also showed that for several q, $SL(2, q)$ has a 2-transitive permutation representation of degree m less than $q+1$. We sum up our results:

(i) $SL(2, 2)$ $m = 2$ (theorem 4.1.4, 4.7.1)
(ii) $SL(2, 3)$ $m = 3$ (theorem 4.1.4, 4.7.1)
(iii) $SL(2, 5)$ $m = 5$ (theorem 4.7.1)
(iv) $SL(2, 7)$ $m = 7$ (theorem 4.7.4)
(v) $SL(2, 9)$ $m = 6$ (theorem 4.7.9)
(vi) $SL(2, 11)$ $m = 11$ (theorem 4.7.12).

It has been known for a long time that the above are the only cases in which $SL(2, q)$ has a permutation representation of degree less than $q+1$ (a theorem of Galois).

5

Finite projective geometry

5.1 Projective planes and affine planes

Projective planes and affine planes furnish the simplest examples of geometric structures. Nevertheless, contrary to the superficial simpleness of their structure, they have complicated properties, and they are presenting interesting problems. Among projective and affine planes, there are certain classes that are most fundamental. They are the projective and affine planes defined over a finite field which are defined in §1.5. The study of the characteristic properties of these special planes is an important part of our subject. The purpose of this chapter is to introduce the work of Ostrom and Wagner (theorems 5.1.32 and 5.1.33 below) who have achieved remarkable results in this direction. The results of this chapter will be a fundamental tool to be used in chapter 6.

In this section, $(v, k, 1)$ will always denote the parameters of a projective plane, unless otherwise stated.

5.1.1 Duality

We first want to reword the definition of a projective plane to gain a better understanding of its properties. Let $(\Omega, \mathfrak{B})$ be a geometric structure and consider the following conditions (the elements of Ω will be called points, those of $\mathfrak{B}$ will be called lines):

(i) If $a_1, a_2 \in \Omega (a_1 \neq a_2)$ then there exists exactly one line containing a_1 and a_2 (this line is denoted by $a_1 a_2$).

(ii) If $l_1, l_2 \in \mathfrak{B} \ (l_1 \neq l_2)$ then $|l_1 \cap l_2| = 1$.

(iii) There exist 4 points, no 3 of which are on one line.

(iii)′ There exist 4 lines, no 3 of which intersect in one point.

Theorem 5.1.1 *For a geometric structure* $(\Omega, \mathfrak{B})$, *the following are equivalent:*

(1) $(\Omega, \mathfrak{B})$ *is a projective plane.*
(2) $(\Omega, \mathfrak{B})$ *satisfies* (i), (ii) *and* (iii).
(3) $(\Omega, \mathfrak{B})$ *satisfies* (i), (ii) *and* (iii)$'$.

Proof (1) → (3): $(\Omega, \mathfrak{B})$ is by definition a block design with parameters $v = b$, $k = r$ and $\lambda = 1$ and $v - 1 > k > 2$. (i) follows from $\lambda = 1$ and (ii) was proved in theorem 1.5.3. Since $k = r > 2$, we can pick two different points a_1 and a_2 on a given line l, and pick two lines l_1 and l_2 through a_1 that are different from l, and two lines l_3 and l_4 through a_2 that are different from l. The four lines l_1, l_2, l_3 and l_4 satisfy (iii)$'$ (see Fig. 1).

(3) → (2): Let l_1, l_2, l_3 and l_4 be four lines satisfying (iii)$'$. Then $l_1 \cap l_2, l_2 \cap l_3, l_3 \cap l_4$ and $l_4 \cap l_1$ satisfy (iii) (see Fig. 2).

(2) → (1): We first show

(*a*) Each line contains the same number of points and this number is greater than or equal to 3. Proof: Let a_1, a_2, a_3, a_4 be 4 points satisfying (iii). Since $a_3, a_4 \notin \overline{a_1 a_2}$, the point $\overline{a_1 a_2} \cap \overline{a_3 a_4} = a_5$ is different from a_1, a_2, a_3 and a_4. Similarly the point $\overline{a_1 a_3} \cap \overline{a_2 a_4} = a_6$ is different from $a_1, \ldots, a_5$ (see Fig. 3). It is clear from the way $a_1, \ldots, a_6$ were chosen that there do not exist two lines containing these 6 points. Hence for every two lines l_1 and l_2 there is a point a that is neither on l_1 nor on l_2. For $x \in l_1$ the point $y = l_2 \cap \overline{ax}$ is a uniquely determined point of l_2. This mapping '$x \to y$' sets up a one-to-one correspondence between the set of points on l_1 and the set of points on l_2, hence the number of points on each line is the same and since $a_5 \in \overline{a_1 a_2}$ this number is not less than 3. In the same way it is proved that:

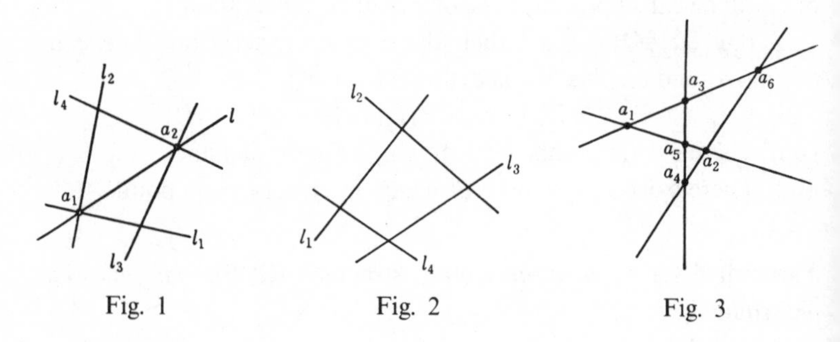

Fig. 1 Fig. 2 Fig. 3

(a') Each point is contained in the same number of lines and this number is greater than or equal to 3.

Next we show

(b) The number of points on a line is equal to the number of lines through a point. Proof: let l be a line and let a be a point not on l. Lines through a intersect l in exactly one point and for each point on l there is exactly one line through a and that point, proving (b).

Now, $(\Omega, \mathfrak{B})$ is a block-design with $\lambda = 1$ because of (i) and (a). Let $(v, b, k, r, 1)$ be the parameter of $(\Omega, \mathfrak{B})$, then $k = r$ because of (b), hence $v = b$ by theorem 1.5.3, i.e. $(\Omega, \mathfrak{B})$ is symmetric. Since $k \geq 3$, this design is not trivial, hence $(\Omega, \mathfrak{B})$ is a projective plane. ∎

Let $\mathbf{P} = (\Omega, \mathfrak{B})$ be a projective plane. For $a \in \Omega$ the subset $\mathfrak{B}_a$ of $\mathfrak{B}$ is defined by

$$\mathfrak{B}_a = \{l \in \mathfrak{B} \mid a \in l\}.$$

If $a_1, a_2 \in \Omega (a_1 \neq a_2)$ then $\mathfrak{B}_{a_1} \neq \mathfrak{B}_{a_2}$ since $k = r > 1$. Now we define a geometric structure $\hat{\mathbf{P}}$ from $\mathbf{P}$ as follows: The set of points of $\hat{\mathbf{P}}$ will be $\mathfrak{B}$ and the set of lines will be the set $\{\mathfrak{B}_a \mid a \in \Omega\}$. Identifying $\mathfrak{B}_a$ with a, this last set can be identified with Ω, i.e. $\hat{\mathbf{P}} = (\mathfrak{B}, \Omega)$. Since $\mathbf{P}$ satisfies conditions (i), (ii) and (iii), $\hat{\mathbf{P}}$ satisfies (i), (ii) and (iii)$'$, i.e. $\hat{\mathbf{P}}$ is also a projective plane by theorem 5.1.1. (This follows also from our original definition of a projective plane and theorem 1.5.3.)

Now, let P denote any statement (or condition) concerning projective planes. By replacing the word 'point' by 'line' and the word 'line' by 'point', whenever they occur in P and by exchanging expressions like 'lies on' and 'goes through', 'connects' and 'intersects', we get a new statement (or condition) $\hat{P}$ concerning projective planes, called the dual statement (or condition) of P. If $P = \hat{P}$ then P is called self-dual. Let Q denote some condition regarding projective planes and let P be a proposition that is true for all projective planes $\mathbf{P} = (\Omega, \mathfrak{B})$ satisfying Q. Then since $\hat{\mathbf{P}}$ are also projective planes, $\hat{P}$ is true for all projective planes satisfying $\hat{Q}$. In particular, if P is true for all projective planes, so is $\hat{P}$. This is called the principle of duality for projective planes.

5.1.2 Coordinate representations for projective and affine planes defined over a field.

Let $\mathbf{P}^{(1)} = (\Omega^{(1)}, \mathfrak{B}^{(1)})$ and $\mathbf{P}^{(2)} = (\Omega^{(2)}, \mathfrak{B}^{(2)})$ be two projective planes. If there exists a one to one mapping f from $\Omega^{(1)}$ to $\Omega^{(2)}$ which induces one-to-one mapping from $\mathfrak{B}^{(1)}$ onto $\mathfrak{B}^{(2)}$, then $\mathbf{P}^{(1)}$ and $\mathbf{P}^{(2)}$ are called isomorphic. Notation: $\mathbf{P}^{(1)} \simeq \mathbf{P}^{(2)}$.

In the consideration of a projective plane $(\Omega, \mathfrak{B})$ we often single out one line from $\mathfrak{B}$ and denote this line by l_∞. The line l_∞ is called the line at infinity, all other lines are called finite lines, the points on l_∞ are called the points at infinity and all points not on l_∞ are called finite points (with regard to l_∞). For a finite line l, we put:

$$\hat{l} = l - (l \cap l_\infty).$$

A pair $(a, l) \in \Omega \times \mathfrak{B}$ such that $a \in l$ is called a flag. If a and l are both finite, a flag (a, l) is called a finite flag. If two finite lines l_1 and l_2 satisfy $l_1 \cap l_2 \in l_\infty$, then l_1 and l_2 are called parallel (with regard to l_∞).

We have already seen, in §1.5, that if we put

$$\tilde{\Omega} = \Omega - l_\infty$$
$$\tilde{\mathfrak{B}} = \{\tilde{l} \mid \tilde{l} = l - (l \cap l_\infty), l \in \mathfrak{B} \text{ and } l \neq l_\infty\},$$

then $\mathbf{A} = (\tilde{\Omega}, \tilde{\mathfrak{B}})$ is an affine plane and that every affine plane can be obtained in this way. The element $\tilde{l} = l - (l \cap l_\infty)$ of $\tilde{\mathfrak{B}}$ is called the line of $\mathbf{A}$ corresponding with the finite line l of $\mathbf{P}$. The line l_∞ is called the line at infinity for $\mathbf{A}$ (note that l_∞ is not a line of $\mathbf{A}$). If $\mathbf{A}^{(1)} = (\tilde{\Omega}^{(1)}, \tilde{\mathfrak{B}}^{(1)})$ and $\mathbf{A}^{(2)} = (\tilde{\Omega}^{(2)}, \tilde{\mathfrak{B}}^{(2)})$ are two affine planes, such that there exists a one-to-one map from $\tilde{\Omega}^{(1)}$ onto $\tilde{\Omega}^{(2)}$ that induces a one-to-one map from $\tilde{\mathfrak{B}}^{(1)}$ onto $\tilde{\mathfrak{B}}^{(2)}$, then $\mathbf{A}^{(1)}$ and $\mathbf{A}^{(2)}$ are called isomorphic. Notation: $\mathbf{A}^{(1)} \simeq \mathbf{A}^{(2)}$.

Theorem 5.1.2 *Let $\mathbf{P}^{(1)}$ and $\mathbf{P}^{(2)}$ be projective planes and let $\mathbf{A}^{(1)}$ and $\mathbf{A}^{(2)}$ be affine planes obtained from $\mathbf{P}^{(1)}$ and $\mathbf{P}^{(2)}$ respectively. If $\mathbf{A}^{(1)} \simeq \mathbf{A}^{(2)}$, then $\mathbf{P}^{(1)} \simeq \mathbf{P}^{(2)}$.*

Proof Let $\mathbf{P}^{(i)} = (\Omega^{(i)}, \mathfrak{B}^{(i)})$ and let $l_\infty^{(i)}$ be the line at infinity for $\mathbf{A}^{(i)} (i = 1, 2)$. Let $\tilde{f}: \tilde{\Omega}^{(1)} \to \tilde{\Omega}^{(2)}$ be a one-to-one mapping that induces a one-to-one mapping between $\mathfrak{B}^{(1)}$ and $\mathfrak{B}^{(2)}$. We define $f:\Omega^{(1)}\to\Omega^{(2)}$ as follows. We let the restriction of f to $\tilde{\Omega}^{(1)}$ coincide with $\tilde{f}$. For $a^{(1)}\in l_\infty^{(1)}$, let $l^{(1)}$ be a finite line through $a^{(1)}$ of $\mathbf{P}^{(1)}$ and let $l^{(2)}$ be the finite line of $\mathbf{P}^{(2)}$ corresponding with $\tilde{l}^{(2)}=\tilde{f}(\tilde{l}^{(1)})$. Now the point $a^{(2)}=l_\infty^{(2)}\cap l^{(2)}\in l_\infty^{(2)}$ is unambiguously defined by $a^{(1)}$. (For, if $l^{(1)\prime}$ is another line satisfying $l_\infty^{(1)}\cap l^{(1)\prime}=a^{(1)}$, then $\tilde{f}(\tilde{l}^{(1)})\cap\tilde{f}(\tilde{l}^{(1)\prime})=\varnothing$. Hence $l^{(2)}\cap l^{(2)\prime}\in l_\infty^{(2)}$, where $l^{(2)\prime}$ denotes the finite line of $\mathbf{P}^2$ that corresponds with $f(\tilde{l}^{(1)\prime})$.) Hence we can define $f(a^{(1)})=a^{(2)}$. Then it follows easily that f has the desired properties. ■

Let $\mathbf{P}=\mathbf{P}_1(2,K)$ be the projective plane defined over a finite field K as defined in §1.5.3. We will introduce coordinates in $\mathbf{P}$ to represent its points and lines. Put $\mathbf{V}=\mathbf{V}(3,K)$. Choosing a basis for $\mathbf{V}$, elements of $\mathbf{V}$ can be represented as elements (row-vectors) of ${}^3K$. Since points of $\mathbf{P}$ are one-dimensional subspaces of $\mathbf{V}$, they can be represented as $\{(\lambda\lambda_1,\lambda\lambda_2,\lambda\lambda_3)\,|\,\lambda\in K\}$ where $\lambda_1,\lambda_2,\lambda_3$ are elements of K, not all of them equal to 0. Putting

$$\{(\lambda\lambda_1,\lambda\lambda_2,\lambda\lambda_3)\,|\,\lambda\in K\}=\langle\lambda_1,\lambda_2,\lambda_3\rangle,$$

each point of $\mathbf{P}$ can be represented in a unique way as:

$$\langle 0,1,0\rangle \text{ or } \langle 1,\lambda,0\rangle \text{ or, } \langle\lambda,\mu,1\rangle \quad \lambda,\mu\in K.$$

We write:

$$\langle 0,1,0\rangle=\infty \quad \langle 1,\lambda,0\rangle=(\lambda) \quad \langle\lambda,\mu,1\rangle=(\lambda,\mu).$$

We call λ and μ the x-coordinate and y-coordinate of the point (λ,μ) respectively.

For $\alpha=(\lambda_1,\lambda_2,\lambda_3)$, $\beta=(\mu_1,\mu_2,\mu_3)\in\mathbf{V}$, the inner product (α,β) is defined by

$$(\alpha,\beta)=\sum_{i=1}^{3}\lambda_i\mu_i.$$

This inner product is symmetric and non-singular. Since there exists a one-to-one correspondence between the lines of $\mathbf{P}$ and the two dimensional subspaces of $\mathbf{V}$, using this inner product each line of **P** can be represented in a unique way as:

$$\langle \lambda, 1, \mu \rangle^{\perp} \text{ or } \langle 1, 0, \mu \rangle^{\perp} \text{ or } \langle 0, 0, 1 \rangle^{\perp}$$

where

$$\langle \lambda, \mu, \nu \rangle^{\perp} = \{(\lambda', \mu', \nu') \in V \mid \lambda\lambda' + \mu\mu' + \nu\nu'\} = 0.$$

This line, corresponding with $\langle 0, 0, 1 \rangle^{\perp}$, is represented as l_{∞}; that corresponding with $\langle 1, 0, \mu \rangle^{\perp}$ as $l_{x+\mu=0}$ (or $x + \mu = 0$); and that corresponding with $\langle \lambda, 1, \mu \rangle^{\perp}$ as $l_{y+\lambda x+\mu=0}$ (or $y + \lambda x + \mu = 0$). It is easy to verify that:

$$\begin{aligned} l_{\infty} &= \{(\lambda) \mid \lambda \in K\} \cup \{(\infty)\} \\ l_{x+\mu=0} &= \{(-\mu, \lambda) \mid \lambda \in K\} \cup \{(\infty)\} \\ l_{y+\lambda x+\mu=0} &= \{(\xi, \eta) \mid \xi, \eta \in K, \eta + \lambda\xi + \mu = 0\} \cup (-\lambda). \end{aligned}$$

Conversely, let K be a field, and let $[\infty]$ be a symbol that represents neither an element from K nor an element from $K \times K$. Put $\Omega = K \times K \cup K \cup \{[\infty]\}$, represent elements of $K \times K$ as $[\lambda, \mu]$ and elements of K as $[\lambda]$. Let $\mathfrak{B}$ be the collection of the following subsets of Ω:

$$\left.\begin{aligned} l_{\infty} &= \{[\xi] \mid \xi \in K\} \cup \{[\infty]\} \\ l_{\lambda} &= \{[-\lambda, \eta] \mid \eta \in K\} \cup \{[\infty]\} \\ l_{\lambda,\mu} &= \{[\xi, \eta] \mid \xi, \eta \in K, \eta + \lambda\xi + \mu = 0\} \cup [-\lambda] \end{aligned}\right\} \quad (5.1.1)$$

where $\lambda, \mu \in K$. It is easy to verify that $\tilde{\mathbf{P}} = (\Omega, \mathfrak{B})$ is a projective plane that is isomorphic to $\mathbf{P}_1(2, K)$.

For the affine plane $\mathbf{A}_1(2, K) = (\tilde{\Omega}, \tilde{\mathfrak{B}})$ we have, using the coordinates introduced into $\mathbf{P}_1(2, K)$:

$$\left.\begin{aligned} \tilde{\Omega} &= \{(\xi, \eta) \mid \xi, \eta \in K\} \\ \tilde{\mathfrak{B}} &= \{\tilde{l}_{(\lambda,\mu)} \mid \lambda, \mu \in K\} \cup \{\tilde{l}_{\lambda} \mid \lambda \in K\} \end{aligned}\right\} \quad (5.1.2)$$

where:

$$\begin{aligned} \tilde{l}_{(\lambda,\mu)} &= \{(\xi, \eta) \mid \xi, \eta \in K, \eta + \lambda\xi + \mu = 0\} \\ \tilde{l}_{\lambda} &= \{(-\lambda, \eta) \mid \eta \in K\}. \end{aligned}$$

Thus we have proved the following theorem:

Theorem 5.1.3 *Let K be a finite field. The geometric structure* $(\Omega, \mathfrak{B})$ *defined by* (5.1.1) *is a projective plane over K and the geometric structure* $(\tilde{\Omega}, \tilde{\mathfrak{B}})$ *defined by* (5.1.2) *is an affine plane over K. Every projective or affine plane defined over K can be obtained in this way.*

5.1.3 Automorphisms

The group of automorphisms of a projective plane $\mathbf{P} = (\Omega, \mathfrak{B})$ is denoted by Aut **P**, i.e.

$$\text{Aut } \mathbf{P} = \{\sigma \in S^{\Omega} \mid l \in \mathfrak{B} \Rightarrow l^{\sigma} \in \mathfrak{B}\}.$$

Let $\mathbf{A} = (\tilde{\Omega}, \tilde{\mathfrak{B}})$ be an affine plane obtained from **P** and let l_{∞} be the line at infinity for **A**. The collection:

$$G = \{\sigma \in \text{Aut } \mathbf{P} \mid l_{\infty}^{\sigma} = l_{\infty}\}$$

is a subgroup of Aut **P**, an element of G induces an automorphism of **A**, different elements of G induce different automorphisms of **A** and every automorphism of **A** can be obtained in this way as we discussed in §1.5. Hence, Aut **A** can be identified with G, whereby Aut **A** becomes a subgroup of Aut **P**.

Let $\sigma \in \text{Aut } \mathbf{P}$. If there exist a point a and a line l such that:

(1) if $x \in l$, then $x^{\sigma} = x$

(2) if l' contains a, then $l'^{\sigma} = l'$,

then σ is called an (a, l)-perspectivity or simply a perspectivity. The point a is called the centre of the perspectivity, the line l its axis. The following facts follow immediately from the definition:

If σ_1 and σ_2 are (a, l)-perspectivities so are σ_1^{-1} and $\sigma_1 \sigma_2$.

If σ is an (a, l)-perspectivity and if $\tau \in \text{Aut } \mathbf{P}$, then the conjugate element σ^{τ} is an (a^{τ}, l^{τ})-perspectivity.

Let σ be an (a, l)-perspectivity. If $a \in l$, then σ is called an (a, l)-elation or sometimes a-elation, l-elation or simply an elation. If $a \in l$, then σ is called an (a, l)-homology, sometimes a-homology, l-homology or simply a homology. The subgroup of Aut **P** generated by the collection of all elations is denoted by $PSL(\mathbf{P})$ and called the little automorphism group (or the little projective group) of **P**. We have $PSL(\mathbf{P}) \trianglelefteq \text{Aut } \mathbf{P}$ by the remarks above.

The conditions (1) and (2) occurring in the definition of a perspectivity are equivalent, as our next theorem shows.

Theorem 5.1.4 *Let* $\mathbf{P} = (\Omega, \mathfrak{B})$ *be a projective plane, let* $\sigma(\neq 1) \in$ Aut $\mathbf{P}$. *Then there exists a point* $a_0 \in \Omega$ *such that* $l^\sigma = l$ *holds for all lines* l *passing through* a_0 *if and only if there exists a line* $l_0 \in \mathfrak{B}$ *such that* $x^\sigma = x$ *holds for all points* x *on* l_0.

Proof First we assume that there exists a point $a_0 \in \Omega$ such that $l^\sigma = l$ for all lines through a_0. If there exists a line l_0 not passing through a_0 such that $l_0^\sigma = l_0$, then every point $x \in l_0$ can be represented as $x = \overline{a_0 x} \cap l_0$, hence $x^\sigma = \overline{a_0 x}^\sigma \cap l_0^\sigma = \overline{a_0 x} \cap l_0 = x$ and the theorem is proved. So we may assume that σ leaves no line, that does not pass through a_0, fixed. Let l denote a line that does not pass through a_0 and put $l^\sigma \cap l = a_1$. We will prove that all points on $l_0 = a_0 a_1$ are left fixed by σ. From $a_1 = l \cap \overline{a_0 a_1}$, we get $a_1^\sigma = l^\sigma \cap \overline{a_0 a_1}^\sigma = l^\sigma \cap \overline{a_0 a_1} = a_1$. Now, let x be an arbitrary point on $l_0 (x \neq a_0)$. Choose l' such that l' passes through x but not through a_0, then: $l'^\sigma \cap l' = x$. (For, suppose: $l'^\sigma \cap l' = c$, where $c \neq x$, then it is proved just as for a_1 that $c^\sigma = c$, hence $\overline{a_1 c}^\sigma = \overline{a_1 c}$ and $\overline{a_1 c}$ does not pass through a_0, contrary to our assumption.) Again it is proved just as for a_1, that $x^\sigma = x$. The converse is clear by duality (see Fig. 4). ■

Theorem 5.1.5 *Let* σ *be a perspectivity. If* $\sigma \neq 1$, *then the centre and axis of* σ *are uniquely determined.*

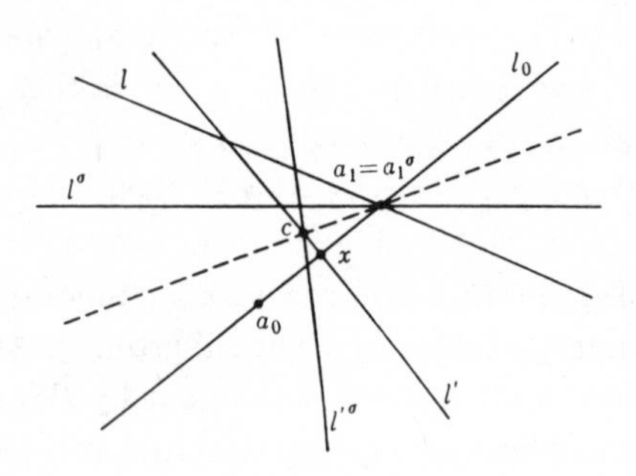

Fig. 4

Proof Suppose that σ has two axes l_1 and l_2. Let l be a line not passing through $l_1 \cap l_2$. Putting $a_1 = l \cap l_1$ and $a_2 = l \cap l_2$, we have $l = \overline{a_1 a_2}$. Therefore $l^\sigma = \overline{a_1 a_2}^\sigma = \overline{a_1 a_2} = l$. For $x \notin l_1 \cap l_2$ there are two lines that pass through x but not through $l_1 \cap l_2$, hence $x^\sigma = x$, i.e. $\sigma = 1$. The proof for the centre follows from the duality principle. ∎

Corollary 5.1.6 *Let σ be a perspectivity.*

(i) *If there is a point a not on the axis of σ that is left fixed by σ, then a is the centre of σ.*

(ii) *If there is a line l not passing through the centre that is left fixed by σ, then l is the axis of σ.*

Corollary 5.1.7 *Let* **P** *be a projective plane of order $n(=k-1)$ and let σ be a perspectivity of* **P** *that has order m (as an element of* Aut **P**). *Then:*

(i) *σ is an elation $\Leftrightarrow m \mid n$*

(ii) *σ is a homology $\Leftrightarrow m \mid n-1$.*

Proof Let a denote the centre and l the axis of σ. If l' is a line that passes through a and is different from l, then σ operates half-regularly on $l' - \{a, l \cap l'\}$ according to corollary 5.1.6, from which the assertion follows. ∎

Theorem 5.1.8 *If σ_1, σ_2 are l-elations, then so is $\sigma_1\sigma_2$. If σ_1 and σ_2 have different centres, then the centre of $\sigma_1\sigma_2$ does not coincide with either the centre of σ_1 or that of σ_2.*

Proof If either σ_1 or σ_2 is the identity, there is nothing to prove, so we may assume $\sigma_1 \neq 1$ and $\sigma_2 \neq 1$. The product $\sigma_1\sigma_2$ leaves all points of l fixed, hence $\sigma_1\sigma_2$ is a perspectivity. If σ_1 and σ_2 have the same centre, then this point is also the centre of $\sigma_1\sigma_2$, so we may assume that σ_1 and σ_2 have different centres. Let a be the centre of $\sigma_1\sigma_2$ and suppose $a \notin l$. Since $a^{\sigma_1\sigma_2} = a$, we have $a^{\sigma_1} = a^{\sigma_2^{-1}}$ and $a^{\sigma_1} \neq a$ by corollary 5.1.6. Putting $b = \overline{aa^{\sigma_1}} \cap l$, we have $\overline{ab}^{\sigma_1} = \overline{a^{\sigma_1}b} = \overline{ab}$. Therefore b is the centre of σ_1 by corollary 5.1.6. From $\overline{aa^{\sigma_1}} = \overline{aa^{\sigma_2^{-1}}}$, we conclude that b is also the centre of σ_2^{-1}, hence of σ_2. Contradiction, therefore $a \in l$. Then it is obvious that a is different from either the centre of σ_1 or of σ_2. ■

The dual theorem of theorem 5.1.8 reads like:

Theorem 5.1.8′ *If σ_1, σ_2 are a-elations, then so is $\sigma_1\sigma_2$. If σ_1 and σ_2 have different axes, then the axis of $\sigma_1\sigma_2$ does not coincide with either the axis of σ_1 or that of σ_2.*

So, the collection of all l-elations forms a subgroup of Aut **P**, denoted by $E(l)$ and called the l-elation group. Similarly, the collection of all a-elations is a subgroup of Aut **P**, denoted by $E(a)$ and called the a-elation group.

Let **A** be an affine plane obtained from the projective plane **P**, and let l_∞ be the line at infinity for **A**. Then $E(l_\infty)$ can be regarded as a subgroup of Aut **A** and is called the translation group of **A**. For a point a and a line l of **P**, let $G(a, l)$ denote the collection of all (a, l)-perspectivities, $G(l)$ the collection of all l-perspectivities and $G(a)$ the collection of all a-perspectivities. $G(a, l)$, $G(l)$ and $G(a)$ are subgroups of Aut **P** satisfying:

$$G(l) = \bigcup_{a \in \Omega} G(a, l), \; G(a) = \bigcup_{l \in \mathfrak{B}} G(a, l)$$

and

$$(\text{Aut } \mathbf{P})_{\langle l \rangle} \geq (\text{Aut } \mathbf{P})_l = G(l) \geq E(l), (\text{Aut } \mathbf{P})_a \geq G(a) \geq E(a).$$

The following corollary follows immediately from corollary 5.1.6.

Corollary 5.1.9 *If $\sigma \in E(l)\,(\sigma \neq 1)$, then σ leaves no point of $\Omega - l$ fixed, hence $E(l)$ operates half-regularly on $\Omega - l$ and $|E(l)|\,|\,n^2$, where $n = k - 1$ denotes the order of* **P**.

Theorem 5.1.10 *Let* **P** *be a projective plane, l a line in* **P**. *If there are two points c_1 and c_2 $(c_1 \neq c_2)$ on l, such that $G(c_1, l) \neq 1$ and $G(c_2, l) \neq 1$, then $E(l)$ is an elementary Abelian group, hence $|E(l)|$ is a prime power.*

Proof Choose $\sigma_i \in G(c_i, l), \sigma_i \neq 1\,(i = 1, 2)$ and a point $a \notin l$. Since

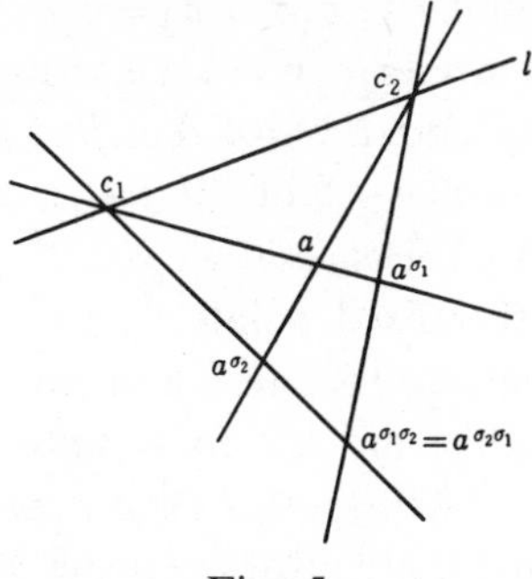

Fig. 5

$\overline{c_i x}^{\sigma_i} = \overline{c_i x}\ \forall x \in \Omega (i = 1, 2)$ and $a = \overline{c_1 a} \cap \overline{c_2 a}$, we have:

$$a^{\sigma_1} = \overline{c_1 a} \cap \overline{c_2 a}^{\sigma_1},\ a^{\sigma_1 \sigma_2} = \overline{c_1 a}^{\sigma_2} \cap \overline{c_2 a}^{\sigma_2}$$

and similarly: $a^{\sigma_2 \sigma_1} = \overline{c_2 a}^{\sigma_1} \cap \overline{c_1 a}^{\sigma_2}$. Therefore, $a^{1\sigma_2} = a^{\sigma_2 \sigma_1}$ for all points of **P**, i.e. $\sigma_1 \sigma_2 = \sigma_2 \sigma_1$, and $\sigma_1 \sigma_2 = \sigma_2 \sigma_1$ is not the identity automorphism and an l-elation with centre different from c_1 and c_2 by theorem 5.1.8 (see Fig. 5).

The above argument has also shown that any two l-elations with different centres commute. For $\tau \in G(c_1, l)$ we have

$$\sigma_2(\sigma_1 \tau) = (\sigma_1 \tau)\sigma_2 = \sigma_1(\tau \sigma_2) = \sigma_1(\sigma_2 \tau) = (\sigma_2 \tau)\sigma_1 = \sigma_2(\tau \sigma_1),$$

since $\sigma_1 \tau \in G(c_1, l)$, hence $\tau \sigma_1 = \sigma_1 \tau$. This shows that any two l-elations with the same centre commute. Therefore $E(l)$ is Abelian. Let σ denote an element of $G(c_1, l)$ of prime order p. If $\rho \neq 1$ denotes an element of $E(l)$ that is not contained in $G(c_1, l)$, then $\sigma\rho$ is an element of $E(l)$ with a centre different from that of σ and ρ. Since $(\sigma\rho)^p = \sigma^p \rho^p = \rho^p$, if $\rho^p \neq 1$ then the centres of ρ and $\sigma\rho$ coïncide. Contradiction. Therefore $\rho^p = 1$, i.e. the pth power of each element of $E(l)$ equals 1, proving the theorem. ■

Theorem 5.1.11 *Let $\sigma_i\ (i = 1, 2)$ be (c_i, l_i)-homologies of order 2 of* **P** *such that $c_1 \in l_2$ and $c_2 \in l_1$, $l_1 \neq l_2$ and $c_1 \neq c_2$. Then $\sigma_1 \sigma_2$ is a (c, l)-homology of order 2, where $c = l_1 \cap l_2$ and $l = \overline{c_1 c_2}$.*

Proof Since $c_1 \in l_2$ and $c_2 \in l_1$, we have $c_2^{\sigma_1} = c_2$ and $l_2^{\sigma_1} = l_2$, hence $\sigma_2^{\sigma_1}$ is a (c_2, l_2)-homology. Therefore $\sigma_1 \sigma_2 \sigma_1 \sigma_2 = \sigma_2^{\sigma_1} \sigma_2$ is a

(c_2, l_2)-homology. Similarly, $\sigma_1\sigma_2\sigma_1\sigma_2 = \sigma_1\sigma_1^{\sigma_2}$ is proved to be a (c_1, l_1)-homology. Hence $(\sigma_1\sigma_2)^2 = 1$, i.e. the order of $\sigma_1\sigma_2$ equals 2. We have $c^{\sigma_1\sigma_2} = c$ and $l^{\sigma_1\sigma_2} = l$. Let $b \notin l, b \neq c$. If $b \in l_1$, then $b^{\sigma_1\sigma_2} = b^{\sigma_2} \neq b$ by corollary 5.1.6. If $b \notin l_1$, then $\overline{c_2 b} \neq \overline{c_2 b^{\sigma_1}} \ni c_2$ again by corollary 5.1.6. Hence $b^{\sigma_1} \notin \overline{c_2 b}$, $b^{\sigma_1\sigma_2} \notin \overline{c_2 b^{\sigma_2}} = \overline{c_2 b}$. Therefore $\sigma_1\sigma_2$ has no other fixed points than c and the points on l. In order to prove that $\sigma_1\sigma_2$ is a (c, l)-homology it suffices to prove that all lines through c are left fixed since $l^{\sigma_1\sigma_2} = l$. Obviously $l_i^{\sigma_1\sigma_2} = l_i\,(i = 1, 2)$. Let $b \notin l \cup l_1 \cup l_2$. Then $c_2 \notin \overline{bb^{\sigma_1\sigma_2}}$ as just shown above. Since $\overline{bb^{\sigma_1\sigma_2}}$ and l_1 are invariant under $\sigma_1\sigma_2$, so is $\overline{bb^{\sigma_1\sigma_2}} \cap l_1$. Since $c_2 = l \cap l_1$, the point $\overline{bb^{\sigma_1\sigma_2}} \cap l_1$ is not a point on l, hence $c \in \overline{bb^{\sigma_1\sigma_2}}$. Therefore $\overline{cb}^{\sigma_1\sigma_2} = \overline{cb}$ (see Fig. 6). ■

Theorem 5.1.12 *Let G be a subgroup of* $(\text{Aut}\,\mathbf{P})_{\langle l_\infty \rangle}$. *Then the following conditions are equivalent*

(i) $(G, \mathfrak{B} - \{l_\infty\})$ *is transitive,*

(ii) $(G, \Omega - l_\infty)$ *and* (G, l_∞) *are both transitive,*

(iii) G *operates transitively on the collection of all finite flags.*

Proof (i) $\Rightarrow$ (ii): Follows from theorem 1.5.8.

(ii) $\Rightarrow$ (iii): Since $|\Omega - l_\infty| = n^2$ and $|l_\infty| = n + 1$, we have $(|\Omega - l_\infty|, |l_\infty|) = 1$. Therefore $|G : G_{a,b}| = n^2(n + 1)$ for $a \in \Omega - l_\infty, b \in l_\infty$. Putting $l = \overline{ab}$, $G_{\langle l \rangle}$ leaves b fixed since $l \cap l_\infty = b$, therefore $G_{a,\langle l \rangle} = G_{a,b}$, hence $|G : G_{a,\langle l \rangle}| = n^2(n + 1)$. Since the number of finite flags also equals $n^2(n + 1)$, the validity of (iii) follows.

(iii) $\Rightarrow$ (i). Obvious. ■

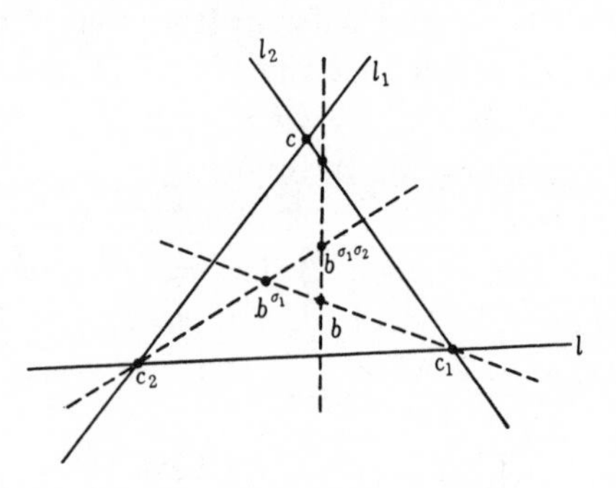

Fig. 6

Corollary 5.1.13 *Let G be a subgroup of* $(\mathrm{Aut}\,\mathbf{P})_{\langle l_\infty \rangle}$ *which satisfies* (i) (*and therefore* (ii) *and* (iii)) *of theorem* 5.1.12. *Let* (a, l) *be a finite flag, put* $|G_{a,\langle l\rangle}| = h$ *and choose* $b \in l_\infty$. *Then:*

(1) $|G| = n^2(n+1)h$
(2) $|G_b| = n^2 h$
(3) $|G_{\langle l\rangle}| = nh$
(4) $|G_a| = (n+1)h$
(5) $|G_{a,b}| = |G_{a,\langle l\rangle}| = h.$

Theorem 5.1.14 *Under the assumptions of theorem* 5.1.12 *we have:* $(G, \Omega - l_\infty)$ *is transitive* $\Leftrightarrow$ $(G_a, \mathfrak{B}_a - \{l_\infty\})$ *is transitive for all* $a \in l_\infty$.

Proof $\Rightarrow$: Let the orbit decomposition of l_∞ under G be given by $l_\infty = T_1 + \ldots + T_r$. If $i \neq j$, then a finite line passing through a point of T_i cannot be mapped by an element of G onto a line passing through a point of T_j. Therefore we see by theorem 1.5.8 that for each i, G operates transitively on the collection of all finite lines passing through some point of T_i. In particular, $(G_a, \mathfrak{B}_a - \{l_\infty\})$ is transitive for $a \in T_i$.

$\Leftarrow$: Let the orbit decomposition of $\mathfrak{B}$ under G be given by $\mathfrak{B} = \{l_\infty\} + \mathfrak{B}_1 + \ldots + \mathfrak{B}_r$, then the number of orbits of (G, l_∞) equals r hence $(G, \Omega - l_\infty)$ is transitive. ∎

Corollary 5.1.15 *Let* $G \leq \mathrm{Aut}\,\mathbf{P}$ *satisfy the assumptions of theorem* 5.1.12 *and the condition* (i) (*and therefore also* (ii) *and* (iii)). *For two points a, b on l_∞ and for two finite lines l, m passing through b we have*

$$|G_{a,b,\langle l\rangle}| = |G_{a,b,\langle m\rangle}|.$$

Proof G operates transitively on $\Omega - l_\infty$ and on l_∞. Since $(|\Omega - l_\infty|, |l_\infty|) = 1$, G_a operates transitively on $\Omega - l_\infty$. Therefore $G_{a,b}$ operates transitively on the collection of all finite lines through b by theorem 5.1.14, hence: $|G_{a,b,\langle l\rangle}| = |G_{a,b,\langle m\rangle}|$. ∎

5.1.4 (a, l)-transitivity

Throughout this section let $\mathbf{P} = (\Omega, \mathfrak{B})$ be a projective plane of order n. Let $a \in \Omega$, $l \in \mathfrak{B}$. If $G(a, l)$ operates transitively on $l' - \{a, l \cap l'\}$ for some line l' through a then **P** is called an (a, l)-transitive plane (or has an (a, l)-transitivity).

Let $H \leq \operatorname{Aut} \mathbf{P}$. If there is a line l such that $l^\sigma = l, \forall \sigma \in H$ and H operates transitively on $\Omega - l$, then **P** is called an H-transitive plane with regard to l. Dually, if there is a point a such that $a^\sigma = a$, $\forall \sigma \in H$ and H operates transitively on $\mathfrak{B} - \mathfrak{B}_a$, then **P** is called an H-transitive plane with regard to a.

We start with the following simple theorem:

Theorem 5.1.16 *The following conditions are equivalent:*

(1) **P** *is an (a, l)-transitive plane, i.e. $G_{a,l}$ operates transitively on $l' - \{a, l \cap l'\}$ for some $l' \in \mathfrak{B}_a$.*

(2) *$G(a, l)$ operates transitively on $l' - \{a, l \cap l'\}$ for all lines $l' \neq l$ that pass through a.*

(3) *$G(a, l)$ operates transitively on $\mathfrak{B}_{a'} - \{l, \overline{aa'}\}$ for some $a' \in l$.*

(4) *$G(a, l)$ operates transitively on $\mathfrak{B}_{a'} - \{l, \overline{aa'}\}$ for all $a' \neq a$ on l.*

Proof (2)$\Rightarrow$(1) and (4)$\Rightarrow$(3) are obvious.

(1)$\Rightarrow$(4) and (3)$\Rightarrow$(2): Assume (1). Choose $a' \in l - \{a, l \cap l'\}$, and $l_1, l_2 \in \mathfrak{B}_{a'} - \{l, \overline{aa'}\}$. Putting $c_i = l_i \cap l'$ $(i = 1, 2)$, there exists an element $\sigma \in G(a, l)$ such that $c_1^\sigma = c_2$, i.e. $l_1^\sigma = l_2$, hence (4) is valid for all $a' \in l - \{a, l \cap l'\}$. Similarly, by the dual argument, if we assume (3) then (2) is valid for all lines $l' \in \mathfrak{B}_a - \{l, \overline{aa'}\}$. ∎

Using corollary 5.1.9 and theorem 5.1.10 we get:

Theorem 5.1.17 *If* **P** *is an $E(l)$-transitive plane with regard to l, then the order n of* **P** *is a power of some prime p and $E(l)$ is an elementary Abelian group whose order is a power of p.*

The dual of this theorem reads as follows:

Theorem 5.1.17′ *If* **P** *is an $E(a)$-transitive plane with regard to a, then the order n of* **P** *is a power of some prime p and $E(a)$ is an elementary Abelian group whose order is a power of p.*

Theorem 5.1.18 *Let σ_1, σ_2 be a non-identity (c_1, l)- and (c_2, l)-homology respectively and let $c_1 \neq c_2$. There is an l-elation $\tau \in \langle \sigma_1, \sigma_2 \rangle$ such that $c_1^\tau = c_2$.*

Proof Put $H = \langle \sigma_1, \sigma_2 \rangle$. Since $l' = \overline{c_1 c_2}$ is an invariant domain of H, $a = l \cap l'$ is a fixed point of H. Let $l' = \{a\} + M_1 + \ldots + M_t$ be the H-orbit decomposition of l' and put $|M_i| = n_i$. Putting $|H_{a_i}| = h_i$ for $a_i \in M_i$ we have $|H| = n_i h_i$. Since all elements of H are l-perspectivities and since a perspectivity has at most one fixed point not on its axis, we have $H_x \cap H_y = \{1\}$ for $x, y \in l' - \{a\}$ $(x \neq y)$. Hence $|H| \geq 1 + \sum_{i=1}^{t} n_i(h_i - 1)$, therefore $h_i > 1$ for at most one i. (For, suppose $h_i > 1$ and $h_j > 1$ with $i \neq j$. Assuming $n_i \geq n_j$ we get: $|H| \geq 1 + n_i(h_i - 1) + n_j(h_j - 1) \gneqq n_j h_j$. Contradiction). Since $\sigma_i \in H_{c_i}$, we have $|H_{c_i}| > 1$ $(i = 1, 2)$, therefore c_1 and c_2 belong to the same H-orbit M_i. Since (H, M_i) is a Frobenius group, there is an element τ in the Frobenius kernel of (H, M_i) such that $c_1^\tau = c_2$. This element τ is a perspectivity with axis l and centre a, since τ leaves l' fixed and has no fixed points on $l' - \{a\}$. Therefore τ is an (a, l)-elation. ∎

Corollary 5.1.19 *Let l, m be two lines of* **P**, *and put $l \cap m = a$. If $G \leq Aut$* **P** *such that for each $x \in m - \{a\}$, G contains at least one non-identity (x, l)-homology, then $G \geq G(a, l)$, and* **P** *is an (a,l)-transitive plane.*

Corollary 5.1.20 *Let l be a line of* **P**. *If there exists a non-identity (a, l)-homology for each $a \notin l$, then* **P** *is an $E(l)$-transitive plane with regard to l.*

The dual statements are:

Corollary 5.1.19′ *Let a, b be two points of* **P**. *If there exists a non-identity (b, m)-homology for each $m \in \mathfrak{B}_a - \{\overline{ab}\}$, then* **P** *is a $(b, \overline{ab})$-transitive plane.*

Corollary 5.1.20′ *Let a be a point of* **P**. *If there exists a non-identity*

(a, l)-homology for each l not passing through a, then **P** *is an $E(a)$-transitive plane with regard to a.*

From now on, instead of '$E(l)$-transitive plane with regard to l' and '$E(a)$-transitive plane with regard to a' we will simply say: '$E(l)$-transitive plane' and '$E(a)$-transitive plane' respectively.

Theorem 5.1.21 *Let l be a line of* **P**, *and let $c_1, c_2 \in l \, (c_1 \neq c_2)$. If* **P** *is a (c_i, l)-transitive plane $(i = 1, 2)$, then* **P** *is an $E(l)$-transitive plane.*

Proof Let a, b be arbitrary points not on l. Putting $d = \overline{c_1 a} \cap \overline{c_2 b}$, there exist $\sigma_1 \in G(c_1, l)$ such that $a^{\sigma_1} = d$ and $\sigma_2 \in G(c_2, l)$ such that $d^{\sigma_2} = b$. Therefore $a^{\sigma_1 \sigma_2} = b$, proving the theorem. ■

The dual of this theorem reads as follows:

Theorem 5.1.21′ *Let a be a point of* **P** *and let l_1 and l_2 be two lines through a $(l_1 \neq l_2)$. If* **P** *is an (a, l_i)-transitive plane $(i = 1, 2)$, then* **P** *is an $E(a)$-transitive plane.*

Theorem 5.1.22 *Let l be a line of* **P**, *$G \leq \operatorname{Aut} \mathbf{P}$. If l is invariant under G, $G \cap E(l) \neq 1$ and G operates transitively on l, then* **P** *is an $E(l)$-transitive plane and $G \geq E(l)$.*

Proof Since (G, l) is transitive and since $G \cap E(l) \neq 1$, $G \cap G(x, l) \neq 1, \forall x \in l$ and $|G \cap G(x, l)|$ is independent from the choice of x. Put $|G \cap G(x, l)| = h \geq 2$. Since $G(a, l) \cap G(b, l) = 1$ for $a, b \in l \, (a \neq b)$, we have $|G \cap E(l)| = (n + 1)(h - 1) + 1$. Putting the number of orbits of $\Omega - l$ under $G \cap E(l)$ equal to m, we have $n^2 = m|G \cap E(l)| = m((n + 1)(h - 1) + 1)$ since $G \cap E(l)$ operates half-regularly on $\Omega - l$. Hence $m | n^2, n^2 \gneq m$ and $n^2 \equiv m \pmod{n + 1}$. Therefore $m = 1$ and $G \cap E(l)$ operates transitively on $\Omega - l$. The fact that $G \cap E(l) = E(l)$ follows from corollary 5.1.9. ■

Since the orders of all presently known projective planes are prime powers, it is conjectured that the order of any projective plane is a prime power. In order to prove the theorem of Ostrom–Wagner, our main aim in this chapter, we need to prove that under certain conditions the order of a projective **P** is a prime power. We have already shown, in theorem 5.1.17, that the order of any $E(l)$-transitive plane is a prime power. The following theorem gives another condition under which the order of **P** is a prime power.

Theorem 5.1.23 *Let l be a line of* **P**, *and let c be a point not on l. If for all $a, b \in l (a \neq b)$, $|G(b, \overline{ac})|$ is even, then the order n of* **P** *is a prime power.*

Proof Pick an element $a \in l$ and put $\overline{ca} = l_\infty$. Since $G(b, \overline{ac})$, where $b \in l$ and $b \neq a$, operates semi-regularly on $l - \{a, b\}$, the order of **P** $(= |l - \{a\}|)$ is odd. We put:

$$H = \langle \sigma | \sigma \text{ is a } (y, \overline{cx})\text{-homology of order 2 for some } y \text{ and } x \in l \rangle$$

$$K = \langle \sigma | \sigma \text{ is a } (y, l_\infty)\text{-homology of order 2 for some } y \in l \rangle.$$

Then $K \leq H$ and the elements of H_a leave l_∞ fixed, hence $H_a \trianglerighteq K$. Since K operates transitively on $l - \{a\}$ by theorem 5.1.18 and since all elements of K are perspectivities with l_∞ as axis, $(K, l - \{a\})$ is a Frobenius group. Now, H operates on l and contains an element that does not leave a fixed. Therefore, H operates 2-transitively on l, K operates faithfully as a Frobenius group on $l - \{a\}$, $|K|$ is even and $|l - \{a\}|$ is odd. Therefore H_a operates primitively on $l - \{a\}$ by lemma 3.4.8. Let N denote the Frobenius-kernel of $(K, l - \{a\})$, then $H_a \trianglerighteq N$, since $H_a \geq K \trianglerighteq N$, and $(N, l - \{a\})$ is regular, hence N is a minimal normal subgroup of H_a by theorem 3.3.6(2). Since N is nilpotent, N has a characteristic subgroup N_0 that is an elementary Abelian group and since $G \trianglerighteq N_0$ we have $N = N_0$. Therefore: the order of $\mathbf{P} = |l - \{a\}| = |N| = |N_0|$ is a prime power. ■

The following theorem plays a crucial role in the proof of the theorem of Ostrom–Wagner.

Theorem 5.1.24 *Let l_∞ be a line of* **P** *and let $G = \{\sigma \in \mathrm{Aut}\,\mathbf{P} \mid l_\infty^\sigma = l_\infty\} \leq \mathrm{Aut}\,\mathbf{P}$ satisfy the following conditions:*

(1) *$G_{a,\langle l\rangle}$ contains a homology of order 2 for every finite flag (a, l).*

(2) *Let $a, b \in l_\infty$ $(a \neq b)$ and let l be a finite line through b. If $G_{a,b,\langle l\rangle}$ contains a homology of order 2, then $G_{a,b,\langle m\rangle}$ contains a homology of order 2 for each finite line m through b.*

Then n is a prime power and at least one of the following holds:

(i) **P** *is an $E(l_\infty)$-transitive plane*

(ii) **P** *is an $E(a)$-transitive plane for some $a \in l_\infty$*

(iii) *There is a point o not on l_∞, such that $|G(a, \overline{bo})|$ is even for all $a, b \in l_\infty$ $(a \neq b)$.*

Proof It follows from (i), (ii) or (iii) that n is a prime power by theorems 5.1.17, 5.1.17′ and 5.1.23. To prove that (i), (ii) or (iii) holds, we distinguish the following cases.

(*a*) Where the centre of a homology of order 2 in G is not a finite point. G contains a homology σ of order 2 by condition (1). The centre a of σ is a point of l_∞, hence the axis l of σ does not coincide with l_∞. Putting $l \cap l_\infty = b$, we have $\sigma \in G_{a,b,\langle l\rangle}$. By condition (2) there is a homology τ of order 2 in each $G_{a,b,\langle m\rangle}$ for all $m \neq l_\infty$ such that $b \in m$. Since the centre of τ is a point of l_∞, it coincides either with a or with b by corollary 5.1.6. Suppose the centre of τ is b, then the axis l' of τ passes through a and G contains a homology of order 2 with the finite point $l' \cap l$ as centre from theorem 5.1.11. Contradiction. Therefore, the centre of τ is a and the axis of τ is m by corollary 5.1.6. Thus G contains a homology with centre a and axis an arbitrary finite line through b. Hence, **P** is an (a, l_∞)-transitive plane by corollary 5.1.19′ (see Fig. 7).

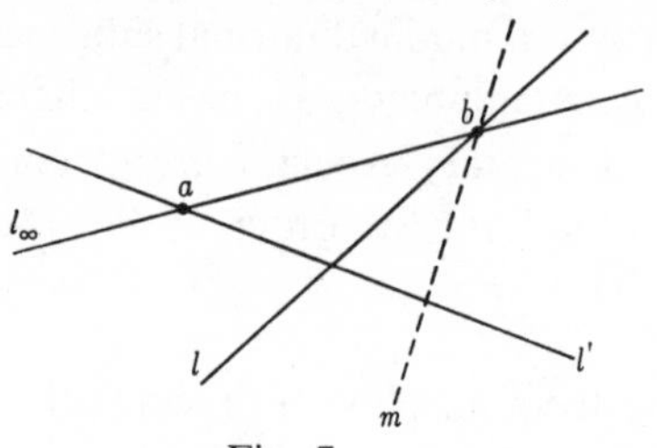

Fig. 7

(a.1) If there exists a point $a' \neq a$, that is the centre of a homology of order 2 in G, then we prove in the same way as above, that **P** is an (a', l_∞)-transitive plane, hence **P** is an $E(l_\infty)$-transitive plane by theorem 5.1.21.

(a.2) If the centre of each homology of order 2 is a, then G contains an (a, l)-homology for each finite line l not through a by condition (1). Hence **P** is an $E(a)$-transitive plane by corollary 5.1.20′.

(b) Where there is only one finite point o that is the centre of a homology of order 2 in G. For $\tau \in G$ the point o^τ is also the centre of a homology of order 2 in G, hence o is a fixed point of G. Choosing $a, b \in l_\infty$ $(a \neq b)$ and $c \in \overline{bo}$ $(c \neq b, c \neq o)$, $G_{c,\langle \overline{ac} \rangle}$ contains a homology σ of order 2 by condition (1). Since σ leaves a, b, c and o fixed, the centre of σ is a and the axis of σ is $\overline{bo}$, so in this case (iii) holds (see Fig. 8).

(c) Where there is more than one point that is the centre of a homology of order 2 in G. Since the axis of any homology that has a finite point as centre is l_∞ by corollary 5.1.6, $E(l_\infty) \neq 1$ by theorem 5.1.18. Now, let σ be a homology of order 2 in G with the finite point o as centre, then $\sigma \in G_{a,b,\langle \overline{bo} \rangle}$ for all $a, b \in l_\infty$. Therefore $G_{a,b,\langle l \rangle}$ contains a homology of order 2 for arbitrary points $a, b \in l_\infty$ and for arbitrary finite lines l through b (by condition (2)).

(c.1) Where there exists a finite line l, such that no finite point on l is the centre of a homology of order 2. Put $l \cap l_\infty = a$, then $G_{a,\langle l \rangle}$ operates transitively on $l_\infty - \{a\}$. (For, let $l_\infty - \{a\} = \Lambda_1 + \ldots + \Lambda_r$ be the orbit decomposition of $l_\infty - \{a\}$ under $G_{a,\langle l \rangle}$. For $b \in \Lambda_1$ let σ be a homology of order 2 contained in $G_{a,b,\langle l \rangle}$. Since the centre of σ is not on l, l_∞ is not the axis of σ. Therefore the only fixed points

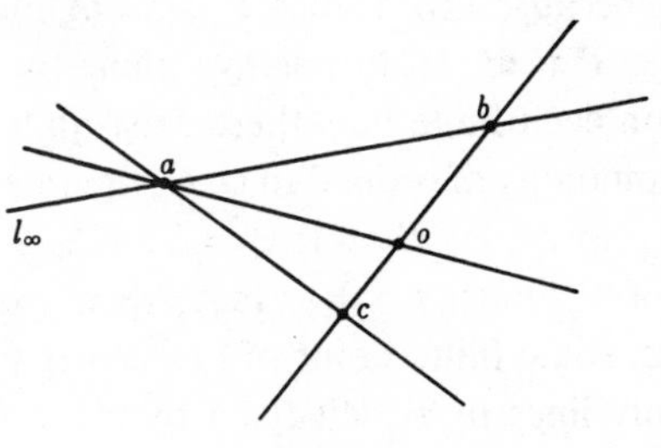

Fig. 8

on l_∞ of σ are a and b, hence $|\Lambda_1|$ is odd and $|\Lambda_2| = \ldots = |\Lambda_r|$ is even. If $r > 1$, then the same argument for some $b' \in \Lambda_2$ would give that $|\Lambda_1|$ is even. Contradiction.) Since l^σ satisfies the same conditions as l for all $\sigma \in G$, $G_{a^\sigma, \langle l^\sigma \rangle}$ operates transitively on $l_\infty - \{a^\sigma\}$. Hence, if there is a $\sigma \in G$ such that $a^\sigma \neq a$, then G is transitive on l_∞ and **P** is $E(l_\infty)$-transitive by theorem 5.1.22, which is condition (i). So we may assume $a^\sigma = a$, $\forall \sigma \in G$. Then $G_{a,c}$ operates transitively on $l_\infty - \{a\}$ for all $c \in l$. (For, let $l_\infty - \{a\} = \Lambda_1 + \ldots + \Lambda_r$ be the orbit decomposition of $l_\infty - \{a\}$ under $G_{a,c}$. For $b \in \Lambda_1$, $G_{c, \langle \overline{bc} \rangle}$ contains a homology σ of order 2 by (1). If σ has other fixed points on l_∞ except a and b, then c is the centre of σ, contrary to assumption (c). Hence the only fixed points on l_∞ of σ are a and b. In particular, if $r > 1$, then $|\Lambda_1|$ is odd and $|\Lambda_2| = \ldots = |\Lambda_r|$ is even. Taking $b' \in \Lambda_2$ we get in the same way that $|\Lambda_1|$ is even. Contradiction, hence $r = 1$.) Next we prove that **P** is an (a, l)-transitive plane. (Since, for $c \in l - \{a\}$ and $b \in l_\infty - \{a\}$, $G_{c, \langle \overline{bc} \rangle}$ contains a homology σ of order 2 by (1) and it is proved in the same way as above that the centre of σ is either a or b. If a is the centre of σ, then $\overline{bc}$ is the axis of σ, therefore G contains a homology with centre a and axis any line through c that does not coincide with l because of the transitivity of $(G_{a,c}, l_\infty - \{a\})$. Hence **P** is an (a, l)-transitive plane by corollary 5.1.19′. If b is the centre of σ, then the axis of σ is l and it is proved in the same way that G contains a homology with axis l and centre any point of $l_\infty - \{a\}$. Therefore, **P** is an (a, l)-transitive plane by corollary 5.1.19.) Hence **P** is an (a^τ, l^τ)-transitive plane for all $\tau \in G$. Now let τ be a homology of order 2 with a finite point o as the centre (and therefore l_∞ as the axis). The existence of such an element is assured by the condition (c) and we note that $a^\tau = a$ and $l^\tau \neq l$ because $l \not\ni o$. Hence **P** is an (a, l^τ)-transitive plane with $l^\tau \neq l$, and so **P** is an $E(a)$-transitive plane by theorem 5.1.21′.

(c.2) Where, on each finite line, there exist finite points that are the centre of a homology of order 2 in G. G_a operates transitively on $\mathfrak{B}_a - \{l_\infty\}$, $\forall a \in l_\infty$. (For, let $\mathfrak{B}_a - \{l_\infty\} = \Lambda_1 + \ldots + \Lambda_r$ be the orbit decomposition of $\mathfrak{B}_a$ under G_a. For $l \in \Lambda_1$ there exists a homology σ of order 2 with some finite point of l as centre (and hence l_∞ as axis) and the only lines in $\mathfrak{B}_a$ left fixed by σ are l_∞ and l. Hence $|\Lambda_1|$ is odd and $|\Lambda_2|$ is even if $r > 1$. If $r > 1$, then we can show in the same way by taking $l' \in \Lambda_2$ that $|\Lambda_1|$ is even, contradiction.) Therefore, G operates transitively on $\Omega - l_\infty$ by theorem 5.1.14. Hence there exists for any finite point a homology of order 2 in G with that point as centre (and hence l_∞ as axis). Therefore $\mathbf{P}$ is an $E(l_\infty)$-transitive plane by corollary 5.1.20. ■

5.1.5 Subplanes and their orders

Let $\mathbf{P} = (\Omega, \mathfrak{B})$ be a projective plane. For $\Omega_1 \subseteq \Omega$ we put

$$\mathfrak{B}_1 = \{l \cap \Omega_1 \,|\, |l \cap \Omega_1| > 1, l \in \mathfrak{B}\}$$

and denote the geometric structure $(\Omega_1, \mathfrak{B}_1)$ by $\mathbf{P}(\Omega_1)$. If $\mathbf{P}(\Omega_1)$ is a projective plane, it is called a subplane of $\mathbf{P}$. If $\Omega_1 \subsetneqq \Omega$, then $\mathbf{P}(\Omega_1)$ is called a proper subplane. For $G \leq \mathrm{Aut}\,\mathbf{P}$, we put:

$$\Omega_1 = F_\Omega(G), \quad \mathfrak{B}_1 = \{l \cap \Omega_1 \,|\, l \in F_{\mathfrak{B}}(G)\}$$

and we denote the geometric structure $(\Omega_1, \mathfrak{B}_1)$ by $F_{\mathbf{P}}(G)$. If $F_{\mathbf{P}}(G)$ is a projective plane, then we have

$$F_{\mathbf{P}}(G) = \mathbf{P}(F_\Omega(G)),$$

namely $F_{\mathbf{P}}(G)$ is a subplane of $\mathbf{P}$. (For, if $l \in \mathfrak{B}$ is such that $|l \cap \Omega_1| > 1$, then $l^G = l$, i.e. $l \in F_{\mathfrak{B}}(G)$. Conversely, let l be an element of $F_{\mathfrak{B}}(G)$. Since $F_{\mathbf{P}}(G)$ is a projective plane, there exist four lines in $F_{\mathfrak{B}}(G)$ no three of which intersect in one point. Hence these lines intersect l in at least two points and the points of intersection are points of Ω_1. Therefore $|l \cap \Omega_1| > 1$, and hence $\{l \cap \Omega_1 \,|\, |l \cap \Omega_1| > 1\} = \{l \cap \Omega \,|\, l \in F_{\mathfrak{B}}(G)\}$, i.e. $F_{\mathbf{P}}(G) = \mathbf{P}(F_\Omega(G))$.

If $F_{\mathbf{P}}(G)$ is a projective plane, it is called the G-subplane of $\mathbf{P}$. If there is no chance of confusion, elements of $\mathfrak{B}_1$ will be represented by their corresponding elements in $F_{\mathfrak{B}}(G)$. If G is a 2-group, the G-subplane $F_{\mathbf{P}}(G)$ is called a 2-subplane. Further, a '2-subplane with regard to $G \leq \mathrm{Aut}\,\mathbf{P}$' will refer to a 2-subplane $F_{\mathbf{P}}(H)$ for some 2-subgroup H of G.

The following theorem is a straightforward consequence of the definition:

Theorem 5.1.25 *Let* $G \leq \mathrm{Aut}\,\mathbf{P}$.

$F_{\mathbf{P}}(G)$ is a projective plane $\Leftrightarrow$ $\Omega_1 = F_\Omega(G)$ contains four points no three of which are on the same line.

Theorem 5.1.26 *Let $\mathbf{P} = (\Omega, \mathfrak{B})$ be a projective plane of order n and let $\mathbf{P}_1 = (\Omega_1, \mathfrak{B}_1)$ be a proper subplane of $\mathbf{P}$. If m denotes the order of $\mathbf{P}_1$, we have:*

(i) $n = m^2$ *or* (ii) $n \geq m^2 + m$.

Proof For $x \in \Omega_1$, the number of lines l such that $x \in l$ and $l \notin \mathfrak{B}_1$ equals $n - m$, hence $|\{l \in \mathfrak{B} \,|\, |l \cap \Omega_1| = 1\}| = (m^2 + m + 1)(n - m)$. Since $|\{l \in \mathfrak{B} \,|\, |l \cap \Omega_1| = 1\}| \leq |\mathfrak{B}| - |\mathfrak{B}_1| = (n^2 + n + 1) - (m^2 + m + 1) = (n - m)(n + m + 1)$, we conclude $m^2 \leq n$. (Note that $m \neq n$, since $\mathbf{P}_1$ is a proper subplane of $\mathbf{P}$.) If $m^2 < n$, then there exists a line that does not contain elements of Ω_1. Let l_0 denote such a line. The intersections of l_0 and the $m^2 + m + 1$ lines $\{l \in \mathfrak{B} \,|\, l \cap \Omega_1 \in \mathfrak{B}_1\}$ (i.e. the lines in $\mathfrak{B}$ that correspond with lines in $\mathfrak{B}_1$) do not belong to Ω_1, hence are all different. Therefore $m^2 + m + 1 \leq n + 1$, i.e. $m^2 + m \leq n$. ∎

Theorem 5.1.27 *Let $\sigma \in \operatorname{Aut} \mathbf{P}$ be of order 2. If σ is not a perspectivity, then $F_{\mathbf{P}}(\langle \sigma \rangle)$ is a subplane of $\mathbf{P}$. If m denotes the order of this subplane, then $n = m^2$.*

Proof Every line of $\mathbf{P}$ contains a point of $\Omega_1 = F_\Omega(\langle \sigma \rangle)$. (For, since the order of σ is 2, we have $l^\sigma \cap l \in \Omega_1$ for those l with $l^\sigma \neq l$. Suppose $l^\sigma = l$. If $a^\sigma = a$ for all $a \notin l$, then $\sigma = 1$. Therefore there exists a point $a \notin l$ such that $a^\sigma \neq a$. Since $\overline{aa^\sigma}$ is invariant under σ, we have $l \cap \overline{aa^\sigma} \in \Omega_1$.) So, if we can prove that $F_{\mathbf{P}}(\langle \sigma \rangle)$ is a projective plane, then we have $n = m^2$, where m denotes the order of $F_{\mathbf{P}}(\langle \sigma \rangle)$, by the proof of theorem 5.1.26. Let us assume that $F_{\mathbf{P}}(\langle \sigma \rangle)$ is not a projective plane. According to theorem 5.1.25, all points of Ω_1 with at most one exception, are on the same line l. If all points of Ω_1 are on l, choose then for $x \in l$ a line $l_x (\neq l)$ through x. Then $\varnothing \neq l_x \cap \Omega_1 \subseteq l_x \cap l = x$, hence $x \in \Omega_1$. Therefore $l = \Omega_1$ and σ is a perspectivity, contradiction. If for some $c \in \Omega_1$ $(c \notin l)$, $\Omega_1 - \{c\} \subseteq l$, again choosing for $x \in l$ a line l through x that does not pass through

c, one proves as above that $x \in \Omega_1$, hence $l \subseteq \Omega_1$, i.e. σ is a perspectivity. Contradiction. ■

Theorem 5.1.28 *Let* $\mathbf{P} = (\Omega, \mathfrak{B})$ *be a projective plane of order* n *and let* $\mathbf{P}_1 = (\Omega_1, \mathfrak{B}_1)$ *be a proper* 2-*subplane of* $\mathbf{P}$ *of order* m. *Then* $n = m^{2^t}$ *for some* $t \in \mathbb{N}$.

Proof The proof is by induction on n. Let $\mathbf{P}_1 = F_{\mathbf{P}}(G)$, where G is a 2-subgroup of Aut $\mathbf{P}$. Since $\mathbf{P}_1$ is a proper subplane, we have $G \neq 1$. Let σ be an element of order 2 of the centre of G. Since $\Omega \supseteq F_\Omega(\langle \sigma \rangle) \supseteq \Omega_1 = F_\Omega(G)$, we have that $\tilde{\mathbf{P}} = (F_\Omega(\langle \sigma \rangle), F_{\mathfrak{B}}(\langle \sigma \rangle))$ is a subplane by theorem 5.1.25 and denoting the order of $\tilde{\mathbf{P}}$ by $\tilde{n}$, we have $n = \tilde{n}^2$ by theorem 5.1.27. Since $\sigma \in Z(G)$, there is a natural homomorphism from G onto a subgroup $\tilde{G}$ of Aut $\tilde{\mathbf{P}}$ and $F_{\mathbf{P}}(G) = F_{\tilde{\mathbf{P}}}(\tilde{G})$. We have $\tilde{n} = m^{2^s}$ for some $s \in \mathbb{N}$ by the induction hypothesis, hence, $n = \tilde{n}^2 = (m^{2^s})^2 = m^{2^{s+1}}$. ■

Theorem 5.1.29 *Let* $\mathbf{P}$ *be a projective plane of odd order* n, *let* l_∞ *be a line of* $\mathbf{P}$ *and assume* $G = (\text{Aut } \mathbf{P})_{\langle l_\infty \rangle}$ *operates transitively on* $\mathfrak{B} - l_\infty$. *Further, let* $\mathbf{P}_1 = (\Omega_1, \mathfrak{B}_1)$ *be a minimal* 2-*subplane of* $\mathbf{P}$ *with regard to* G *and put* $H = G^{\Omega_1}_{\langle \Omega_1 \rangle} \leq \text{Aut } \mathbf{P}_1$. *Then the following conditions hold*:

(1) $H_{c,\langle l \rangle}$ *contains a homology of order* 2 *for every finite flag* (c, l) *of* $\mathbf{P}_1$.

(2) *Let* $a, b \in \mathbf{P}_1$ *be two points on* l_∞. *If for some finite line* l *of* $\mathbf{P}_1$ *through* b, $H_{a,b,\langle l \rangle}$ *contains a homology of order* 2, *then* $H_{a,b,\langle m \rangle}$ *contains a homology of order* 2 *for any finite line* m *of* $\mathbf{P}_1$ *through* b.

Proof From corollary 5.1.13 we have that $|G| = n^2(n+1)h$ with $h = |G_{c,a}| = |G_{c,\langle l \rangle}|$, where $a \in l_\infty$ and (c, l) is an arbitrary finite flag of $\mathbf{P}$. Putting $2^u \top n+1$ and $2^v \top h$, we have $u \geq 1$. Pick a maximal 2-subgroup P of G such that $\mathbf{P}_1 = F_{\mathbf{P}}(P)$ and put $|P| = 2^w$. Choosing (c, l) to be a finite flag in $\mathbf{P}_1$, we have $G_{c,\langle l \rangle} \geq P$, hence $w \leq v$. Therefore P is not a Sylow 2-subgroup of G, and hence there exists a 2-subgroup $\tilde{P}$ of G such that $[\tilde{P} : P] = 2$. Since $P \lhd \tilde{P}$, the elements of $\tilde{P}$ induce automorphisms of $\mathbf{P}_1$ and because of the minimality of $\mathbf{P}_1$ an element $\tau \in \tilde{P} - P$ induces a perspectivity of order 2 of $\mathbf{P}_1$ by theorem 5.1.27. Since the order of $\mathbf{P}_1$ is odd by theorem 5.1.28, τ induces a homology by corollary 5.1.7. Hence τ leaves some finite flag (c', l') of $\mathbf{P}_1$ fixed and $\tilde{P} \leq G_{c',\langle l' \rangle}$. Since $|G_{c',\langle l' \rangle}| = |G_{c,\langle l \rangle}|$ by theorem 5.1.12, hence $w \lneq v$ and there exists a 2-group $\tilde{\tilde{P}}$ such that $G_{c,\langle l \rangle} \geq \tilde{\tilde{P}} \rhd P$ and $[\tilde{\tilde{P}} : P] = 2$. An element $\sigma \in \tilde{\tilde{P}} - P$ induces a homology of order 2 of $\mathbf{P}_1$, proving (1). Let $a, b \in \mathbf{P}_1$, such that $a, b \in l_\infty$. If l and m are finite lines of $\mathbf{P}_1$ through b, then $|G_{a,b,\langle l \rangle}| = |G_{a,b,\langle m \rangle}|$ by corollary 5.1.15. If $H_{a,b,\langle l \rangle}$ contains a homology of order 2, then some element of $G_{a,b,\langle l \rangle}$ induces a homology of order 2 of $\mathbf{P}_1$, hence P is no Sylow 2-subgroup of $G_{a,b,\langle l \rangle}$. Therefore P is no Sylow 2-subgroup of $G_{a,b,\langle m \rangle}$ and there exists a 2-subgroup $\tilde{P}$ such that $P \lhd \tilde{P} \leq G_{a,b,\langle m \rangle}$ and $[\tilde{P} : P] = 2$. Then an element $\tau \in \tilde{P} - P$ induces a homology of order 2 of $\mathbf{P}_1$ because of the definition of $\mathbf{P}_1$. ■

Using theorem 5.1.24 and the previous theorem, we get.

Theorem 5.1.30 *Let* $\mathbf{P}$ *be a projective plane of odd order* n. *If for some line* l $(\mathrm{Aut}\ \mathbf{P})_{\langle l \rangle}$ *operates transitively on* $\mathfrak{B} - \{l\}$, *then* n *is a prime power.*

5.1.6 $E(l)$-transitivity and Desarguean planes

In this section we will prove a fundamental result on projective planes over a finite field. If a certain number of points of a projective plane are on one line l, these points are called collinear (in l). Dually, if a certain number of lines of a projective plane intersect in one point a, these lines are called concurrent (in a). Three different points that are not collinear are said to form a triangle. A projective plane $\mathbf{P}$ satisfying the following condition (i) is called a Desarguean plane:

(i) If $\{a_1, a_2, a_3\}$ and $\{b_1, b_2, b_3\}$ are two triangles in $\mathbf{P}$, such that $\overline{a_1 b_1}$, $\overline{a_2 b_2}$ and $\overline{a_3 b_3}$ are concurrent, then $\overline{a_1 a_2} \cap \overline{b_1 b_2}$, $\overline{a_1 a_3} \cap \overline{b_1 b_3}$ and $\overline{a_2 a_3} \cap \overline{b_2 b_3}$ are collinear. Again, if $\mathbf{P}$ satisfies the following condition (ii), then $\mathbf{P}$ is called weakly Desarguean.

(ii) If $\{a_1, a_2, a_3\}$ and $\{b_1, b_2, b_3\}$ are two triangles in $\mathbf{P}$ such that $\overline{a_1b_1}, \overline{a_2b_2}$ and $\overline{a_3b_3}$ are concurrent in d, and if d, $\overline{a_1a_2} \cap \overline{b_1b_2}$ and $\overline{a_1a_3} \cap \overline{b_1b_3}$ are collinear, then $a_2a_3 \cap b_2b_3$ is also on the line determined by these three points.

Theorem 5.1.31 *For a projective plane* **P**, *the following conditions are equivalent:*

(1) $\mathbf{P} \simeq \mathbf{P}_1(2, K)$, *where K is a finite field.*
(2) **P** *is Desarguean.*
(3) **P** *is weakly Desarguean.*
(4) *For any line l of* **P**, **P** *is an $E(l)$-transitive plane.*

Proof (1)$\Rightarrow$(2): Let $\mathbf{P} = \mathbf{P}_1(2, K)$. Let $\{a_1, a_2, a_3\}$ and $\{b_1, b_2, b_3\}$ be two triangles, such that $\overline{a_1b_1}, \overline{a_2b_2}$ and $\overline{a_3b_3}$ are concurrent in O. We have to show that $d_1 = \overline{a_2a_3} \cap \overline{b_2b_3}$, $d_2 = \overline{a_1a_3} \cap \overline{b_1b_3}$ and $d_3 = \overline{a_1a_2} \cap \overline{b_1b_2}$ are collinear. If O coincides with one of the points a_i, b_j $(1 \le i \le 3)$, if some vertex of $\{a_1, a_2, a_3\}$ coincides with a vertex of $\{b_1, b_2, b_3\}$ or if $\overline{a_ia_j} = \overline{b_ib_j}$ for some i, j $(i \neq j)$, then (2) is almost obvious. So we assume that the 7 points $O, a_1, \ldots, b_3$ are all different, and that $\overline{a_ia_j} \neq \overline{b_ib_j}$ for $i \neq j$.

(*a*) For $O \notin \overline{d_1d_2}$: Defining $\mathbf{P}_1(2, K)$ via $V = V(3, K)$, we put: $d_2 = \langle e_1 \rangle, O = \langle e_2 \rangle$ and $d_1 = \langle e_3 \rangle$. Then e_1, e_2, e_3 form a basis for V, and points and lines of $\mathbf{P}_1(2, K)$ have a coordinate representation with respect to this base. Using the notation of §5.1.2, we have: $d_1 = (0, 0)$, $O = (\infty)$, $d_2 = (0)$, $\overline{d_1d_2} = l_{y=0}$. Further $(0, 0) \in \overline{a_2a_3}$, $O \notin \overline{a_2a_3}$, and $a_i, b_i \notin \overline{Od_2} = l_\infty$ for $i = 1, 3$.

(*a*.1) For $a_2 \notin \overline{Od_2}$: We can write $\overline{a_2a_3} = l_{y=\lambda_1 x}$ for some $\lambda_1 (\neq 0) \in K$, hence we can represent a_2 and a_3 as $a_2 = (\mu, \lambda_1\mu)$, $a_3 = (\nu, \lambda_1\nu)$ for some $\mu, \nu \in K$ $(\mu \neq \nu)$. Similarly, since $(0, 0) \in \overline{b_2b_3}$ and $(\infty) \in (\overline{a_ib_i})$ $(i = 2, 3)$, we can write $\overline{b_2b_3} = l_{y=\lambda_2 x}$ for some $\lambda_2 (\neq 0) \in K$, $b_2 = (\mu, \lambda_2\mu)$ and $b_3 = (\nu, \lambda_2\nu)$. Since $(0) \in \overline{a_1a_3}$, $(0) \in \overline{b_1b_3}$, $(\infty) \in \overline{a_1b_1}$ and $a_1, b_1 \notin l_\infty$, we can write $a_1 = (\varepsilon, \lambda_1\nu)$, $b_1 = (\varepsilon, \lambda_2\nu)$ (see Fig. 9). Hence:

$$\overline{a_1a_2} = l_{\lambda_1(\mu-\nu)x + (\varepsilon-\mu)y = \lambda_1\mu(\varepsilon-\nu)},$$

$$\overline{b_1b_2} = l_{\lambda_2(\mu-\nu)x + (\varepsilon-\mu)y = \lambda_2\mu(\varepsilon-\nu)},$$

therefore: $\overline{a_1a_2} \cap \overline{b_1b_2} = d_3 = (\mu(\varepsilon-\nu)/(\mu-\nu), 0) \in \overline{d_1d_2}$.

(*a*.2) For $a_2 \in \overline{Od_2}$: Since $a_2, b_2 \in l_\infty$ $(a_2 \neq b_2)$ we can write $a_2 = (\lambda_1)$,

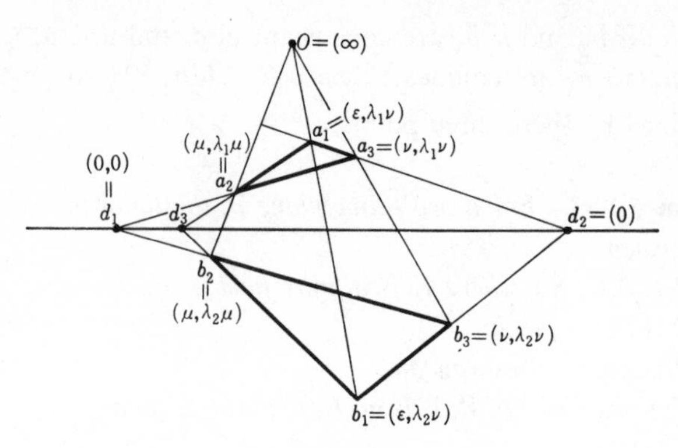

Fig. 9

$b_2 = (\lambda_2)(\lambda_1 \neq \lambda_2)$. Therefore $a_3 = (\mu, \lambda_1\mu), b_3 = (\mu, \lambda_2\mu), a_1 = (v, \lambda_1\mu)$, $b_1 = (v, \lambda_2\mu)$ for some μ, $v \in K$ and

$$\overline{a_1a_2} = l_{y=\lambda_1 x+\lambda_1(\mu-v)}, \quad \overline{b_1b_2} = l_{y=\lambda_2 x+\lambda_2(\mu-v)},$$

hence: $\overline{a_1a_2} \cap \overline{b_1b_2} = d_3 = (v-\mu, 0) \in \overline{d_1d_2}$ (see Fig. 10).

(*b*) For $O \in \overline{d_1}d_2$: Put $d_2 = \langle e_1 \rangle, b_3 = \langle e_2 \rangle$ and $O = \langle e_3 \rangle$. Then e_1, e_2, e_3 form a basis for V and we consider coordinate representations with respect to this basis. We have $O = (0, 0)$, $b_3 = (\infty)$ and $d_2 = (0)$. By multiplying e_1, e_2 and e_3 by suitably chosen scalars we can achieve that $a_1 = (1, 1)$. Now, we can write $b_1 = (1)$, $a_3 = (0, 1)$, $b_2 = (\lambda, m\lambda)$ $(\lambda m \neq 0)$. Hence $d_1 = (\lambda, 0)$, $\overline{b_1b_2} = l_{y=x+(m\lambda-\lambda)}$, $\overline{d_1a_3} = l_{y=-\lambda^{-1}x+1}$ and, $\overline{b_2O} = l_{y=mx}$. Since $a_2 =$

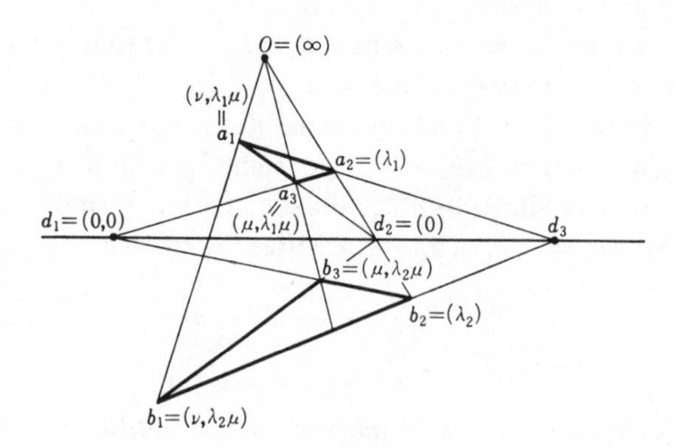

Fig. 10

$\overline{d_1a_3} \cap \overline{b_2O}$, we have:

$$a_2 = \begin{cases} (m) & \text{if } m\lambda + 1 = 0 \\ \left(\dfrac{\lambda}{m\lambda+1}, \dfrac{m\lambda}{m\lambda+1}\right) & \text{if } m\lambda + 1 \neq 0. \end{cases}$$

Hence:

$$\overline{a_1a_2} = \begin{cases} l_{y=\frac{1}{m\lambda-\lambda+1}x+\frac{m\lambda-\lambda}{m\lambda-\lambda+1}} & \text{if } m\lambda+1 \neq \lambda \\ l_{x=1} & \text{if } m\lambda+1=\lambda. \end{cases}$$

and

$$d_3 = \overline{a_1a_2} \cap \overline{b_1b_2} = \begin{cases} (\lambda - m\lambda, 0) & \text{if } m\lambda+1 \neq \lambda \\ (1, 0) & \text{if } m\lambda+1=\lambda. \end{cases}$$

Therefore $d_3 \in \overline{d_1d_2}$ (see Fig. 11).

(2) ⇒ (3): Obvious from the fact that (i) implies (ii).

(3) ⇒ (4): Put $\mathbf{P} = (\Omega, \mathfrak{B})$. If l is a line, and if $a_1, b_1 \notin l\,(a_1 \neq b_1)$, then we will prove that there exists an l-elation that maps a_1 onto b_1. First, we define a permutation σ_{a_1,b_1} of $\Omega - \overline{a_1b_1}$ as follows:

(i) $c^{\sigma_{a_1,b_1}} = c$ for $c \in l$.

(ii) If $c \notin l$, put $\overline{a_1c} \cap l = d, \overline{a_1b_1} \cap l = O, \overline{db_1} \cap \overline{cO} = b$ and define $c^{\sigma_{a_1,b_1}} = b$.

To continue the proof we must first introduce the following lemma.

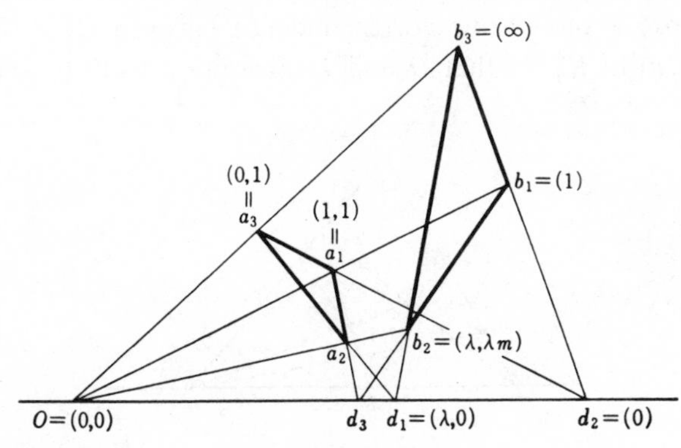

Fig. 11

Lemma 5.1.32 *Pick* $a_2 \in \Omega - (\overline{a_1 b_1} \cup l)$ *and put* $a_2^{\sigma_{a_1,b_1}} = b_2$. *Then* $a_2 \neq b_2$, *and* σ_{a_1,b_1} *and* σ_{a_2,b_2} *agree on* $\Omega - (\overline{a_1 b_1} \cup \overline{a_2 b_2})$.

Proof Obviously $a_2 \neq b_2$. Pick $a_3 \in \Omega - (\overline{a_1 b_1} \cup \overline{a_2 b_2})$ and put $a_3^{\sigma_{a_1,b_1}} = b_3$. If $a_3 \in \overline{a_1 a_2}$, then obviously $a_3^{\sigma_{a_2,b_2}} = b_3$, so assume $a_3 \notin \overline{a_1 a_2}$. Using the weak Desarguean property for $\{a_1, a_2, a_3\}$ and $\{b_1, b_2, b_3\}$, we have $\overline{a_2 a_3} \cap \overline{b_2 b_3} \in l$, hence $a_3^{\sigma_{a_2,b_2}} = b_3$ (see Fig. 12). (End of proof of lemma 5.1.32.) ■

Using this lemma we can extend σ_{a_1,b_1} to become a permutation σ of Ω. Pick $a_2 \in \Omega - (\overline{a_1 b_1} \cup l)$ and put $a_2^{\sigma_{a_1,b_1}} = b_2$. Now we define:

$$a^\sigma = \begin{cases} a^{\sigma_{a_1,b_1}} & a \notin \overline{a_1 b_1} \\ a^{\sigma_{a_2,b_2}} & a \in \overline{a_1 b_1}, \end{cases}$$

and this definition is supported by the lemma. The permutation σ has no fixed points in $\Omega - l$, but σ leaves all points of l and all lines through O fixed. Further, σ maps all lines through a_1 into lines, hence σ maps arbitrary lines into lines by the lemma. Therefore σ is an l-elation satisfying $a_1^\sigma = b_1$.

(4) ⇒ (1): (*a*) Introduction of coordinates: Choose four points O, I, X, Y in $\mathbf{P}$ no three of which are colinear. Put $l_\infty = \overline{XY}$ and $l_\infty \cap \overline{OI} = Z$. Let n denote the order of $\mathbf{P}$ and let K be a set consisting of n elements. Pick and fix two elements from K and call them 0 and 1. $|\Omega - l_\infty| = n^2$, so we can try to identify the elements of $\Omega - l_\infty$ with the elements of $K \times K$. We put $O = (0, 0)$ and $I = (1, 1)$ and we choose a one-to-one correspondence between $\overline{OI} - \{Z\}$ and $\tilde{K} = \{(a, a) | a \in K\}$ so that O and I correspond to $(0, 0)$ and $(1, 1)$

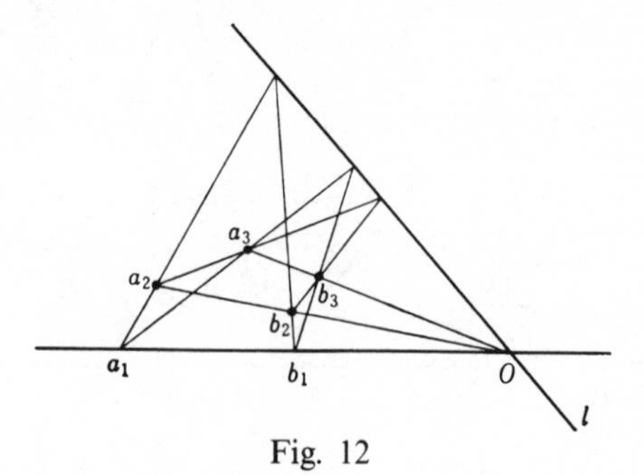

Fig. 12

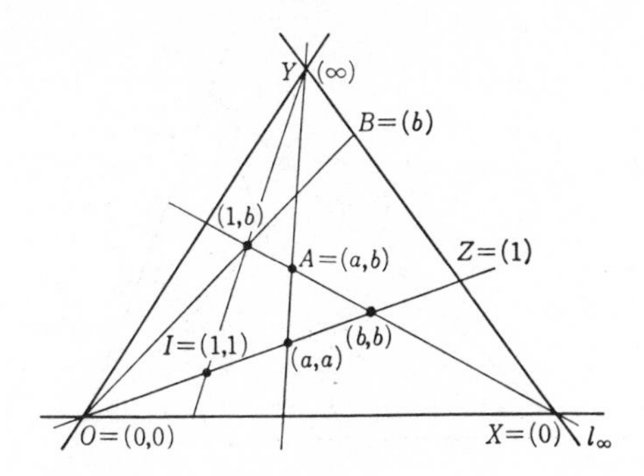

Fig. 13

respectively. We denote elements in $\overline{OI} - \{Z\}$ by the corresponding elements in $\tilde{K}$ under this correspondence. If for $A \notin l_\infty$, $\overline{AY} \cap \overline{OI} = (a, a)$ and $\overline{AX} \cap \overline{OI} = (b, b)$, then we identify A with (a, b). Let ∞ be a symbol that is not an element of K and put $Y = (\infty)$. If $B \in l_\infty$ $(B \neq Y)$ and if $\overline{OB} \cap \overline{YI} = (1, b)$, then we write $B = (b)$. In particular, we have $Z = (1)$ and $X = (0)$ (see Fig. 13).

Now, we have established a one-to-one correspondence between the points of Ω and $K \times K \cup K \cup \{\infty\}$. If $p \in \Omega$ corresponds with (a, b), (a) or (∞), then we will write $p_{(a,b)}$, $p_{(a)}$ or $p_{(\infty)}$, or simply $p = (a, b)$, (a) or (∞) respectively and (a, b), (a) or (∞) is called the coordinate representation of p. In particular, if $p = (a, b)$, then a is called the x-coordinate and b is called the y-coordinate of p.

(*b*) Definition of operations. We define addition and multiplication on K in the following way. For $x, b, y \in K$ we define $y = x + b$ if and only if $\overline{p_{(0,b)}Z} \cap \overline{p_{(x,x)}Y} = p_{(x,y)}$, and $y = xb$ if and only if $\overline{p_{(b)}O} \cap \overline{p_{(x,x)}Y} = p_{(x,y)}$ (see Figs. 14 and 15). If $y = x + b$, then $\overline{p_{(x,y)}Z} \cap \overline{OY} = p_{(0,b)}$ and $\overline{p_{(0,b)}Z} \cap \overline{Xp_{(y,y)}} = p_{(x,y)}$, therefore two elements from among y, x and b determine the third unambiguously. Similarly, if $y = xb$, then $\overline{Op_{(x,y)}} \cap l_\infty = p_{(b)}$ and $\overline{Op_{(b)}} \cap \overline{Xp_{(y,y)}} = p_{(x,y)}$, therefore two elements from among y, x and b determine the third unambiguously provided none of them is 0. Also if x and b are not zero, then xb is not zero.

(*c*) Coordinate representation of lines. From the definition we have

$$l_\infty = \{(a) \mid a \in K \text{ or } a = \infty\}.$$

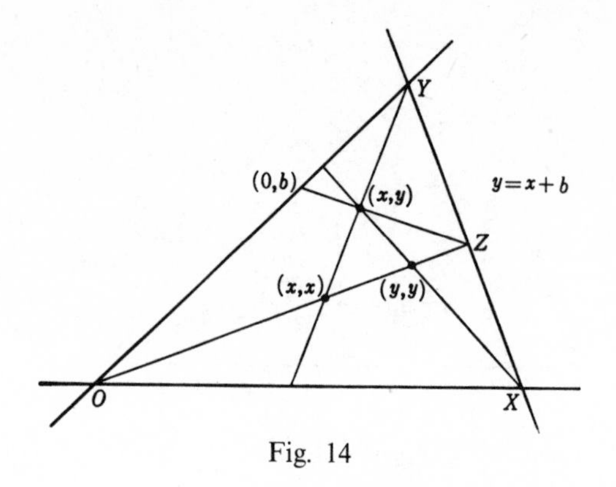

Fig. 14

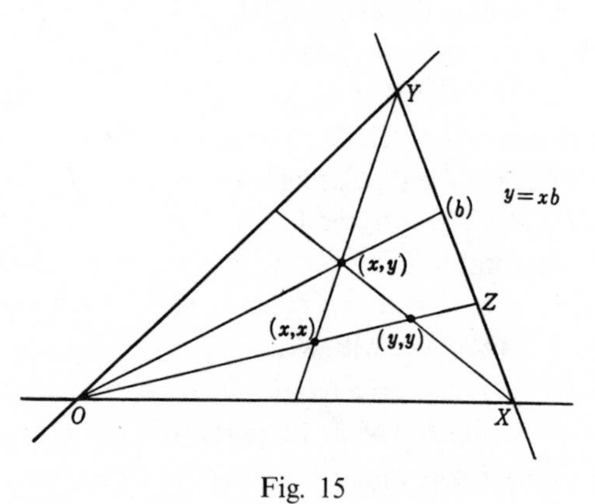

Fig. 15

Let $l \neq l_\infty$ denote a line through Y. Putting $l \cap \overline{OI} = (a, a)$, we have

$$l = \{(a, b) \mid b \in K\} \cup \{(\infty)\}.$$

We represent this line by $x = a$. Finally, let l denote an arbitrary line not through Y. Putting $l \cap l_\infty = p_{(m)}$ and $l \cap \overline{OY} = p_{(0,b)}$, we have

$$l = \{(x, xm + b) \mid x \in K\} \cup \{m\}.$$

(For, if $p_{(x,y)} = A \in l$, then:

$$\overline{Op_{(m)}} \cap \overline{YA} = B = p_{(x,xm)}$$

and

$$\overline{BX} \cap \overline{OZ} = C = p_{(xm,xm)}.$$

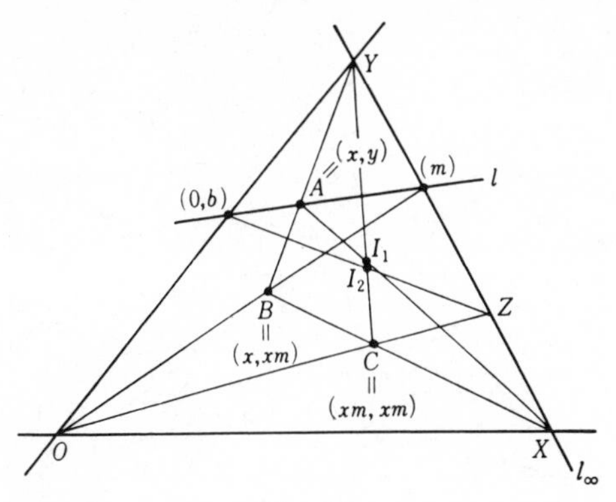

Fig. 16

Therefore, all points on $\overline{YC}$, except Y, can be represented as $(xm, *)$. In particular:

$$\overline{CY} \cap \overline{AX} = I_1 = (xm, y),$$
$$\overline{CY} \cap \overline{p_{(0,b)}Z} = I_2 = (xm, xm + b),$$

hence it suffices to prove $I_1 = I_2$. Since **P** is an $E(l_\infty)$-transitive plane, we can choose $\sigma_1 \in G(Z, l_\infty)$, $\sigma_2 \in G(p_{(m)}, l_\infty)$ and $\sigma_3 \in G(X, l_\infty)$ such that $O^{\sigma_1} = C$, $O^{\sigma_2} = B$ and $B^{\sigma_3} = C$. Since $O^{\sigma_2\sigma_3\sigma_1^{-1}} = O$ and since $\sigma_2\sigma_3\sigma_1^{-1} \in E(l_\infty)$, we have $\sigma_2\sigma_3 = \sigma_1$ by corollary 5.1.9, hence $p_{(0,b)}^{\sigma_2\sigma_3} = p_{(0,b)}^{\sigma_1}$. Now, we easily get $p_{(0,b)}^{\sigma_1} = I_2$ and $p_{(0,b)}^{\sigma_2\sigma_3} = A^{\sigma_3} = I_1$, hence $I_1 = I_2$ (see Fig. 16).) This line is represented by $y = xm + b$. Therefore, a line of **P** can be represented unambiguously as

$$l_\infty, x = a \text{ or } y = xm + b$$

for some $a, b, m \in K$.

If two lines $y = xm_1 + b_1$ and $y = xm_2 + b_2$ are such that $m_1 \neq m_2$, then the two lines are different and their intersection is not on l_∞. Hence, the elements $a, b \in K$ satisfying $b = am_1 + b_1$ and $b = am_2 + b_2$ are uniquely determined and (a, b) represents the point of intersection of the two lines.

Next, we will prove that K with the addition and multiplication defined in (b) is an alternating field.

(d) K is an Abelian group under addition: We first prove the associativity. For $b \in K$ choose $\sigma \in G(Y, l_\infty)$ such that $O^\sigma = p_{(0,b)}$, then, $\forall u \in K$, $p_{(u,u)}^\sigma = p_{(u,u+b)}$ and $l_{y=u}^\sigma = l_{y=u+b}$, hence $p_{(v,u)}^\sigma = p_{(v,u+b)}$,

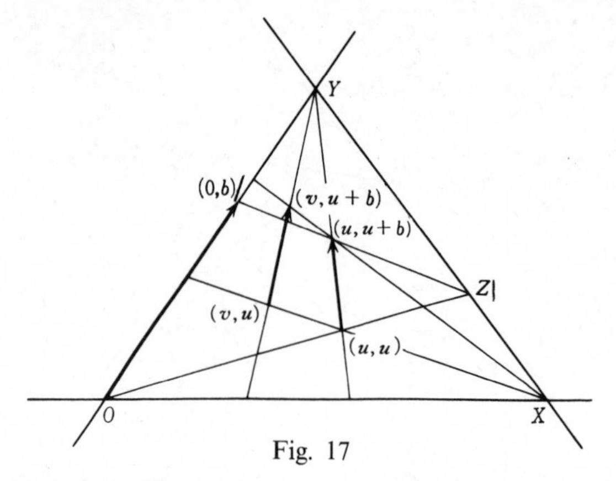

Fig. 17

$\forall u, v \in K$ (see Fig. 17). For $c \in K$ choose $\tau \in G(Y, l_\infty)$ such that $O^\tau = p_{(0,c)}$, then $O^{\sigma\tau} = p^\tau_{(0,b)} = p_{(0,b+c)}$. Therefore, $p^{\sigma\tau}_{(0,a)} = p_{(0,a+(b+c))}$, $\forall a \in K$. On the other hand, $p^{\sigma\tau}_{(0,a)} = p^\tau_{(0,a+b)} = p_{(0,(a+b)+c)}$. Therefore $a + (b + c) = (a + b) + c$.

Commutativity: follows from $\sigma\tau = \tau\sigma$ which was proved in theorem 5.1.10. Zero element: $0 + a = a$ since $\overline{p_{(0,a)}Z} \cap \overline{OY} = p_{(0,a)}$ and $a + 0 = a$ since $\overline{OZ} \cap \overline{p_{(a,a)}Y} = p_{(a,a)}$. Existence of inverse: already shown in (*b*).

(*e*) The unit element for the multiplication is 1 and $0a = a0 = 0$, $\forall a \in K$: Since $\overline{ZO} \cap \overline{p_{(a,a)}Y} = p_{(a,a)}$ we have $a = a1$ and since $p_{(1,a)} \in \overline{p_{(1,a)}O} = l_{y=xa}$ we have $1a = a$. The last assertion is obvious.

(*f*) $(a + b)c = ac + bc$ and $c(a + b) = ca + cb$, $\forall a, b, c \in K$: Since $\overline{p_{(c)}p_{(b,0)}} = l_{y=xm+t}$ passes through $p_{(c)}$ we have $m = c$ and since it passes also through $p_{(b,0)}$ we have $0 = bc + t$. Therefore this line is represented by $y = xc - bc$. Choose $\sigma \in G(X, l_\infty)$ such that $O^\sigma = p_{(b,0)}$. Then

$$p^\sigma_{(a,a)} = l^\sigma_{y=x} \cap l^\sigma_{y=a} = l_{y=x-b} \cap l_{y=a} = p_{(a+b,a)},$$

hence $l^\sigma_{x=a} = l_{x=a+b}$. Further:

$$p^\sigma_{(a,ac)} = l^\sigma_{y=xc} \cap l^\sigma_{x=a} = l_{y=xc-bc} \cap l_{x=a+b} = p_{(a+b,(a+b)c-bc)}.$$

On the other hand:

$$p^\sigma_{(a,ac)} = l^\sigma_{x=a} \cap l^\sigma_{y=ac} = l_{x=a+b} \cap l_{y=ac} = p_{(a+b,ac)}.$$

Therefore $(a + b)c = ac + bc$ (see Fig. 18).

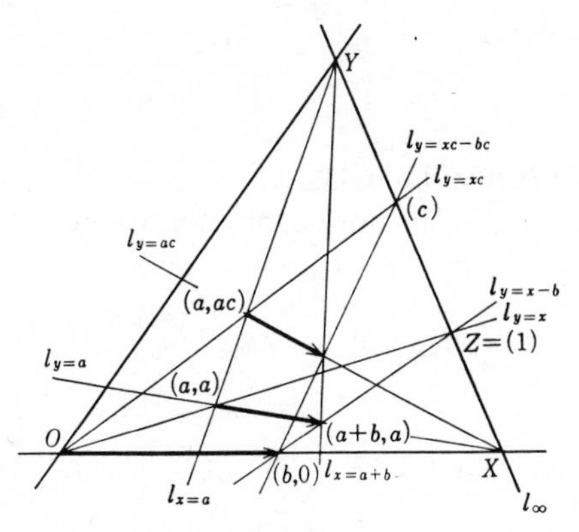

Fig. 18

Next, choose $\sigma \in G(Y, \overline{OY})$ such that $X^\sigma = p_{(a)}$. Since $\overline{p_{(0,d)}X^\sigma} = \overline{p_{(0,d)}p_{(a)}}$, we have $l^\sigma_{y=d} = l_{y=xa+d}$. Therefore,

$$p^\sigma_{(c,cb)} = l^\sigma_{y=cb} \cap l^\sigma_{x=c} = l_{y=xa+cb} \cap l_{x=c} = p_{(c,ca+cb)}.$$

In particular, $p^\sigma_{(1,b)} = p_{(1,a+b)}$, hence $l^\sigma_{y=xb} = l_{y=x(a+b)}$. Therefore,

$$p^\sigma_{(c,cb)} = l^\sigma_{y=xb} \cap l^\sigma_{x=c} = l_{y=x(a+b)} \cap l_{x=c} = p_{(c,c(a+b))}.$$

Therefore, $c(a+b) = ca + cb$ (see Fig. 19).

(*g*) If $a \in K$ and $a \neq 0$, then there exists an inverse element a^{-1} for a (i.e. $a^{-1}a = aa^{-1} = 1$) and moreover $a^{-1}(ab) = (ba)a^{-1} = b$, $\forall b \in K$. Choose $\sigma \in G(O, \overline{OY})$ such that $X^\sigma = p_{(-1-a,0)}$. Since $\overline{p_{(0,b+ab)}X} = l_{y=b+ab}$, $\overline{p_{(0,b+ab)}p_{(-1-a,0)}} = l_{y=xb+b+ab}$ and $p^\sigma_{(0,b+ab)} =$

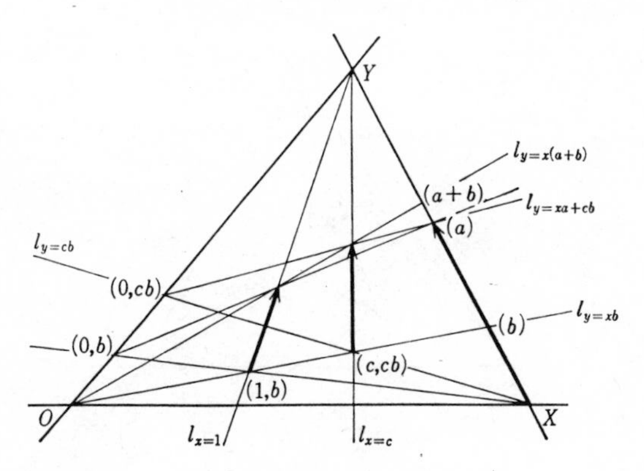

Fig. 19

$p_{(0,b+ab)}$, $\forall b \in K$, we have $l^{\sigma}_{y=b+ab} = l_{y=xb+b+ab}$. Since $p_{(1,b+ab)} = l_{x=1} \cap l_{y=b+ab} \cap l_{y=x(b+ab)}$ and $l^{\sigma}_{y=x(b+ab)} = l_{y=x(b+ab)}$ we have: $p^{\sigma}_{(1,b+ab)} = l^{\sigma}_{x=1} \cap l_{y=xb+b+ab} \cap l_{y=x(b+ab)}$. If we take $b \neq 0$, then we have $b \neq b + ab$ from (*b*), and hence $l_{y=xb+b+ab} \cap l_{y=x(b+ab)} \notin l_{\infty}$ and so we can write: $l^{\sigma}_{x=1} = l_{x=c}$ for some $c \in K$ where $c \neq 0, 1$. Then we have

$$p^{\sigma}_{(1,b+ab)} = p_{(c,cb+b+ab)} = p_{(c,c(b+ab))},$$

and hence

$$cb + b + ab = c(b + ab) \quad \forall b \in K.$$

Putting $c = 1 + a'$ we get $a'(ab) = b$, $\forall b \in K$. Since $a' \neq 0$, applying the above argument for a' instead of a we can find an a'' such that $a''(a'd) = d, \forall d \in K$. Putting $b = d = 1$, we get $a'a = 1 = a''a'$ and putting $d = a$, we get $a'' = a$. Hence a' is the inverse element of a, and $a^{-1}(ab) = b, \forall b \in K$ (see Fig. 20).

Next we prove the relation $(ba)a^{-1} = b, \forall b \in K$. Choose $\sigma \in G(O, \overline{OX})$ such that $Y^{\sigma} = (0, -1)$. For $c \in K (c \neq 0)$ we have

$$l^{\sigma}_{x=c} = \overline{p_{(c,0)} Y^{\sigma}} = \overline{p_{(c,0)} p_{(0,-1)}} = l_{y=xc^{-1}-1}.$$

For $b \in K$ $(b \neq 0)$ we have: $p_{(1,1-ab)} = l_{x=1} \cap l_{y=x(1-ab)}$, hence:

$$p^{\sigma}_{(1,1-ab)} = l_{y=x-1} \cap l_{y=x(1-ab)} = p_{((ab)^{-1},(ab)^{-1}-1)}.$$

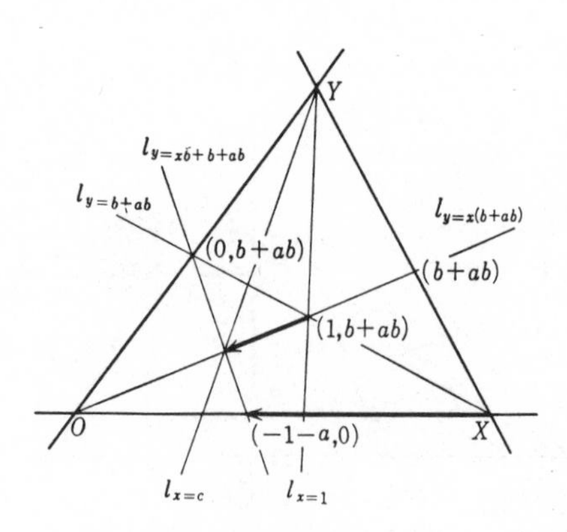

Fig. 20

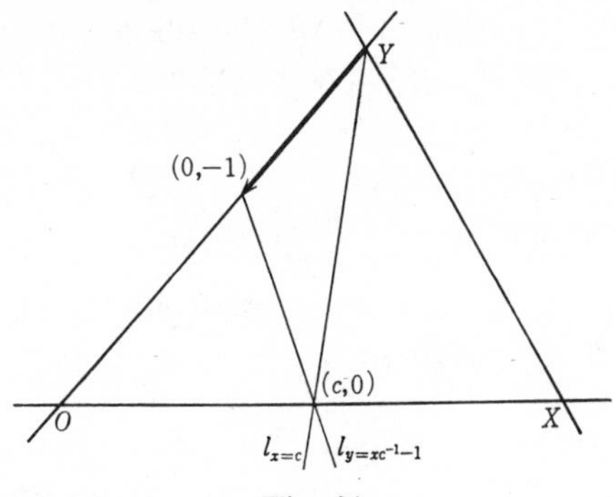

Fig. 21

Also, $p_{(a,1-ab)} = l_{x=a} \cap l_{y=x(a^{-1}-b)}$, hence

$$p^{\sigma}_{(a,1-ab)} = l_{y=xa^{-1}-1} \cap l_{y=x(a^{-1}-b)} = p_{(b^{-1},b^{-1}a^{-1}-1)}.$$

Since $p_{(a,1-ab)} \in \overline{p_{(1,1-ab)}X}$, we have $p^{\sigma}_{(a,1-ab)} \in \overline{p^{\sigma}_{(1,1-ab)}X}$, hence $(ab)^{-1} = b^{-1}a^{-1}$. We know already that $a(a^{-1}b^{-1}) = b^{-1}$, $\forall b \in K$, hence,

$$(ba)a^{-1} = (a(a^{-1}b^{-1}))^{-1} = (b^{-1})^{-1} = b \quad \forall b \in K \text{ (see Fig. 21).}$$

(*h*) We prove the relations $a(ab) = (aa)b$ and $(ba)a = b(aa)$: If $a = \pm 1$, these relations are obvious, so we may assume $a \neq \pm 1$. Putting $t = (a^{-1} - (a+1)^{-1})(a^2 + a)$, we have

$$\begin{aligned}(a+1)t &= (a+1)(a+1-(a+1)^{-1}a^2 - (a+1)^{-1}a) \\ &= (a+1)^2 - a^2 - a = a+1.\end{aligned}$$

Hence $(a+1)^{-1}((a+1)t) = 1$, therefore $t = 1$ by (*g*), i.e. $a^{-1} - (a+1)^{-1} = (a^2+a)^{-1}$ and $b = (a^{-1} - (a+1)^{-1})((a^2+a)b)$, $\forall b \in K$. Putting $w = (a^{-1} - (a+1)^{-1})(a(ab) + ab)$ we have

$$\begin{aligned}(a+1)w &= (a+1)(ab + b - (a+1)^{-1}(a(ab)) - (a+1)^{-1}(ab)) \\ &= (a+1)(ab+b) - a(ab) - ab = (a+1)b,\end{aligned}$$

hence $w = b$, i.e. $(a^2 + a)b = a(ab) + ab$, hence $(aa)b = a(ab)$. The second relation $(ba)a = b(aa)$ is proved in a similar way.

(i) Now we come to the end of the proof. We have proved that K is an alternating field and since $|K| < \infty$ we conclude that K is a field by the theorem of Artin–Zorn (theorem 1.3.18), hence the coordinate representation described in (*c*) is that over the finite field K and we have $\mathbf{P} \simeq \mathbf{P}_1(2, K)$ by theorem 5.1.3. (End of proof of theorem 5.1.31.) ■

5.1.7 The theorem of Ostrom–Wagner

Now let us prove the theorem of Ostrom–Wagner, one of the main objectives of this chapter.

Theorem 5.1.33 (Ostrom–Wagner) *Let* $\mathbf{P} = (\Omega, \mathfrak{B})$ *be a projective plane. If* Aut $\mathbf{P}$ *operates 2-transitively on* Ω, *then* $\mathbf{P} \simeq \mathbf{P}_1(2, K)$ *for some finite field* K.

This theorem is a consequence of the following slightly more general result:

Theorem 5.1.34 (Wagner) *Let* $\mathbf{P} = (\Omega, \mathfrak{B})$ *be a projective plane, let* l_∞ *be a line of* $\mathbf{P}$ *and let* $G \leq \text{Aut}\,\mathbf{P}$ be such that $l_\infty^G = l_\infty$. *If* G *operates transitively on* $\mathfrak{B} - \{l_\infty\}$, *then* $\mathbf{P}$ *is an* $E(l_\infty)$*-transitive plane and* $G \geq E(l_\infty)$.

We first show how theorem 5.1.33 is derived from theorem 5.1.34. Since $G = \text{Aut}\,\mathbf{P}$ operates 2-transitively on Ω, $(G, \mathfrak{B})$ is 2-transitive by theorem 1.5.12. Therefore, $G_{\langle l \rangle}$ operates transitively on $\mathfrak{B} - \{l\}$ for any line l of $\mathbf{P}$, hence $\mathbf{P}$ is an $E(l)$-transitive plane by theorem 5.1.34. Since $\mathbf{P}$ is an $E(l)$-transitive plane for all lines l of $\mathbf{P}$, the truth of theorem 5.1.33 follows from theorem 5.1.31.

Proof of theorem 5.1.34 Let n denote the order of $\mathbf{P}$, then from theorem 5.1.13, $|G| = n^2(n+1)h$ with $h = |G_{a,b}| = |G_{b,\langle l \rangle}|$, where $a \in l_\infty$ and (b, l) is a finite flag.

(1) Case that n is even: Putting $2^u \top n$ and $2^v \top h$, we have $u \geq 1$ and $2^{2u+v} \top |G|$. If S is a Sylow 2-subgroup of G, we have $|S| = 2^{2u+v} \neq 1$. Let $\sigma \in Z(S)$ be an element of order 2. By theorems 5.1.12 and 5.1.22 it suffices to show that $\sigma \in E(l_\infty)$. Let us assume that $\sigma \notin E(l_\infty)$. If σ is not a perspectivity, then $F_{\mathbf{P}}(\langle \sigma \rangle)$ is a subplane of order $\sqrt{n}$ of $\mathbf{P}$ by theorem 5.1.27, hence, $|F_\Omega(\sigma)| = n + \sqrt{n} + 1$.

Since $l_\infty^\sigma = l_\infty$, we have $|F_{l_\infty}(\sigma)| = \sqrt{n}+1$. Therefore $|F_{\Omega - l_\infty}(\sigma)| = n$. If σ is a perspectivity, then σ is an elation since n is even and l_∞ is not the axis of σ since $\sigma \notin E(l_\infty)$. Therefore $|F_{\Omega - l_\infty}(\sigma)| = n$. So in both cases $|F_{\Omega - l_\infty}(\sigma)| = n$. Since $\sigma \in Z(S)$, S operates on $F_{\Omega - l_\infty}(\sigma)$. Let $F_{\Omega - l_\infty}(\sigma) = T_1 + \ldots + T_r$ be the S-orbit decomposition of $F_{\Omega - l_\infty}(\sigma)$. Pick one point a_i from each $T_i\,(i = 1, \ldots, r)$. Since $(G, \Omega - l_\infty)$ is transitive, $|G : G_{a_i}| = n^2$, hence $2^{2u} \,\|\, |G : S_{a_i}|$. Therefore: $2^{2u} \,\|\, |S : S_{a_i}|$, hence $2^{2u} \,\|\, |T_i|$. So: $2^{2u} \,\|\, |F_{\Omega - l_\infty}(\sigma)| = n$. From $2^u \nmid n$, we conclude $u = 0$. Contradiction.

(2) Case that n is odd: Again it suffices to prove that G contains an l_∞-elation as in the case (1), i.e. that $G \cap E(l_\infty) \neq \varnothing$. According to theorem 5.1.30 we have $n = p^s$, where p is an odd prime. Put $p^t \nmid h$. Take a point $a \in l_\infty$ and let S be a Sylow p-subgroup of G_a. Since S is also a Sylow p-subgroup of G, we have $|S| = p^{2s+t}$. For $c \notin l_\infty$ put $|c^S| = p^u (\leq n^2 = p^{2s})$, then $|S_c| = p^{2s+t-u}$. From $|G_{a,c}| = h$ and $G_{a,c} \geq S_c$ we conclude $p^{2s+t-u} | h$, hence $p^{2s} \leq p^u$ and therefore $p^{2s} = p^u$. So $(S, \Omega - l_\infty)$ is transitive. We prove next that $(S_x, \mathfrak{B}_x - \{l_\infty\})$ is transitive for all $x \in l_\infty$. (For, let $l_\infty = T_1 + \ldots + T_r$ be the S-orbit decomposition of l_∞, then the number of orbits of $(S, \Omega) = r + 1$ and this number equals the number of orbits of $(S, \mathfrak{B})$ by theorem 1.5.8. Hence, putting $\mathfrak{B}_{T_i} = \{l \neq l_\infty \,|\, l \cap T_i \neq \varnothing\}$ we arrive at the S-orbit decomposition of $\mathfrak{B} : \mathfrak{B} = \{l_\infty\} + \mathfrak{B}_{t_1} + \ldots + \mathfrak{B}_{T_r}$. Hence, in particular $(S_x, \mathfrak{B}_x - \{l_\infty\})$ is transitive.) Let b denote an element such that $b \in l_\infty$ $(b \neq a)$ and $|S_b|$ is maximal, i.e. $|S_x| \leq |S_b|$ for all $x \in l_\infty$ $(x \neq a)$. Since $S_b \leq S_c$ for $c \in F_{l_\infty}(S_b) - \{a, b\}$, we conclude $S_b = S_c$ from the definition of b, and so $S_b\,(= S_c)$ operates transitively on $\mathfrak{B}_c - \{l_\infty\}$. Let $\sigma \in Z(S_b)$ be an element of order p, then $\sigma \in E(l_\infty)$. (For, $|F_\Omega(\sigma)| \geq 2$ since $a, b \in F_\Omega(\sigma)$, hence $|F_{\mathfrak{B}}(\sigma)| \geq 2$ by theorem 1.5.10. Pick $l \in F_{\mathfrak{B}}(\sigma)\,(l \neq l_\infty)$ and put $c = l \cap l_\infty$. If $c \in F_{l_\infty}(S_b)$, then S_b operates transitively on $\mathfrak{B}_c - \{l_\infty\}$ and since $\sigma \in Z(S_b)$ leaves the element $l(\neq l_\infty)$ of $\mathfrak{B}_c$ fixed, σ leaves all elements of $\mathfrak{B}_c$ fixed. Therefore σ is a perspectivity with centre c. Since $|\sigma| = p \geq 3$, $|l_\infty| = 1 + p^s$, $l_\infty^\sigma = l_\infty$ and $|F_{l_\infty}(\sigma)| \geq 2$, we have $|F_{l_\infty}(\sigma)| \geq 3$. Therefore l_∞ is the axis of σ by corollary 5.1.6 and $\sigma \in E(l_\infty)$. If $c \notin F_{l_\infty}(S_b)$, then $c^\tau = d \neq c$ for some $\tau \in S_b$. Hence putting $e = l \cap l^\tau$, we have $e \notin l_\infty$ and $e^\sigma = l^\sigma \cap l^{\tau\sigma} = l \cap l^\tau = e$. Since S_b operates transitively on $\mathfrak{B}_b - \{l_\infty\}$ and since $\sigma \in Z(S_b)$ leaves $\overline{be} \in \mathfrak{B}_b\,(\overline{be} \neq l_\infty)$ fixed, σ

leaves all elements of $\mathfrak{B}_b$ fixed. Therefore, σ is a perspectivity with centre b. But l and l^τ are two lines not passing through b that are invariant under σ, hence $\sigma = 1$ by corollary 5.1.6. Contradiction.) Hence we have $G \cap E(l_\infty) \neq \varnothing$ which proves our theorem. ■

5.2 Higher-dimensional projective geometry

A geometric structure $\mathbf{P} = (\Omega, \mathfrak{B})$ satisfying the following three conditions, is called a (finite) projective geometry. Elements of Ω are called points, elements of $\mathfrak{B}$ are called lines.

(1) For $p_1, p_2 \in \Omega$ $(p_1 \neq p_2)$ there is exactly one line containing p_1 and p_2 (this line is denoted by $\overline{p_1 p_2}$).

(2) $|l| \geq 3$ for all $l \in \mathfrak{B}$.

(3) If p_1, p_2, p_3, p_4, p_5 are different points of Ω such that $p_4 \in \overline{p_1 p_2}$ and $p_5 \in \overline{p_1 p_3}$, then $\overline{p_4 p_5} \cap \overline{p_2 p_3} \neq \varnothing$.

Excluding the case $|\mathfrak{B}| \leq 1$, projective geometries are $(v, k, 1)$-designs. (For, it suffices to prove that each line contains the same number of points. Let l_1, l_2 be different lines. If $l_1 \cap l_2 \neq \varnothing$, then put $p = l_1 \cap l_2$ and choose different points $p_1, p_2, p_3 \in \Omega$ such that $l_1 = \overline{p_1 p}$, $l_2 = \overline{p_2 p}$ and $p_3 \in \overline{p_1 p_2}$. Then the mapping from l_1 into l_2 defined by

$$q \rightarrow \overline{p_3 q} \cap l_2 \in l_2 \quad (q \in l_1)$$

is one-to-one and onto by (3), hence $|l_1| = |l_2|$. If $l_1 \cap l_2 = \varnothing$, then pick $p_1 \in l_1, p_2 \in l_2$ arbitrarily. Then: $|l_1| = |\overline{p_1 p_2}| = |l_2|$.)

The number $k - 1 = |l| - 1$ (i.e. the order of $\mathbf{P}$ as a design) is called the order of the projective geometry.

As we saw in §1.5.3, a projective design $\mathbf{P}_1(n, K)$ satisfies the conditions for a projective geometry. $\mathbf{P}_1(n, K)$ is called a projective geometry defined over a finite field K. The order of $\mathbf{P}_1(n, K)$ equals $|K|$.

Let $\mathbf{P} = (\Omega, \mathfrak{B})$ be a projective geometry. If $\tilde{\Omega} \subseteq \Omega$ and $\tilde{\mathfrak{B}} \subseteq \mathfrak{B}$ satisfy:

(i) if $a, b \in \tilde{\Omega}$ $(a \neq b)$, then $\overline{ab} \subset \tilde{\Omega}$

(ii) if $\tilde{\mathfrak{B}} = \{l \in \mathfrak{B} \mid l = ab$ for some $a, b \in \tilde{\Omega}\}$ then $(\tilde{\Omega}, \tilde{\mathfrak{B}})$ is called a subspace of $\mathbf{P}$. A subspace of $\mathbf{P}$ is a projective geometry again. If $(\Omega_1, \mathfrak{B}_1)$ and $(\Omega_2, \mathfrak{B}_2)$ are subspaces of $\mathbf{P}$, so is $(\Omega_1 \cap \Omega_2, \mathfrak{B}_1 \cap \mathfrak{B}_2)$.

Therefore, if $\Lambda = \{a_0, \dots, a_r\} \subseteq \Omega$, then there exists a minimal subspace containing Λ. This minimal subspace is denoted by $\langle \Lambda \rangle$ or $\langle a_0, \dots, a_r \rangle$ and called the subspace spanned by $a_0, \dots, a_r$. If Λ is such that $\langle \Lambda \rangle \neq \langle \Lambda - \{a_i\} \rangle$ for all i, then Λ is called independent. If Λ is independent and such that $\mathbf{P} = \langle \Lambda \rangle$, then Λ is called a basis of $\mathbf{P}$. If $\mathbf{P}_1 = (\Omega_1, \mathfrak{B}_1)$ and $\mathbf{P}_2 = (\Omega_2, \mathfrak{B}_2)$ are subspaces of $\mathbf{P}$, then the subspaces $\langle \mathbf{P}_1, \mathbf{P}_2 \rangle = \langle \Omega_1 \cup \Omega_2 \rangle$ and $\mathbf{P}_1 \cap \mathbf{P}_2 = \langle \Omega_1 \cap \Omega_2 \rangle$ are called the join and the intersection of $\mathbf{P}_1$ and $\mathbf{P}_2$ respectively.

Lemma 5.2.1 *Let $\mathbf{P} = (\Omega, \mathfrak{B})$ be a projective geometry of order m.*

(1) *Let $\Lambda = \{a_0, \dots, a_r\}$ be independent and put $\mathbf{P}_1 = \langle \Lambda \rangle = (\Omega_1, \mathfrak{B}_1)$. Let b be an arbitrary element of $\Omega - \Omega_1$ and put $\mathbf{P}_2 = \langle \Lambda \cup \{b\} \rangle = (\Omega_2, \mathfrak{B}_2)$. Then $\{a_0, a_1, \dots, a_r, b\}$ is independent and $|\Omega_2| = m|\Omega_1| + 1$.*

(2) *If Λ is a basis for $\mathbf{P}$, then $|\Omega| = 1 + m + \dots + m^r$ where $r = |\Lambda| - 1$. Hence this number r is independent from the choice of a basis of $\mathbf{P}$. It is called the dimension of $\mathbf{P}$ and denoted by* dim $\mathbf{P}$.

Proof (1) Put $\tilde{\Omega} = \{x \mid x \in \overline{by}, y \in \Omega_1\}$. If $x_1, x_2 \in \tilde{\Omega}$, then $\overline{x_1 x_2} \subset \tilde{\Omega}$ (For, pick $y_i \in \Omega_1$ $(i = 1, 2)$ such that $x_i \in \overline{by_i}$, then $\emptyset \neq \overline{bx} \cap \overline{y_1 y_2} \in \Omega_1$ for all $x \in \overline{x_1 x_2}$ by (3) of the definition of a projective geometry.) Therefore $\tilde{\Omega} = \Omega_2$ and $|\Omega_2| = m|\Omega_1| + 1$. Hence $\{a_0, a_1, \dots, a_r, b\}$ is independent.

(2) Follows from (1). ∎

The dimension of a subspace is defined to be the dimension of that subspace, regarded as a projective geometry. If $\Lambda = \emptyset$, then we write $\langle \Lambda \rangle = \emptyset$ and we define dim $\emptyset = -1$. If $\Lambda = \{a\}$, then $\langle \Lambda \rangle = \{a\}$ is 0-dimensional. In this case Ω has only one point. If $\Lambda = \{a, b\}$ $(a \neq b)$, then $\langle \Lambda \rangle$ is 1-dimensional and $\langle \Lambda \rangle = \overline{ab}$. Two-dimensional subspaces of $\mathbf{P}$ are called planes. If $\mathbf{P}$ is n-dimensional, then an $(n-1)$-dimensional subspace of $\mathbf{P}$ is called a hyperplane. A one-dimensional projective geometry is called a projective line. It is easy to see that 2-dimensional projective geometries are projective planes in the sense of §5.1.1.

Theorem 5.2.2 *Let $\mathbf{P}$ be a projective geometry. If $\mathbf{P}_1$ and $\mathbf{P}_2$ are* *subspaces of* $\mathbf{P}$, *then:*

$$\dim \mathbf{P}_1 + \dim \mathbf{P}_2 = \dim\langle \mathbf{P}_1, \mathbf{P}_2 \rangle + \dim \mathbf{P}_1 \cap \mathbf{P}_2.$$

Proof We apply induction on $r = \dim\langle \mathbf{P}_1, \mathbf{P}_2\rangle - \dim(\mathbf{P}_1 \cap \mathbf{P}_2)$. If $r = 0$, then $\mathbf{P}_1 = \mathbf{P}_2$ and the assertion is obviously true. So assume $r > 0$. If $\mathbf{P}_1 \cap P_2 = \mathbf{P}_1$ or $\mathbf{P}_2$, then the above equality obviously holds, so we may assume $\mathbf{P}_i \neq \mathbf{P}_1 \cap \mathbf{P}_2$ $(i = 1, 2)$. Putting $\mathbf{P}_i = (\Omega_i, \mathfrak{B}_i)$ $(i = 1, 2)$ and $\mathbf{P}_1 \cap \mathbf{P}_2 = (\Omega_0, \mathfrak{B}_0)$, we have $\Omega_i \supsetneqq \Omega_0$ $(i = 1, 2)$ and $\Omega_1 \cap \Omega_2 = \Omega_0$. Pick an arbitrary $\beta \in \Omega_2 - \Omega_1$ and put $\tilde{\mathbf{P}} = \langle \Omega_1, \beta \rangle$ and $\tilde{\mathbf{P}}_2 = \tilde{\mathbf{P}} \cap \mathbf{P}_2$. If $\tilde{\mathbf{P}} \neq \langle \mathbf{P}_1, \mathbf{P}_2 \rangle$, then

$$\dim \tilde{\mathbf{P}} + \dim \mathbf{P}_1 \cap \mathbf{P}_2 = \dim \mathbf{P}_1 + \dim \tilde{\mathbf{P}}_2$$

and

$$\dim\langle \mathbf{P}_1, \mathbf{P}_2 \rangle + \dim \tilde{\mathbf{P}}_2 = \dim \tilde{\mathbf{P}} + \dim \mathbf{P}_2$$

by the induction hypothesis and the truth of the equality of the theorem follows from these two equalities. If $\tilde{\mathbf{P}} = \langle \mathbf{P}_1, \mathbf{P}_2 \rangle$, then for all $\gamma(\neq \beta) \in \Omega_2$, $\overline{\beta\gamma} \cap \Omega_1 \neq \varnothing$ and $\overline{\beta\gamma} \cap \Omega_1 \subseteq \Omega_2 \cap \Omega_1 = \Omega_0$, hence $\langle \Omega_0, \beta \rangle = \mathbf{P}_2$. Therefore, $\dim \mathbf{P}_2 - \dim \mathbf{P}_1 \cap \mathbf{P}_2 = 1 = \dim\langle \mathbf{P}_1, \mathbf{P}_2 \rangle - \dim \mathbf{P}_1$ by lemma 5.2.1 from which the theorem follows. ■

As we saw in §5.1, it was quite complicated to give conditions for a projective plane to be the projective plane over some finite field. For higher dimensions, the situation is very simple: every n-dimensional $(n \geq 3)$ projective geometry is isomorphic with a projective geometry $\mathbf{P}_1(n, K)$ for some finite field K. To prove this, some preparation is needed.

Let $V = V(n + 1, K)$ be an $(n + 1)$-dimensional K-module, let $\mathbf{P}(n, K)$ be the projective space obtained from V and let $\mathbf{P}_1(n, K) = (\Omega, \mathfrak{B})$ be the projective geometry defined over the field K. Choose a basis $a_0, a_1, \ldots, a_n$ for V, then every element of Ω can be represented as $\langle a \rangle = \{\lambda a \mid \lambda \in K\}$ for some $a(\neq 0) \in V$. If $a = \lambda_0 a_0 + \ldots + \lambda_n a_n$, we put $\langle a \rangle = [\lambda_0, \ldots, \lambda_n]$ and we call $\lambda_0, \ldots, \lambda_n$ the coordinates of $\langle a \rangle$ with respect to the base $a_0, \ldots, a_n$. It is easy to verify that:

$$[\lambda_0, \ldots, \lambda_n] = [\mu_0, \ldots, \mu_n] \Leftrightarrow \mu_i = \lambda\lambda_i \ (i = 1, \ldots, n) \text{ for some } \lambda(\neq 0) \in K.$$

If $l \in \mathfrak{B}$, and if $[\lambda_0, \ldots, \lambda_n]$ and $[\mu_0, \ldots, \mu_n]$ are different points of l, then l can be represented as:

$$l = \{[\lambda\lambda_0 + \mu\mu_0, \ldots, \lambda\lambda_n + \mu\mu_n] \mid (\lambda, \mu) \in K \times K \text{ and } (\lambda, \mu) \neq (0, 0)\}.$$

Theorem 5.2.3 *Let* $\mathbf{P} = (\Omega, \mathfrak{B})$ *be an n-dimensional projective geometry* ($n \geq 3$), *then all planes of* $\mathbf{P}$ *are Desarguean.*

Proof Let π be a plane of $\mathbf{P}$. Let π also denote the set of points of that plane. According to theorem 5.1.31 it suffices to show that π is $E(l)$-transitive for all lines l of π. Let $\mathbf{P}_0$ denote a 3-dimensional subspace of $\mathbf{P}$ containing π, let $\pi' \neq \pi$ denote a plane of $\mathbf{P}_0$ such that $\pi \cap \pi' = l$, and finally let $p \in \mathbf{P}_0$ be a point that is not contained in π or π' (the existence of p follows from lemma 5.2.1). We define a mapping $f_p : \pi \to \pi'$ in the following way: for $X \in \pi$, we have $\dim(\overline{pX} \cap \pi') = 0$ by theorem 5.2.2, so we can define:

$$f_p(X) = \overline{pX} \cap \pi'.$$

This mapping is clearly one-to-one and onto and maps lines into lines by (3) of the definitions of a projective geometry, i.e. f_p is an isomorphism from π onto π'. Let g_p represent the inverse map of f_p. For $p, q \in \mathbf{P}_0$ not on π or π' the mapping $f_p g_q$ is an automorphism of π, that leaves the points on l fixed, hence $f_p g_q$ is a perspectivity. Now, let A, B be two different points of $\pi - l$. Put $\overline{AB} \cap l = C$, pick $p \in \mathbf{P}_0$ such that p is not on π or π' and put $\overline{pA} \cap \pi' = A'$, and $BA' \cap \overline{pC} = q$. (It is possible to pick points like this by theorem 5.2.2 and (3) of the definition of a projective geometry (see Fig. 22).)

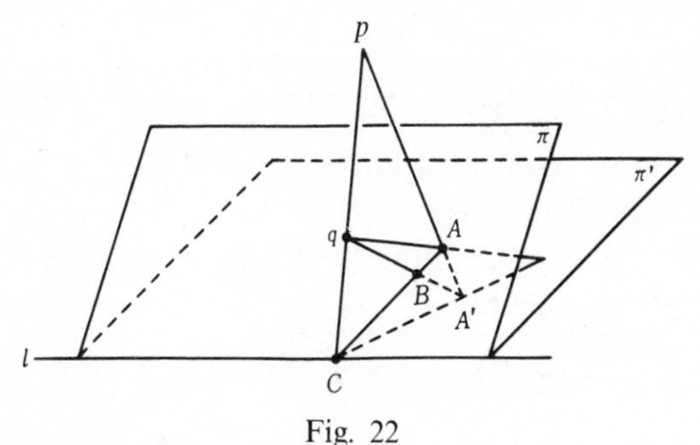

Fig. 22

Now it is easily shown that $\sigma = f_p g_q \in E(l)$ and $A^\sigma = B$, proving the theorem. ■

Theorem 5.2.4 *Let* $\mathbf{P}$ *be an* $n(\geq 3)$*-dimensional projective geometry, then* $\mathbf{P} \simeq \mathbf{P}_1(n, K)$ *for some finite field* K.

Proof Since each line of $\mathbf{P}$ contains the same number of points, there exists a finite field K such that every plane of $\mathbf{P}$ is isomorphic with $\mathbf{P}_1(2, K)$, by theorem 5.1.31 and 5.2.3. We proceed with the proof of the theorem by induction on n, namely, we will prove $\mathbf{P} \simeq \mathbf{P}_1(n, K)$ under the assumption that all $(n-1)$-dimensional subspaces of $\mathbf{P}$ are isomorphic with $\mathbf{P}_1(n-1, K)$. Let $\mathbf{P}_0$ be an $(n-2)$-dimensional subspace of $\mathbf{P}$ and let $\mathbf{P}_1$ and $\mathbf{P}_2$ be two different $(n-1)$-dimensional subspaces of $\mathbf{P}$ containing $\mathbf{P}_0$ (existence of such subspaces follows from lemma 5.2.1). Choose $a \in \mathbf{P}_1, b \in \mathbf{P}_2$ such that $a, b \notin \mathbf{P}_0$. Since $\mathbf{P}_0 \simeq \mathbf{P}_1(n-2, K)$, we can introduce coordinates in $\mathbf{P}_0$. Now, we fix a system of coordinates in $\mathbf{P}_0$ and we will show that this system of coordinates can be extended to both $\mathbf{P}_1$ and $\mathbf{P}_2$ consistently. If $x \in \mathbf{P}_0$ is represented as $x = [\lambda_1, \ldots, \lambda_{n-1}]$, then we represent x as a point of $\mathbf{P}$ as $x = [0, \lambda_1, \ldots, \lambda_{n-1}, 0]$. It is possible to define a coordinate system on $\mathbf{P}_1$ that extends this representation of points of $\mathbf{P}_0$ and such that $a = [1, 0, \ldots, 0]$. In the same way, it is possible to introduce a coordinate system in $\mathbf{P}_2$ such that $b = [0, \ldots, 0, 1]$. Then points of $\mathbf{P}_1$ are represented as $[\lambda_0, \lambda_1, \ldots, \lambda_{n-1}, 0]$ and points of $\mathbf{P}_2$ are represented as $[0, \lambda_1, \ldots, \lambda_n]$ and we have, for $\lambda(\neq 0) \in K$,

$$[\lambda\lambda_0, \ldots, \lambda\lambda_{n-1}, 0] = [\lambda_0, \ldots, \lambda_{n-1}, 0]$$

and

$$[0, \lambda\lambda_1, \ldots, \lambda\lambda_n] = [0, \lambda_1, \ldots, \lambda_n]$$

Next we want to define coordinates for a point $x \in \mathbf{P}$ not in $\mathbf{P}_1$ or $\mathbf{P}_2$. We distinguish two cases:

(*a*) Where $x \notin \overline{ab}$: According to theorem 5.2.2 we can define three points c, d and e by:

$$\overline{ax} \cap \mathbf{P}_2 = c, \quad \overline{bx} \cap \mathbf{P}_1 = d$$
$$\overline{bc} \cap \overline{ad} = \langle a, b, x \rangle \cap \mathbf{P}_0 = e.$$

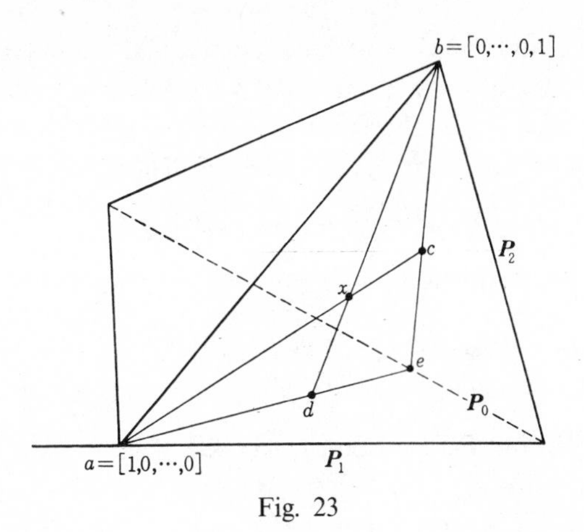

Fig. 23

Putting $e=[0, v_1, \ldots, v_{n-1}, 0]$, d and c are represented as: $d=[v_0, v_1, \ldots, v_{n-1}, 0]$ and $c=[0, v_1, \ldots, v_n]$. We put

$$x=[v_0, v_1, \ldots, v_n]$$

(see Fig. 23).

(*b*) Where $x \in \overline{ab}$: Pick $c \in \mathbf{P}_0$ and pick $d \in \overline{xc}$ such that $d \neq x$ and $d \neq c$. If

$$c=[0, v_1, \ldots, v_{n-1}, 0]$$

and

$$d=[v_0, v_1, \ldots, v_{n-1}, v_n]$$

then we put:

$$x=[v_0, 0, \ldots, 0, v_n].$$

This definition is independent from the choice of c and d, as we proceed to prove. Put $e_1=\overline{bd} \cap \overline{ac}$ and $e_2=\overline{ad} \cap \overline{bc}$. Then e_1 and e_2 are written as

$$e_1=[v_0, v_1, \ldots, v_{n-1}, 0] \text{ and } e_2=[0, v_1, \ldots, v_n].$$

Now pick $c' \in \mathbf{P}_0$ and $d' \in \overline{xc'}$ such that $d' \neq x$ and $d' \neq c'$ and put $c'=[0, \mu_1, \ldots, \mu_{n-1}, 0]$ and $d'=[\mu_0, \mu_1, \ldots, \mu_{n-1}, \mu_n]$. Then

$$e_1'=\overline{bd'} \cap \overline{ac'}=[\mu_0, \ldots, \mu_{n-1}, 0]$$

and

$$e_2'=\overline{ad'} \cap \overline{bc'}=[0, \mu_1, \ldots, \mu_{n-1}, \mu_n].$$

Since $e_1 \in \overline{ac}$ and $e'_1 \in \overline{ac'}$, we have $\overline{cc'} \cap \overline{e_1 e'_1} \neq \emptyset$ and since $d \in \overline{xc}$ and $d' \in \overline{xc'}$, we also have: $\overline{cc'} \cap \overline{dd'} \neq \emptyset$. Further, $d \in \overline{be_1}, d' \in \overline{be'_1}$, hence $\overline{dd'} \cap \overline{e_1 e'_1} \neq \emptyset$, therefore: $\overline{e_1 e'_1} \cap \overline{cc'} = \overline{dd'} \cap \overline{cc'} \neq \emptyset$ and in the same way $\overline{e_2 e'_2} \cap \overline{cc'} = \overline{dd'} \cap \overline{cc'} \neq \emptyset$. Therefore:

$$\overline{e_1 e'_1} \cap \overline{cc'} = \overline{e_2 e'_2} \cap \overline{cc'} \neq \emptyset.$$

Denoting this point by y, we have

$$\begin{aligned} y &= [\nu\nu_0 + \mu\mu_0, \dots, \nu\nu_{n-1} + \mu\mu_{n-1}, 0] \\ &= [0, \nu'\nu_1 + \mu'\mu_1, \dots, \nu'\nu_n + \mu'\mu_n] \quad \text{for some } \nu, \mu, \nu', \mu' \in K \end{aligned}$$

since both $\mathbf{P}_1$ and $\mathbf{P}_2$ are isomorphic with $\mathbf{P}_1(n-1, K)$. Therefore: $\nu_0 : \mu_0 = \mu : -\nu$ and $\nu_n : \mu_n = \mu' : -\nu'$. Since $c \neq c'$ we must have $\nu_i : \nu_j \neq \mu_i : \mu_j$ for some i, j with $1 \leq i, j \leq n-1$. Therefore $\mu : -\nu = \mu' : -\nu'$, hence $\nu_0 : \mu_0 = \nu_n : \mu_n$ (see Fig. 24).

Now, every point of $\mathbf{P}$ can be represented as:

$$[\nu_0, \nu_1, \dots, \nu_n] \quad \text{(where } \nu_0, \dots, \nu_n \in K \text{ and are not all 0)}$$

and it is easy to check that:

$$[\nu_0, \nu_1, \dots, \nu_n] = [\mu_0, \mu_1, \dots, \mu_n] \Leftrightarrow \nu_i = \lambda\mu_i \quad (i = 0, \dots, n)$$

for some $\lambda \in K$ $(\lambda \neq 0)$. To finish the proof of the theorem it suffices to prove that the line through $x = [\nu_0, \dots, \nu_n]$ and $y = [\mu_0, \dots, \mu_n]$

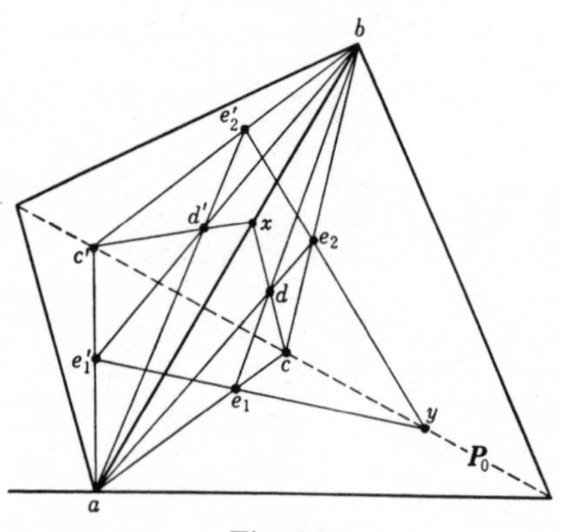

Fig. 24

$(x \neq y)$ can be represented as

$$\{[\nu\nu_0 + \mu\mu_0, \ldots, \nu\nu_n + \mu\mu_n] | (\nu, \mu) \in K \times K, (\nu, \mu) \neq (0, 0)\}.$$

We distinguish the following cases

(*c*) If $x, y \in \mathbf{P}_1$ or if $x, y \in \mathbf{P}_2$, then the assertion is obviously true.

(*d*) If $x \in \mathbf{P}_1$ and $y \in \mathbf{P}_2$, then we can put

$$x = [\nu_0, \nu_1, \ldots, \nu_{n-1}, 0]$$

and

$$y = [0, \mu_1, \ldots, \mu_n].$$

Since $\overline{ax} \cap \mathbf{P}_2 = x_0 = [0, \nu_1, \ldots, \nu_{n-1}, 0]$, we have:

$$\overline{yx_0} = \{[0, \nu\nu_1 + \mu\mu_1, \ldots, \nu\nu_{n-1} + \mu\mu_{n-1}, \mu\mu_n] | \nu, \mu \in K\}.$$

Therefore, all points of the plane $\langle a, y, x_0 \rangle$ can be represented as:

$$[\xi, \nu\nu_1 + \mu\mu_1, \ldots, \nu\nu_{n-1} + \mu\mu_{n-1}, \mu\mu_n] \quad \nu, \mu, \xi \in K$$

and in a similar way, putting $y_0 = \overline{by} \cap \mathbf{P}_1$, the points of the plane $\langle b, y_0, x \rangle$ can be represented as

$$[\nu\nu_0, \nu\nu_1 + \mu\mu_1, \ldots, \nu\nu_{n-1} + \mu\mu_{n-1}, \eta] \quad \nu, \mu, \eta \in K.$$

Since $\overline{yx}$ is the line of intersection of the two planes $\langle a, y, x_0 \rangle$ and $\langle b, y_0, x \rangle$, we have $\overline{yx} = \{[\nu\nu_0, \nu\nu_1 + \mu\mu_1, \ldots, \nu\nu_{n-1} + \mu\mu_{n-1}, \mu\mu_n] | \nu, \mu \in K\}$ (see Fig. 25).

(*e*) The remaining case: if $\overline{xy} \cap \mathbf{P}_1 \neq \overline{xy} \cap \mathbf{P}_2$, we are reduced to case (*d*). If $\overline{xy} \cap \mathbf{P}_1 = \overline{xy} \cap \mathbf{P}_2 = z$, there are two possibilities. If $\overline{xy} \cap \overline{ab} = \emptyset$, then $\overline{xy}$ is the line of intersection of the two planes

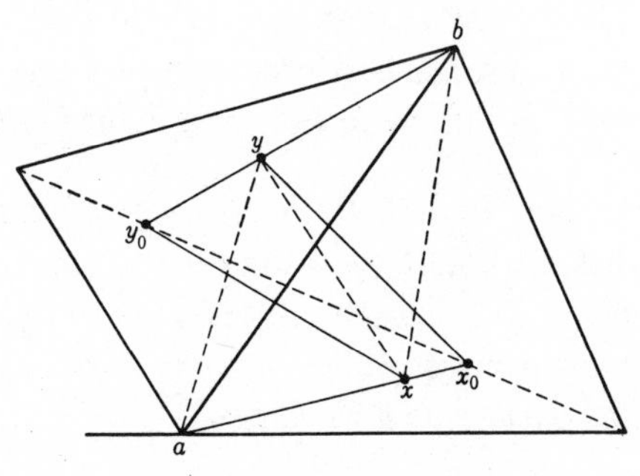

Fig. 25

$\langle a, z, x \rangle$ and $\langle b, z, y \rangle$ and the proof is similar to the proof in case (*d*). If $\overline{xy} \cap \overline{ab} \neq \varnothing$, the proof is easy from the definition of the coordinates. ■

5.3 Characterization of projective geometries

The $P_r(n, K)$ are special but important examples of block designs. One often needs to show that block designs arising from a study of permutation groups can be represented by some $P_r(n, K)$. Theorems 5.1.31 and 5.2.4 are fundamental results in this direction. In this section we will use these theorems to prove the results of Dembowski and Wagner.

Let $\mathbf{D} = (\Omega, \mathfrak{B})$ be a block design with parameters (v, b, k, r, λ). For $a_1, \ldots, a_n \in \Omega$ we call $\bigcap_{\{a_1,\ldots,a_n\} \subseteq B \in \mathfrak{B}} B$ the subspace of $\mathbf{D}$ spanned by $a_1, \ldots, a_n$ and we denote this subspace by $\langle a_1, \ldots, a_n \rangle$. If there is no element $B \in \mathfrak{B}$ such that $\{a_1, \ldots, a_n\} \subseteq B$, then we put $\langle a_1, \ldots, a_n \rangle = \Omega$. In particular, if $a, b \in \Omega \, (a \neq b)$, then $\langle a, b \rangle$ is called the line connecting a and b and is denoted by $\overline{ab}$. Lines are also called one-dimensional subspaces. From the definition of block designs a line is completely determined by any two different points on that line, i.e. if $c, d \in \overline{ab} \, (c \neq d)$, then $\overline{ab} = \overline{cd}$. The points $a_1, \ldots, a_n \in \Omega$ are called collinear if $\langle a_1, \ldots, a_n \rangle$ is a line. If $a, b, c \in \Omega$ are not collinear, then $\langle a, b, c \rangle$ is called a plane, or a two-dimensional subspace. If a', b', c' are three non-collinear points in $\langle a, b, c \rangle$, then $\langle a', b', c' \rangle \subseteq \langle a, b, c \rangle$ but not necessarily $\langle a', b', c' \rangle = \langle a, b, c \rangle$. If each plane of $\mathbf{D}$ is contained in the same number of elements of $\mathfrak{B}$, $\mathbf{D}$ is called a smooth block design. If $\mathbf{D}$ is a smooth block design then a plane of $\mathbf{D}$ is uniquely determined by any three non-collinear points of that plane. Moreover, every line of $\mathbf{D}$ contains the same number of points, as we will prove in the following theorem.

Theorem 5.3.1 *Let* $\mathbf{D} = (\Omega, \mathfrak{B})$ *be a smooth block-design with parameters* (v, b, k, r, λ). *Let* t *be the number of blocks which contain an arbitrary plane (this number is independent of the choice of planes). Then the number of points on a line is the same for all lines, and denoting this number by* h *we have:*

$$\lambda(k - h) = (v - h)t. \tag{5.3.1}$$

Proof Choose a line l and count the number of elements of $\{(B, x) | B \in \mathfrak{B}, l \subset B, x \in B - l\}$ in two different ways as the proof of theorem 1.5.1. Then $\lambda(k - |l|) = (v - |l|)t$, proving the theorem. ■

This theorem shows that it is possible to construct some other block designs from a smooth block design. Namely we have the following theorem.

Theorem 5.3.2 *Let $\mathbf{D} = (\Omega, \mathfrak{B})$ be a smooth block-design with parameters (v, b, k, r, λ), let t denote the number of blocks which contain an arbitrary plane and let h denote the number of points on an arbitrary line.*

(1) *If $\mathfrak{B}^*$ denotes the collection of all lines of $\mathbf{D}$, then $\mathbf{D}^* = (\Omega, \mathfrak{B}^*)$ is a block design with parameters:*

$$\left(v, \frac{v(v-1)}{h(h-1)}, h, \frac{v-1}{h-1}, 1\right).$$

(2) *Let $a \in \Omega$. Put $\Omega_a = \{l | l$ is a line of $\mathbf{D}$ and $a \in l\}$ and $\mathfrak{B}_a = \{[B] | B \in \mathfrak{B}$ and $a \in B\}$, where $[B] = \{l | l$ is a line of $\mathbf{D}, l \subset B$ and $a \in l\}$, then $\mathbf{D}_a = (\Omega_a, \mathfrak{B}_a)$ is a block-design with parameters:*

$$\left(\frac{v-1}{h-1}, r, \frac{k-1}{h-1}, \lambda, t\right) \quad \text{provided } \lambda > 1.$$

(Notice that it is possible that $\mathbf{D}^*$ or $\mathbf{D}_a$ is trivial.)

(3) $(v-1) \le r(h-1)$ *and* $(r-1)t \le \lambda(\lambda - 1)$. *If $\mathbf{D}$ is symmetric the equality signs hold.*

Proof (1), (2): Left to the reader.

(3) The first inequality follows from theorem 1.5.1 if $\lambda = 1$, and from Fisher's inequality applied to $\mathbf{D}_a$ if $\lambda > 1$. Using (5.3.1) we get:

$$h - 1 = \frac{\lambda(k-1) - t(v-1)}{\lambda - t} = \frac{(\lambda^2 - tr)(v-1)}{(\lambda - t)r}.$$

Substituting this in the first inequality, yields the second. If $\mathbf{D}$ is symmetric, then for $\mathfrak{B}_a \ni [\mathrm{B}_1], [\mathrm{B}_2] (\neq)$ we have $|\mathrm{B}_1 \cap \mathrm{B}_2| =$ a constant and therefore $|[\mathrm{B}_1] \cap [\mathrm{B}_2]| =$ a constant. Thus $\mathbf{D}_a$ is a symmetric design. Therefore the first inequality becomes an equality, hence the second inequality also becomes an equality. ■

Theorem 5.3.3 (Dembowski–Wagner) *Let* $\mathbf{D} = (\Omega, \mathfrak{B})$ *be a non-trivial design with parameters* (v, b, k, r, λ). *The following conditions are all equivalent:*

(1) $\mathbf{D}$ *is the design consisting of the points and hyperplanes of some projective geometry of dimension* $n \geq 2$.

(2) $\mathbf{D}$ *is a smooth, symmetric design.*

(3) *Every plane is contained in exactly* $\lambda(\lambda - 1)/(k - 1)$ *blocks.*

(4) *Every line contains exactly* $(b - \lambda)/(r - \lambda)$ *points.*

(5) *Each line intersects each block.*

Proof (1)⇒(2). If the dimension of the projective geometry in question equals 2, then $\mathbf{D}$ is a projective plane, obviously satisfying (2). (The number of blocks containing a plane equals 0.) If the dimension ≥ 3, then $\mathbf{D} = P_{n-1}(n, K)$ for some finite field K by theorem 5.2.4, from which (2) follows.

(2)⇒(3). Follows from theorem 5.3.2(3). (Note that $k = r$, since $\mathbf{D}$ is symmetric.)

(3)⇒(4). Since

$$\frac{\lambda(\lambda - 1)}{k - 1} \leq \frac{\lambda(\lambda - 1)}{r - 1}$$

by theorem 5.3.2(3), we have $k \geq r$. On the other hand since $k \leq r$ by Fisher's inequality, $k = r$, i.e. $\mathbf{D}$ is symmetric. From (5.3.1) and $t = \lambda(\lambda - 1)/(k - 1)$ we have

$$\frac{\lambda(k - h)}{v - h} = \frac{\lambda(\lambda - 1)}{k - 1}.$$

Using $v = b$, $k = r$ and $(v - 1)\lambda = k(k - 1)$ we get (4).

(4)⇒(5): The number of blocks containing a line l equals λ, and the number of blocks intersecting l in one point is

$$\frac{(r - \lambda)(b - \lambda)}{r - \lambda} = b - \lambda,$$

therefore l intersects each block.

(5) ⇒ (1): If $\lambda = 1$, then **D** is a projective geometry of dimension 2 (i.e. a projective plane) by theorem 1.5.5. So we may assume $\lambda > 1$. Let l be an arbitrary line. Since l intersects every block, we have $|l|(r-\lambda)+\lambda = b$, i.e. the number of points on an arbitrary line equals $(b-\lambda)/(r-\lambda)$. Let a be a point not on l and let t be the number of blocks containing the plane $\langle a, l\rangle$, then we find by computing the number of blocks containing a:

$$r = t + \frac{(\lambda - t)(b-\lambda)}{(r-\lambda)},$$

hence **D** is smooth. If $\mathfrak{B}^*$ denotes the collection of all lines of **D**, then we will prove that $\mathbf{P} = (\Omega, \mathfrak{B}^*)$ is a projective geometry:

(1) Since all lines contain the same number of points every line is completely determined by two of its points.

(2) Each line contains at least 3 points. (For, suppose that each line contains just 2 points. Let a be a point and let B be a block not containing a, then every line through a intersects B, hence $|B| = v-1$ since these lines contain only 2 points. Therefore **D** is trivial, contrary to our assumption.)

(3) Let $\pi = \langle a, b, c\rangle$ be a plane, choose $d \in \overline{ab}$ and $e \in \overline{ac}$ such that a, b, c, d, e are all different. Let B be a block that contains $\overline{bc}$ but not a. Let f be the point of intersection of $\overline{de}$ and B, then: $\langle f, b, c\rangle \subsetneq \langle a, b, c\rangle$. Therefore f, b, c are collinear from the smoothness of **D** and $\overline{de} \cap \overline{bc} \neq \varnothing$.

Hence $\mathbf{P} = (\Omega, \mathfrak{B}^*)$ is a projective geometry. Since all blocks B of **D** are clearly subspaces of **P** and intersect each line, they are all hyperplanes of **P**. (This follows, for example, from the fact $\mathbf{P} \simeq \mathbf{P}_1(n, K)$ by theorem 5.2.4 or from theorem 5.2.2.) Since $b \geq v$, $\mathfrak{B}$ is just the collection of hyperplanes of **P**. ■

6

Finite groups and finite geometries

For each finite geometry $\mathbf{D} = (\Omega, \mathfrak{B})$, we obtain a permutation group of Ω as its automorphism group. Conversely, given a permutation group (G, Ω), it is often possible to construct a finite geometry $\mathbf{D} = (\Omega, \mathfrak{B})$ such that G is considered as a subgroup of Aut $\mathbf{D}$ and use this geometry to study the properties of (G, Ω). Using this approach, one can ask 'What relation exists between the properties of finite geometries and the properties of permutation groups?' For example, one would like to know which permutation groups give rise to designs of the form $\mathbf{P}_i(n, K)$. In this chapter we consider this problem for 2-transitive permutation groups.

6.1 Designs constructed from 2-transitive groups

Let (G, Ω) be a transitive permutation group of degree n. For a natural number k, such that $2 \leq k \leq n-1$, let $\Omega^{(k)}$ denote the collection of all subsets of k elements of Ω and let us consider G to be a permutation group on $\Omega^{(k)}$ in an obvious way. If $\mathfrak{B}$ is an invariant domain of $(G, \Omega^{(k)})$, then $\mathbf{D} = (\Omega, \mathfrak{B})$ is a geometric structure and G is considered to be a subgroup of Aut $\mathbf{D}$. Since (G, Ω) is transitive, Aut $\mathbf{D}$ ($\geq G$) operates transitively on Ω. Moreover if we assume that (G, Ω) is 2-transitive, then the number of elements of $\mathfrak{B}$ containing two elements of Ω is independent of the choice of those two elements, hence $\mathbf{D}$ is a design. Conversely, each design, the automorphism group of which operates 2-transitively on the set of its points, can be obtained in this way. In particular, if $(G, \Omega^{(k)})$ is not transitive and if $\mathfrak{B}$ is an arbitrary non-trivial invariant domain of $(G, \Omega^{(k)})$ (i.e. $\mathfrak{B} \subsetneq \Omega^{(k)}$), then and only then $\mathbf{D} = (\Omega, \mathfrak{B})$ is a non-trivial design. (Note that we have $2 \lneq k \lneq n-1$ by our assumption.)

Let (G, Ω) be a 2-transitive group. For $B \in \Omega^{(k)}$ let $\mathfrak{B}(B)$ denote the orbit of $(G, \Omega^{(k)})$ containing B, i.e.:

$$\mathfrak{B}(B) = B^G = \{X \in \Omega^{(k)} \mid X = B^\sigma, \sigma \in G\}.$$

We denote the resulting design $(\Omega, \mathfrak{B}(B))$ by $\mathbf{D}(B)$. The parameters of $\mathbf{D}(B)$ are denoted by $(v_B, b_B, k_B, r_B, \lambda_B)$. Then $v_B = |\Omega|, b_B = |\mathfrak{B}(B)| = |G : G_{\langle B \rangle}|$ and the automorphism group of $\mathbf{D}(B)\,(\geq G)$ operates 2-transitively on the set of points Ω and transitively on the set of the blocks $\mathfrak{B}(B)$. The structure of $\mathbf{D}(B)$ depends on not only (G, Ω), but also choice of $B \subset \Omega$. In this section we will study some relations between structures of (G, Ω) and $\mathbf{D}(B)$ under suitable choice of B. First we shall give some conditions for $\mathbf{D}(B)$ to be a non-trivial symmetric design:

Theorem 6.1.1 *Let (G, Ω) be a 2-transitive group of degree n.*

(1) *There exists a subset B of Ω such that $\mathbf{D}(B)$ is a non-trivial, symmetric design if and only if there exists a subgroup H of G satisfying the following two conditions:*

(i) $|G : H| = n$.

(ii) *The permutation group (H, Ω) is not transitive and has no fixed points.*

(2) *More precisely the following holds: If $B \subseteq \Omega$ is such that $\mathbf{D}(B)$ is a non-trivial symmetric design, then $H = G_{\langle B \rangle}$ satisfies conditions* (i) *and* (ii). *Conversely, if H is a subgroup of G satisfying* (i) *and* (ii), *then $\mathbf{D}(B)$ is a non-trivial symmetric design for every orbit B of (H, Ω). In this case (H, Ω) has exactly two orbits.*

Proof It suffices to prove (2). Let $\mathbf{D}(B)$ be a non-trivial, symmetric design. Since $\mathbf{D}(B)$ is symmetric, we have $n = v_B = b_B = |G : G_{\langle B \rangle}|$, i.e. $H = G_{\langle B \rangle}$ satisfies (i). If $G_{\langle B \rangle}$ has a fixed point a, then $G_{\langle B \rangle} = G_a$ since $G_a \geq G_{\langle B \rangle}$ and $|G : G_{\langle B \rangle}| = n = |G : G_a|$. Therefore $B = \{a\}$ or $B = \Omega - \{a\}$ since (G, Ω) is 2-transitive, contrary to the fact that $\mathbf{D}(B)$ is non-trivial. Hence H satisfies (ii).

Conversely, let $H \leq G$ satisfy (i) and (ii) and let B be an orbit of (H, Ω). Putting $|B| = k$, we have $|\Omega^{(k)}| > n$ since $1 \lneqq |B| \lneqq n - 1$. Since $|B^G| = |G : G_{\langle B \rangle}| \leq |G : H| = n$, $(G, \Omega^{(k)})$ is not transitive and $\mathbf{D}(B)$ is a non-trivial design. From $b_B = |B^G| \leq n = v_B$ and Fisher's inequality, we conclude $b_B = v_B$, i.e. $\mathbf{D}(B)$ is symmetric. Let the number of orbits of (H, Ω) equal r, then $r \geq 2$ by (ii). Let χ be a permutation character of (G, Ω), then χ can be written as:

$$\chi = 1_G + \chi_0,$$

where 1_G denotes the identity character and χ_0 denotes an irreducible character (theorem 3.2.6). According to theorem 3.2.4, we have:

$$(1_H, \chi_H) = \frac{1}{|H|} \sum_{\sigma \in H} \chi(\sigma) = r,$$

hence $(1_H, \chi_{0|H}) = r - 1$. Therefore $(1_H^G, \chi_0) = r - 1$ by theorem 2.10.13 and letting e denote the identity of G, we get

$$n = (1_H^G)(e) \geq (r-1)\chi_0(e) = (r-1)(n-1),$$

hence $r \leq 2$. Combining this with $r \geq 2$, we get $r = 2$. ■

Next we want to find conditions under which $\mathbf{D}(B)$ becomes a non-trivial design with $\lambda_B = 1$. First, suppose that $\mathbf{D}(B)$ is such that $\lambda_B = 1$. Choose $a \in B$ and put $\Delta = B - \{a\}$. If $\sigma \in G_a$ is such that $\Delta^\sigma \cap \Delta \neq \varnothing$, then $|B^\sigma \cap B| \geq 2$, hence $B^\sigma = B$ since $\lambda_B = 1$ and hence we have $\Delta^\sigma = \Delta$, i.e. Δ is a set of imprimitivity of $(G_a, \Omega - \{a\})$. In particular, if $|B| > 2$, then $(G_a, \Omega - \{a\})$ is not primitive. This gives the following lemma:

Lemma 6.1.2 *Let (G, Ω) be a 2-transitive group, and let B be a subset of Ω with $|B| > 2$. If $\mathbf{D}(B)$ is a non-trivial design with $\lambda_B = 1$, then (G, Ω) is not 2-primitive. In fact, $B - \{a\}$ is a non-trivial set of imprimitivity of $(G_a, \Omega - \{a\})$.*

Conversely, let (G, Ω) be 2-transitive, but not 2-primitive. Let $a \in \Omega$, let Δ be a non-trivial set of imprimitivity of $(G_a, \Omega - \{a\})$ and put $B = \Delta \cup \{a\}$. Since $B^G \neq \Omega^{(|B|)}$ (assuming $n > 6$), $\mathbf{D}(B)$ is non-trivial, but it is not necessarily true that $\lambda_B = 1$. To find sufficient conditions for $\lambda_B = 1$ to be true, we need some preparations.

Let (G, Ω) be 2-transitive. Let $a, b \in \Omega$, let D denote a set of imprimitivity of $(G_a, \Omega - \{a\})$ such that $b \in D$ and put $B = \Delta(a, b) = \{a\} \cup D$. For x, y arbitrary elements of Ω, we can choose $\sigma \in G$ such that

$a^\sigma = x$ and $b^\sigma = y$, then we define $\Delta(x, y)$ by

$$\Delta(x, y) = B^\sigma = \{x\} \cup D^\sigma.$$

Since D is a set of imprimitivity of $(G_a, \Omega - \{a\})$, the subset $\Delta(x, y)$ of Ω is independent of the choice of σ, but completely determined by x and y. Moreover it is easy to check that Δ satisfies:

$$\Delta(x, y) \supseteq \{x, y\} \tag{6.1.1}$$

$$\Delta(x, y)^\sigma = \Delta(x^\sigma, y^\sigma) \tag{6.1.2}$$

$$\Delta(x, y) = \Delta(x, z) \quad \forall z \in \Delta(x, y) - \{x\} \tag{6.1.3}$$

for all $x, y \in \Omega$. Generally, a map $\Delta : \Omega \times \Omega - \{(x, x) | x \in \Omega\} \rightarrow 2^\Omega$ satisfying (6.1.1), (6.1.2) and (6.1.3) is called a pre-design function. If $2 \lneq |\Delta(x, y)| \lneq n$, then Δ is called a non-trivial pre-design function. We have shown that a system of sets of imprimitivity of $(G_a, \Omega - \{a\})$ determines a pre-design function. Conversely, it is easily seen that if Δ is a pre-design function for (G, Ω), then $\Delta(a, b) - \{a\}$ is a set of imprimitivity of $(G_a, \Omega - \{a\})$. Therefore, there exists a one-to-one correspondence between the systems of sets of imprimitivity of $(G_a, \Omega - \{a\})$ and the set of pre-design functions of (G, Ω). Under this correspondence, a non-trivial system of set of imprimitivity corresponds with a non-trivial pre-design function and conversely. Hence:

Lemma 6.1.3 *Let (G, Ω) be a 2-transitive group. (G, Ω) is not 2-primitive if and only if (G, Ω) has a non-trivial pre-design function.*

Let Δ be a pre-design function, then the design $\mathbf{D}(\Delta(a, b))$ is denoted by $\mathbf{D}(\Delta)$ and the parameters of $\mathbf{D}(\Delta)$ are denoted by $(v_\Delta, b_\Delta, k_\Delta, r_\Delta, \lambda_\Delta)$. If Δ satisfies in addition to (6.1.1), (6.1.2) and (6.1.3) also:

$$\Delta(x, y) = \Delta(y, x) \quad \text{for some } x, y \in \Omega, \tag{6.1.4}$$

then Δ is called symmetric. Since (G, Ω) is 2-transitive, (6.1.4) is equivalent with:

$$\Delta(x, y) = \Delta(y, x) \quad \text{for all } x, y \in \Omega. \tag{6.1.5}$$

We are now in a position to formulate and prove the following theorem:

Theorem 6.1.4 *Let G be a 2-transitive group.*

(1) *If Δ is a (non-trivial) symmetric pre-design function, then* $\mathbf{D}(\Delta)$ *is a (non-trivial) design with $\lambda_\Delta = 1$.*

(2) *If $B \subseteq \Omega$ is such that $\mathbf{D}(B)$ is a (non-trivial) design with $\lambda_B = 1$, then there exists a (non-trivial) symmetric pre-design function Δ such that: $\mathbf{D}(B) = \mathbf{D}(\Delta)$. Actually, for $x, y \in \Omega (x \neq y)$ we define: $\Delta(x, y) = B^\sigma$, where $\sigma \in G$ is such that $x, y \in B^\sigma$.*

Proof (1) For $a, b \in \Delta(x, y)$ $(a \neq b)$, it suffices to show that $\Delta(x, y) = \Delta(a, b)$. If $x = a$, this is true by (6.1.3). If $x \neq a$, then $\Delta(x, y) = \Delta(x, a) = \Delta(a, x) = \Delta(a, b)$.

(2) Let x and y be arbitrary elements of Ω. Since $\lambda_B = 1$, the set $B^\sigma, \sigma \in G$ satisfying $x, y \in B^\sigma$, is unambiguously determined by x and y and this defines a symmetric pre-design function Δ such that $\mathbf{D}(B) = \mathbf{D}(\Delta)$.

Next, we want to derive a group theoretic condition under which $\mathbf{D}(B)$ becomes a design with $\lambda_B = 1$. Let $H \leq G$ and $K \leq H$. Then K is called a (G, H)-weakly closed subgroup if the following condition is satisfied:

(i) if $K^\sigma \leq H$ $(\sigma \in G)$, then $K^\sigma = K$.

If the following stronger condition:

(ii) $K^\sigma \cap H \leq K \quad \forall \sigma \in G$

is satisfied, then K is called a (G, H)-strongly closed subgroup. It is easy to verify that in both cases $K \trianglelefteq H$.

Theorem 6.1.5 *Let (G, Ω) be a 2-transitive group, let $a, b \in \Omega$, let $W(\neq 1)$ be a subgroup of $G_{a,b}$ and put $B = F_\Omega(W)$. If W is either*

(i) *a $(G, G_{a,b})$-weakly closed subgroup, or*

(ii) *a $(G_a, G_{a,b})$-strongly closed subgroup,*

then the design $\mathbf{D}(B)$ is such that $\lambda_B = 1$.

Proof Let x and y be two arbitrary elements of Ω and choose

$\sigma \in G$ such that $a^\sigma = x$ and $b^\sigma = y$. Since in both cases (i) and (ii) $W \trianglelefteq G_{a,b}$, it is easy to check that W^σ is independent of the choice of σ, but completely determined by x and y. We put $W^\sigma = W(x, y)$ and $\Delta(x, y) = F_\Omega(W(x, y))$. If we can prove that Δ is a symmetric pre-design function of (G, Ω), then the assertion of the theorem follows by theorem 6.1.4. The first condition $(x, y) \in \Delta(x, y)$ is obviously satisfied. For $\sigma \in G$, we have

$$\Delta(x, y)^\sigma = F_\Omega(W(x, y))^\sigma = F_\Omega(W(x^\sigma, y^\sigma)) = \Delta(x^\sigma, y^\sigma),$$

proving the second condition. To prove conditions (6.1.3) and (6.1.4) we have to consider the two cases (i) and (ii) separately.

(i) Where W is $(G, G_{a,b})$-weakly closed. For $x, y \in \Omega$, $W(x, y)$ is clearly $(G, G_{x,y})$-weakly closed. Since $W(x, y) \leq G_{x,z}$ for $z \in \Delta(x, y)$ $(z \neq x)$, we have $W(x, y) = W(x, z)$, hence $\Delta(x, y) = \Delta(x, z)$. Similarly we conclude from $W(y, x) \leq G_{x,y}$ that $W(y, x) = W(x, y)$, hence $\Delta(y, x) = \Delta(x, y)$. This proves that Δ is a symmetric pre-design function.

(ii) Where W is $(G_a, G_{a,b})$-strongly closed. For $x, y \in \Omega$, $W(x, y)$ is clearly $(G_x, G_{x,y})$-strongly closed. Hence:

$$W(x, y) \cap G_z \leq W(x, z) \quad \forall z \in \Omega - \{x\}.$$

In particular, we see that $W(x, y) = W(x, z)$ holds for $z \in \Delta(x, y)$ $(z \neq x)$, hence $\Delta(x, y) = \Delta(x, z)$. Now, the only thing that remains to be proved is that Δ is symmetric. Since $|\Delta(x, y)| = |\Delta(y, x)|$, it suffices to prove that $F_\Omega(W(y, x)) \subseteq F_\Omega(W(x, y))$ and to prove this, it suffices to prove the following proposition.

Proposition *Let P be a Sylow p-subgroup of $W(y, x)$ for some prime p. If there exists an element $z \in \Omega$ such that $P \leq G_z$, then there exists a Sylow p-subgroup Q of $W(x, y)$ such that $Q \leq G_z$.*

(In fact, let us assume the proposition is proved. Take $z \in F_\Omega(W(y, x))$. Then from the proposition G_z contains a Sylow p-subgroup of $W(x, y)$ for each prime p, hence G_z contains $W(x, y)$ and we can conclude $F_\Omega(W(y, x)) \subseteq F_\Omega(W(x, y))$.)

To prove this proposition, we need the following lemma:

Lemma 6.1.6 *Let H be a group, let L and K be subgroups of H such that $L \trianglelefteq H$ and $H = LK$. If P and Q are Sylow p-subgroups of K and L respectively, such that $P \ltimes Q$ (i.e. $P \le \mathcal{N}_H(Q)$), then PQ is a Sylow p-subgroup of LK and $P \cap Q$ is a Sylow p-subgroup of $L \cap K$ and moreover $P \cap Q = P \cap L = K \cap Q$.*

Proof Since PQ is a p-subgroup of LK, $PQ \cap L$ is a p-subgroup of L containing Q. Since Q is a Sylow p-subgroup of L, we have $PQ \cap L = Q$. We have

$$\begin{aligned}|KL:PQ| &= |KL:PL|\,|PL:PQ| = |K:K\cap PL|\,|L:PQ\cap L|\\ &= |K:P|\,|L:Q|/|K\cap PL:P|,\end{aligned}$$

and this is relatively prime with p, hence PQ is a Sylow p-subgroup of KL. Since $(P \cap L) \ltimes Q$, $(P \cap L)Q$ is a p-subgroup of L. Therefore $P \cap L = P \cap Q$ using the fact that Q is a Sylow p-subgroup of L. Therefore

$$|K\cap L:P\cap Q| = |K\cap L:P\cap L| = |P(K\cap L):P| \not\equiv 0 \pmod p,$$

hence $P \cap Q$ is a Sylow p-subgroup of $K \cap L$. The equality $K \cap Q = P \cap Q$ follows easily from this. (End of proof of lemma 6.1.6.) ■

Now, let us prove the proposition. Let P be a Sylow p-subgroup of $W(y, x)$ satisfying the condition stated in the proposition. If $z = x$, the proposition is obviously true, so we may assume $z \neq x$. If $P \le W(x, y)$, we just put $Q = P$, so we may assume $P \not\le W(x, y)$. Since $W(x, z) \trianglelefteq G_{x,z}$ and $P \le G_{x,z}$, there exists a Sylow p-subgroup Q of $W(x, z)$ such that $P \ltimes Q$. We will prove that this Q has the desired properties, i.e. that Q is a Sylow p-subgroup of $W(x, y)$ and $Q \le G_z$. Since $Q \le W(x, z)$, we clearly have $Q \le G_z$. Since $W(x, y) \underset{G}{\sim} W(x, z)$, the order of Q equals the order of the Sylow p-subgroups of $W(x, y)$, hence it suffices to prove $Q \le W(x, y)$. Since $Q \le W(x, z)$ and since $W(x, z) \cap G_y \le W(x, y)$, it suffices to prove $Q \le G_y$, namely $Q = Q_y$.

We first prove $\mathcal{N}_G(P) \le G_y$. (Proof: Suppose there is an element $\sigma \in \mathcal{N}_G(P)$ such that $y^\sigma \neq y$. Then:

$$P \le W(y,x) \cap W(y^\sigma, x^\sigma) = W(y,x)_{y^\sigma} \cap W(y^\sigma, x^\sigma)_y \le W(y, y^\sigma) \cap W(y^\sigma, y),$$

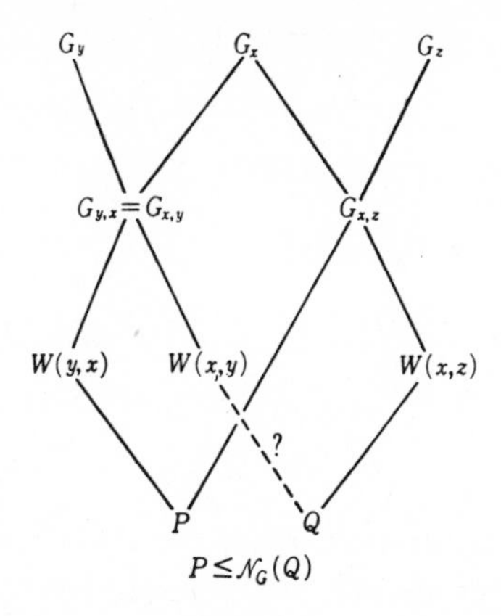

hence P is a Sylow p-subgroup of $W(y, y^\sigma) \cap W(y^\sigma, y)$. Since $P \cap W(x, y)$ is a Sylow p-subgroup of $W(y, x) \cap W(x, y)$ by lemma 6.1.6, we conclude using the 2-transitivity of (G, Ω) that $|P| = |P \cap W(x, y)|$, hence $P \leq W(x, y)$, contrary to our assumption.)

Next we prove: $\mathcal{N}_Q(P) = Q_y$. (Proof: Since $P \ltimes Q \cap G_y = Q_y$, we see that PQ_y is a p-group. From $PQ_y \leq G_{y,x}$ we get $PQ_y \ltimes W(y, x) \cap PQ_y = P$, hence $\mathcal{N}_Q(P) \geq Q_y$. On the other hand, $\mathcal{N}_Q(P) = \mathcal{N}_G(P) \cap Q \leq G_y \cap Q = Q_y$, hence $\mathcal{N}_Q(P) = Q_y$.)

Now, let us assume $Q_y \lneqq Q$. Since $P \trianglelefteq PQ_y$ and $P \ntrianglelefteq PQ$, we have, $PQ \gneqq PQ_y$ and there exists an element $\sigma \in PQ, \sigma \notin PQ_y$ such that $(PQ_y)^\sigma = PQ_y$. $P^\sigma \trianglelefteq (PQ_y)^\sigma = PQ_y$ since $P \trianglelefteq PQ_y$, hence $PP^\sigma \leq PQ_y$ and $P \leq \mathcal{N}_G(P^\sigma) = \mathcal{N}_G(P)^\sigma \leq G_{y^\sigma}$. However, $\sigma \notin PQ_y = (PQ)_y = PQ \cap G_y$, hence $\sigma \notin G_y$, i.e. $y^\sigma \neq y$. Therefore;

$$P \leq W(y, x) \cap G_{y^\sigma} \leq W(y, y^\sigma).$$

Similarly we conclude from $P^\sigma \leq \mathcal{N}_G(P) \leq G_y$ that

$$P^\sigma \leq W(y^\sigma, x^\sigma) \cap G_y \leq W(y^\sigma, y).$$

Hence PP^σ is a Sylow p-subgroup of $W(y, y^\sigma)W(y^\sigma, y)$ by lemma 6.1.6. Using the 2-transitivity of (G, Ω) we conclude that the order of the Sylow p-subgroups of $W(y, x)W(x, y)$ equals $|PP^\sigma|$. Since $P \ltimes W(x, y)$ it is possible to choose a Sylow p-subgroup Q_1 of $W(x, y)$ such that $P \ltimes Q_1$ and according to lemma 6.1.6, PQ_1 is a Sylow p-subgroup of $W(y, x)W(x, y)$ and $P \cap Q_1 = P \cap W(x, y)$

is a Sylow p-subgroup of $W(y,x)\cap W(x,y)$. We also have

$$\begin{aligned}P\cap Q_1 &= P\cap W(x,y) = P\cap W(x,y)\cap G_z \le P\cap W(x,z)\\ &= P\cap Q \le W(y,x)\cap W(x,z)\cap G_y \le W(y,x)\cap W(x,y)\end{aligned}$$

hence $P\cap Q_1 = P\cap Q$. Therefore

$$|PP^\sigma| = |PQ_1| = |P||Q_1|/|P\cap Q_1| = |P||Q|/|P\cap Q| = |PQ|,$$

which is contrary to $PP^\sigma \le PQ_y \lneqq PQ$. (End of proof of theorem 6.1.5.) ■

Let (G,Ω) be a 2-transitive group, let a be a point of Ω and let N be a normal subgroup of G_a. We are going to construct a design using this N. For $x\in\Omega$, choose $\sigma\in G$ such that $a^\sigma = x$. Since $N \trianglelefteq G_a$, N^σ is independent of the choice of σ, we can define $N(x) = N^\sigma$. Then from the definition of $N(x)$, for all $\tau\in G$, we have $N(x)^\tau = N(x^\tau)$ and $(N(x)_y)^\tau = N(x^\tau)_{y\tau}$. Since for $y\in\Omega-\{x\}$ both $y^{N(X)}$ and $F_{\Omega-\{x\}}(N(x)_y)$ are sets of imprimitivity of $(G_x, \Omega-\{x\})$, we have that Δ_N and Γ_N defined by

$$\Delta_N(x,y) = \{x\}\cup y^{N(x)} \quad\text{and}\quad \Gamma_N(x,y) = F_\Omega(N(x)_y) \tag{6.1.6}$$

are both pre-design functions of (G,Ω), giving rise to designs $\mathbf{D}(\Delta_N)$ and $\mathbf{D}(\Gamma_N)$. Since

$$(N(x)_y)^\sigma \cap G_{x,y} \le N(x)\cap G_{x,y} = N(x)_y \quad \text{for } \sigma\in G_x,$$

$N(x)_y$ is a $(G_x, G_{x,y})$-strongly closed subgroup, hence Γ_N is symmetric and $\mathbf{D}(\Gamma_N)$ is a design with $\lambda = 1$ by theorem 6.1.5.

Lemma 6.1.7 *For $x, y\in\Omega$ we have:*

$$\mathcal{N}_{N(x)}(N(x)_y N(y)_x) = \mathcal{N}_{N(x)}(N(x)_y).$$

Proof For $\sigma\in\mathcal{N}_{N(x)}(N(x)_y N(y)_x)$ we have $N(x)_y N(y)_x = (N(x)_y N(y)_x)^\sigma = N(x)_{y\sigma} N(y^\sigma)_x$. In particular $N(x)_{y\sigma} \le G_{x,y}$ and $(N(x)_y)^\sigma = N(x)_{y\sigma} \le N(x)\cap G_{x,y} = N(x)_y$. Hence we have $\mathcal{N}_{N(x)}(N(x)_y N(y)_x) \le \mathcal{N}_{N(x)}(N(x)_y)$. To prove the opposite inclusion, since $\mathcal{N}_{N(x)}(N(x)_y N(y)_x) \ge N(x)_y$, it suffices to prove that

$$|\mathcal{N}_{N(x)}(N(x)_y N(y)_x) : N(x)_y| \ge |\mathcal{N}_{N(x)}(N(x)_y) : N(x)_y|.$$

$N(x)$ and $N(x)N(y)_x$ both operate transitively on $\Delta_N(x, y) - \{x\} = y^{N(x)}$ and the subgroups leaving y fixed are $N(x)_y$ and $N(x)_y N(y)_x$ respectively. Since Γ_N is symmetric, the sets of fixed points of $N(x)_y$ and $N(y)_x$ coincide, hence the sets of fixed points in $\Delta_N(x, y) - \{x\}$ under $N(x)_y$ and $N(x)_y N(y)_x$ coincide and contain y. Hence if we denote this set of fixed points by Ω_0, then $\mathcal{N}_{N(x)}(N(x)_y)$ and $\mathcal{N}_{N(x)N(y)_x}(N(x)_y N(y)_x)$ operate transitively on Ω_0 by Witt's theorem (theorem 3.4.3) and their subgroups leaving y fixed are $N(x)_y$ and $N(x)_y N(y)_x$ respectively. Therefore:

$$\begin{aligned}
&|\mathcal{N}_{N(x)}(N(x)_y) : N(x)_y| \\
&\quad = |\mathcal{N}_{N(x)N(y)_x}(N(x)_y N(y)_x) : N(x)_y N(y)_x| \\
&\quad = |\mathcal{N}_{N(x)}(N(x)_y N(y)_x) N(y)_x : N(x)_y N(y)_x| \\
&\quad = |\mathcal{N}_{N(x)}(N(x)_y N(y)_x) : \mathcal{N}_{N(x)}(N(x)_y N(y)_x) \cap N(x)_y N(y)_x| \\
&\quad \leq |\mathcal{N}_{N(x)}(N(x)_y N(y)_x) : N(x)_y|.
\end{aligned}$$

■

6.2 Characterization of projective transformation groups

In the previous chapters we have learned some results on the relation between finite groups and finite geometries. As an application of those results we will prove here some results connected with the characterization in terms of permutation groups of projective transformation groups. A fundamental result in this direction is:

Theorem 6.2.1 (O'Nan) *Let (G, Ω) be a 2-transitive group. If for some $a \in \Omega$ the group G_a contains an Abelian, normal subgroup A $(A \neq 1)$, such that $(A, \Omega - \{a\})$ is not semi-regular, then*

$$PSL(m+1, q) \leq G \leq P\Gamma L(m+1, q)$$

for some natural number $m \geq 2$ and some prime power q.

Proof Since $(A, \Omega - \{a\})$ is not semi-regular, there exists an element σ of prime order p such that σ has fixed points in $\Omega - \{a\}$. Hence the subgroup of A generated by all elements of order p is not $\{1\}$. Since this subgroup is a characteristic subgroup of A, it is a normal subgroup of G_a and it is not semi-regular as a permutation group on $\Omega - \{a\}$. Therefore, we may assume that A is an elementary Abelian p-group for some prime p from the beginning. Since A is a p-group and since A operates not semi-regularly on $\Omega - \{a\}$, the pre-design functions Δ_A and Γ_A, as defined in (6.1.6), are not trivial. Therefore, they give rise to non-trivial designs $\mathbf{D}(\Delta_A) = (\Omega, \mathfrak{B}_1)$ and $\mathbf{D}(\Delta_A) = (\Omega, \mathfrak{B}_2)$, where:

$$\mathfrak{B}_1 = \{X \subset \Omega \mid X = \Delta_A(x, y), x, y \in \Omega, x \neq y\}$$
$$\mathfrak{B}_2 = \{X \subset \Omega \mid X = \Gamma_A(x, y), x, y \in \Omega, x \neq y\}.$$

As we saw in §6.1, $\mathbf{D}(\Gamma_A)$ is such that $\lambda_{\Gamma_A} = 1$. From now on, we will write Δ and Γ instead of Δ_A and Γ_A.

(*a*) $\Delta(x, y) \subseteq \Gamma(x, y)$. (Proof: Since $\Delta(x, y) = \{x\} \cup y^{A(x)}$, $A(x)$ operates transitively on $y^{A(x)}$. Since $A(x)$ is Abelian, $A(x)_y$ leaves all points of $y^{A(x)}$ fixed, i.e. $\Delta(x, y) \subseteq F_\Omega(A(x)_y) = \Gamma(x, y)$.) For $z \in \Gamma(x, y)$ $(z \neq x)$ we have $\Gamma(x, y) = \Gamma(x, z) \supseteq \Delta(x, z)$ and $\Delta(x, y) \cap \Delta(x, z) = \{x\}$ or $\Delta(x, y) = \Delta(x, z)$. Therefore, $|\Gamma(x, y)| = 1 + h(|\Delta(x, y)| - 1)$ for some integer h. Since $A(x)$ operates transitively on $\Delta(x, y) - \{x\}$, we have $|\Delta(x, y)| = 1 + p^n$, where we have put $|A(x) : A(x)_y| = p^n$. Therefore $|\Gamma(x, y)| = 1 + hp^n$.

(*b*) $\Gamma(x, y)^{A(x)} = \Gamma(x, y)$. (Proof: $\Delta(x, y)^\sigma = \Delta(x, y)$ for $\sigma \in A(x)$. Hence $\Gamma(x, y) \cap \Gamma(x, y)^\sigma \supseteq \Delta(x, y)$ by (*a*) and $\Gamma(x, y) = \Gamma(x, y)^\sigma$ since $\lambda_\Gamma = 1$.) Therefore, $\Gamma(x, y)^{A(z)} = \Gamma(x, y)$ for $z \in \Gamma(x, y)$, i.e. $A(x)$ leaves all blocks of $\mathbf{D}(\Gamma)$ that contain x fixed.

(*c*) $A(x) \cap A(y) = \{1\}$ for $x, y \in \Omega$ $(x \neq y)$. (Proof: Pick $\sigma \in A(x) \cap A(y)$, then $z^\sigma = z$ for $z \in \Gamma(x, y)$ since $\Gamma(x, y) = F_\Omega(A(x)_y)$. For $z \notin \Gamma(x, y)$ we have $z = \Gamma(z, y) \cap \Gamma(x, z)$, hence $z^\sigma = \Gamma(z, y)^\sigma \cap \Gamma(x, z)^\sigma = \Gamma(z, y) \cap \Gamma(x, z) = z$. Therefore $\sigma = 1$.)

(*d*) If $B^\sigma = B$ for $B \in \mathfrak{B}_2$ and $\sigma \in A(x)$, then $x \in B$ or σ leaves all points of B fixed. (Proof: Suppose there is a $y \in B$ with $y^\sigma \neq y$. Since $y, y^\sigma \in \Gamma(x, y) = \Gamma(x, y)^\sigma$, we have $|\Gamma(x, y) \cap B| \geq 2$, hence $\Gamma(x, y) = B$ since $\lambda_\Gamma = 1$.)

(*e*) If $\sigma \in A(x)$ leaves two points y_1 and y_2 of $B \in \mathfrak{B}_2$ fixed, then σ leaves all points of B fixed. (Proof: Since $B^\sigma = B$, σ leaves all points of B fixed or $x \in B$ by (*d*). If $x \in B$, then $B = \Gamma(x, z)$ where $z = y_1$ or y_2 and so we have $\sigma \in A(x)_z$ and $B = F_\Omega(A(x)_z)$, from which the assertion follows.) In particular, if $H \leq A(x)$ is such that $|F_\Omega(H)| \geq 2$, then $\Gamma(a, b) \subseteq F_\Omega(H)$ for $a, b \in F_\Omega(H)$ $(a \neq b)$.

Now, let Y be a subset of Ω, satisfying:

$$\text{if } a, b \in Y, \text{ then } \Gamma(a, b) \subset Y$$

and put $\mathfrak{B} = \{X \subset Y \mid X = \Gamma(a, b) \text{ for some } a, b \in Y, a \neq b\}$. The geometric structure $(Y, \mathfrak{B})$ (sometimes denoted simply by Y, the structure being understood) is called a subspace of $\mathbf{D}(\Gamma)$ and denoted by $\mathbf{D}(\Gamma)_{|Y}$. If $|\Gamma(a, b)| \lneq |Y|$, then Y is called a non-trivial subspace. In this case, $\mathbf{D}(\Gamma)_{|Y}$ is a non-trivial design with $\lambda = 1$, as the reader may prove.

(f) Let Y be a subspace of $\mathbf{D}(\Gamma)$. Then,

(i) $Y^{A(x)} = Y$ for $x \in Y$,

(ii) $\langle A(x) \mid x \in Y \rangle$ operates transitively on Y.

(Proof: (i) follows from $Y = \bigcup_{y \in Y - \{x\}} \Gamma(x, y)$ and (b). To prove (ii), pick $a, b \in Y$ $(a \neq b)$. Since $\Gamma(a, b)^{A(x)} = \Gamma(a, b)$ for $x \in \Gamma(a, b) (\subseteq Y)$, $B = \Gamma(a, b)$ is an invariant domain of $\langle A(x) \mid x \in B \rangle$ and it suffices to prove that $\langle A(x) \mid x \in B \rangle$ operates transitively on B. Since $A(x)$ operates non-trivially on B for $x \in B$, $\langle A(x) \mid x \in B \rangle$ operates non-trivially on B. Also, $G_{\langle B \rangle}$ operates 2-transitively on B, because (G, Ω) is 2-transitive and $\lambda_\Gamma = 1$. Since $\langle A(x) \mid x \in B \rangle \trianglelefteq G_{\langle B \rangle}$, $\langle A(x) \mid x \in B \rangle$ operates transitively on B by theorem 3.3.6).

(g) Let Y be a non-trivial subspace of $\mathbf{D}(\Gamma)$ and put $H = G_{\langle Y \rangle}$. For $\sigma \in H$ and $K \leq H$ their restrictions to Y are denoted by $\bar{\sigma} = \sigma_{|Y}$ and $\bar{K} = K_{|Y}$ respectively. We have for $x, y \in Y$ $(x \neq y)$:

(i) $\mathscr{C}_{\bar{H}}(\bar{\sigma}) \leq \bar{H}_x$ for $\bar{\sigma} \in \overline{A(x)}$, $(\bar{\sigma} \neq 1)$.

(ii) $|\overline{A(x)} : \overline{A(x)}_y| \leq |\overline{A(x)}_y|$.

(Proof: (i) For $\bar{\tau} \in \mathscr{C}_{\bar{H}}(\bar{\sigma})$ we have $\bar{\sigma} = \bar{\sigma}^\tau \in \overline{A(x)}^\tau = \overline{A(x)^\tau} = \overline{A(x^\tau)}$. Therefore $\bar{\sigma} \in \overline{A(x)} \cap \overline{A(x^\tau)}$. If $x^\tau \neq x$, then $A(x) \cap A(x^\tau) = 1$ by (c), contrary to $\bar{\sigma} \neq 1$, hence $\tau \in H_x$.

(ii) $\overline{A(y)}_x \neq 1$ since Y is non-trivial. Let $\bar{\sigma}$ be a non-identity element of $\overline{A(y)}_x$. Define $f : \overline{A(x)} \to \bar{H}$ by $f(\bar{\tau}) = [\bar{\tau}, \bar{\sigma}]$ for $\bar{\tau} \in A(x)$. Since

$$\mathscr{N}_{A(x)}(A(x)_y A(y)_x) = \mathscr{N}_{A(x)}(A(x)_y) = A(x)$$

by lemma 6.1.7, hence

$$[A(x), A(y)_x] \leq [A(x), A(x)_y A(y)_x] \leq A(x)_y A(y)_x \leq G_{x,y}.$$

On the other hand we have $[A(x), A(y)_x] \leq A(x)$ since $A(x) \trianglelefteq G_x$.

Therefore

$$[A(x), A(y)_x] \le A(x)_y$$

and f is a map from $\overline{A(x)}$ into $\overline{A(x)}_y$. Therefore if $\tau_1, \tau_2 \in A(x)$, then

$$f(\bar{\tau}_1\bar{\tau}_2) = [\bar{\tau}_1\bar{\tau}_2, \bar{\sigma}] = \bar{\tau}_2^{-1}[\bar{\tau}_1, \bar{\sigma}]\bar{\tau}_2[\bar{\tau}_2, \bar{\sigma}]$$
$$= [\bar{\tau}_1, \bar{\sigma}][\bar{\tau}_2, \bar{\sigma}] = f(\bar{\tau}_1)f(\bar{\tau}_2)$$

(the third equality follows from the commutativity of $\overline{A(x)}$). So f is a homomorphism and $|\overline{A(x)} : \mathrm{Ker}\, f| \le |\overline{A(x)}_y|$. For $\bar{\tau} \in \mathrm{Ker}\, f$ we have $[\bar{\tau}, \bar{\sigma}] = 1$, hence

$$\bar{\tau} \in \overline{A(x)} \cap \mathscr{C}_{\bar{H}}(\bar{\sigma}) \le \overline{A(x)} \cap \bar{H}_y = \overline{A(x)}_y,$$

i.e. $\mathrm{Ker}\, f \le \overline{A(x)}_y$, hence $|\overline{A(x)} : \overline{A(x)}_y| \le |\overline{A(x)}_y|$.

(*h*) $|A(x)_y : A(x)_{y,z}| = p^n$ for $z \in \Omega - \Gamma(x, y)$. (Proof: $Y = F_\Omega(A(x)_{y,z})$ is a subspace of $\mathbf{D}(\Gamma)$ and non-trivial, because $\Gamma(x, y) \subsetneq Y$. Considering restrictions to Y, we have $\overline{A(x)} = A(x)/A(x)_{y,z}$ and $\overline{A(x)}_y = A(x)_y/A(x)_{y,z}$, hence $|\overline{A(x)} : \overline{A(x)}_y| = |A(x) : A(x)_y| = p^n$. Therefore, $p^n \le |\overline{A(x)}_y|$ by (*g*). On the other hand:

$$|\overline{A(x)}_y| = |A(x)_y/A(x)_{y,z}| = |z^{A(x)_y}| \le |z^{A(x)}| = p^n.)$$

So, we get $|\overline{A(x)}| = p^{2n}$ and $|\overline{A(x)}_y| = p^n$.

(*i*) Putting $Y = F_\Omega(A(x)_{y,z})$ for $z \in \Omega - \Gamma(x, y)$ we will show that $\mathbf{D}(\Gamma)_{|Y}$ is a Desarguean projective plane.

We first compute the number of blocks of $\mathbf{D}(\Gamma)_{|Y}$ containing $x \in Y$. We denote this number by r. For $u, v \in Y - \{x\}$ we have

$$\overline{A(x)}_u = \overline{A(x)}_v \quad \text{or} \quad \overline{A(x)}_u \cap \overline{A(x)}_v = 1$$

according to whether $A(x)_u = A(x)_v$ or $A(x)_u \neq A(x)_v$ respectively. (For, if $v \in \Gamma(x, u)$, then $A(x)_u = A(x)_v$ since $\Gamma(x, u) = \Gamma(x, v)$ and if $v \notin \Gamma(x, u)$, then $|A(x)_u : A(x)_{u,v}| = p^n$ by (*h*), hence $|A(x)_{u,v}| = |A(x)_{y,z}|$ From $A(x)_{y,z} \le A(x)_{u,v}$, we conclude $A(x)_{u,v} = A(x)_{y,z}$, hence $\overline{A(x)}_u \cap \overline{A(x)}_v = 1$.) Therefore, noticing that

$$\Gamma(x, u) = \Gamma(x, v) \Leftrightarrow A(x)_u = A(x)_v,$$

we conclude

$$r \le \frac{|\overline{A(x)}| - 1}{|\overline{A(x)}_y| - 1} = \frac{p^{2n} - 1}{p^n - 1} = p^n + 1.$$

On the other hand we have $|\Gamma(x, y)| = hp^n + 1 \leq r$ by Fisher's inequality (theorem 1.5.2). Therefore, $h = 1$ and $|\Gamma(x, y)| = 1 + p^n = r$. Hence $\mathbf{D}(\Gamma)_{|Y}$ is a non-trivial, symmetric design with $\lambda = 1$, i.e. a projective plane. Let $\Gamma(a, b)$ be a block of $\mathbf{D}(\Gamma)_{|Y}$, then $\langle A(x) | x \in \Gamma(a, b) \rangle$ operates transitively on $\Gamma(a, b)$ by (f). Since $\overline{A(u)}_v \neq 1$ for $u, v \in \Gamma(a, b)(u \neq v)$, we can take an element $\bar{\sigma} \neq 1$ from $A(u)_v$. Then $\bar{\sigma}$ is a $(u, \Gamma(a, b))$-elation by (b). Since $\langle \overline{A(x)} | x \in \Gamma(a, b) \rangle$ operates transitively on $\Gamma(a, b)$, $\mathbf{D}(\Gamma)_{|Y}$ is Desarguean by theorems 5.1.22 and 5.1.31.

(j) $\mathbf{D}(\Gamma) \simeq \mathbf{P}_1(m, q)$ for some $m \geq 2$ and $q = p^n$. (Proof: If $\Omega = F_\Omega(A(x)_{y,z})$ for $z \in \Omega - \Gamma(x, y)$, then $\mathbf{D}(\Gamma) \simeq \mathbf{P}_1(2, q)$ by (i) and theorem 5.1.31 (or 5.1.33). If $\Omega \supsetneqq F_\Omega(A(x)_{y,z})$, then $\mathbf{D}(\Gamma)$ is a projective geometry of dimension 3 or higher, hence $\mathbf{D}(\Gamma) \simeq \mathbf{P}_1(m, q)$ $(m \geq 3)$ by theorem 5.2.4.

(k) Since $G \leq \operatorname{Aut} \mathbf{P}_1(m, q) (= \mathbf{D}(\Gamma))$ we have $G \leq P\Gamma L(m+1, q)$ by the fundamental theorem of projective geometry (theorem 1.5.15). To prove that $PSL(m+1, q) \leq G$, it suffices to show that G contains all elations of the projective geometry $\mathbf{D}(\Gamma)$, since the elations generate $PSL(m+1, q)$ by theorem 4.6.1. According to (b), the elements of $A(x)$ are elations with centre $x \in \mathbf{P}(m, q)$ (theorem 4.6.1(4)). Since the order of the group determined by the collection of all elations with centre $x \in \mathbf{P}(m, q)$ equals q^m (theorem 4.6.1(3)), it suffices to show that $|A(x)| = q^m$ in order to prove that $PSL(m+1, q) \leq G$. We prove this by induction on m.

If $m = 2$, then $|A(x)| = q^2$ by (h) (noting that $A(x)_{y,z} = 1$). So we assume $m \geq 3$. Let Y be a hyperplane of $\mathbf{D}(\Gamma) = \mathbf{P}_1(m, q)$, then $H = G_{\langle Y \rangle}$ operates 2-transitively on Y. (For, since Y is a subspace in the sense of (e), H operates transitively on Y by (f). Therefore it suffices to show that $(H_x, Y - \{x\})$ is transitive for $x \in Y$. Pick $u, v \in Y - \{x\} (u \neq v)$. Since Y is a subspace, we have $\Gamma(u, v) \subseteq Y$. If $x \in \Gamma(u, v)$, then $A(x)$ operates transitively on $\Gamma(u, v) - \{x\} = \Gamma(x, u) - \{x\} = \Delta(x, u) - \{x\}$ (note that $h = 1$, i.e. $\Gamma(x, y) = \Delta(x, y)$ as we saw in the proof of (i)). Since $x \in Y$, we have $Y^{A(x)} = Y$ by (f). This proves the assertion in the case that $x \in \Gamma(u, v)$. Let us assume $x \notin \Gamma(u, v)$. $\langle A(t)_x | t \in \Gamma(u, v) \rangle$ operates on $\Gamma(u, v)$ by (f). Since $\Gamma(u, v) \neq \Gamma(t, x)$ for $t \in \Gamma(u, v)$, $A(t)_x$ operates non-trivially on $\Gamma(u, v)$, hence $A(t)_x$ operates semi-regularly on $\Gamma(u, v) - \{t\}$ by (e). Now, if $\Gamma(u, v)$ is divided into two invariant domains T_1 and T_2 under $\langle A(t)_x | t \in \Gamma(u, v) \rangle$, then

$$|A(t_1)_x| \,|\, |T_1| - 1 \quad \text{and} \quad |A(t_2)_x| \,|\, |T_1|$$

for $t_1 \in T_1$ and $t_2 \in T_2$, hence $|T_1| \equiv 1 \pmod p$ and $|T_1| \equiv 0 \pmod p$, contradiction. Therefore $(H_x \geq) \langle A(t)_x | t \in \Gamma(u, v) \rangle$ operates transitively on $\Gamma(u, v)$, so there is an element of H_x mapping u into v.)

Now, for $x \in Y$, $A(x)_{|Y} (\neq 1)$ is an Abelian, normal subgroup of $H_{x|Y}$ and does not operate semi-regularly on $Y - \{x\}$ since $m \geq 3$. Therefore, $H_{|Y} \geq PSL(m, q)$ and $|A(x) : A(x)_Y| = q^{m-1}$ by the inductive assumption. Since G operates transitively on the point set of $\mathbf{P}(m, q)$, the number $|A(x)|$ is independent from x, therefore the number $|A(x)_Y|$ is independent from the choice of the hyperplane Y and the point $x \in Y$.

Since $\forall \sigma \in A(x)\,(\sigma \neq 1)$ is an elation with centre x, σ leaves all the points of some hyperplane through x fixed. Therefore

$$A(x) = \bigcup_{Y \text{ hyperplane through } x} A(x)_Y.$$

Since $A(x)_Y \cap A(x)_{Y'} = 1$ if $Y' \neq Y$ and since the number of hyperplanes through x of $\mathbf{P}(m, q)$ equals $(q^m - 1)/(q - 1)$, we have

$$|A(x)| - 1 = (|A(x)_Y| - 1)\frac{q^m - 1}{q - 1}.$$

Putting $|A(x)_Y| = p^\beta$, we have $|A(x)| = p^\beta q^{m-1}$, hence

$$p^\beta q^{m-1} - 1 = (p^\beta - 1)\frac{q^m - 1}{q - 1},$$

namely $q + q^{m-1}p^\beta = p^\beta + q^m$. Therefore $p^\beta = q$, hence $|A(x)| = q^{m-1}p^\beta = q^m$. ■

Using this theorem we can prove the following theorem:

Theorem 6.2.2 *Let (G, Ω) be a 2-transitive group of degree n. Let $H \leq G$ be such that (H, Ω) is not transitive, $|G : H| = n$ and $F_\Omega(H) = \varnothing$. If H contains a solvable normal subgroup $N \neq 1$ such that $F_\Omega(N) \neq \varnothing$, then $PSL(m+1, q) \leq G \leq P\Gamma L(m+1, q)$ for some natural number $m \geq 2$ and some prime power q.*

Proof (H, Ω) has exactly two orbits by theorem 6.1.1. Let Λ denote one of these orbits. Putting $\mathfrak{B} = \{X \subseteq \Omega | X = \Lambda^\sigma, \sigma \in G\}$, $\mathbf{D} = (\Omega, \mathfrak{B})$ is a symmetric design such that $G \leq \mathrm{Aut}\,\mathbf{D}$. According to theorem 1.5.9, $(G, \mathfrak{B})$ is 2-transitive and H is the subgroup leaving the element Λ of $\mathfrak{B}$ fixed. Since $N \trianglelefteq H$ and $F_\Omega(N) \neq \emptyset$, N leaves all elements of some orbit of (H, Ω) fixed, and hence an arbitrary $\sigma \in N$ has at least two fixed points in Ω. Therefore an arbitrary $\sigma \in N$ also has, regarded as a permutation of $\mathfrak{B}$, at least two fixed points by theorem 1.5.10. Since N is solvable, N contains an Abelian characteristic subgroup $N_0 \neq 1$ and N_0 is a normal subgroup of H and does not operate semi-regularly on $\mathfrak{B} - \{\Lambda\}$. Since $(G, \mathfrak{B})$ is a faithful permutation group by theorem 1.5.10, we conclude $PSL(m+1, q) \leq G \leq P\Gamma L(m+1, q)$ by theorem 6.2.1. ∎

This theorem can be extended as follows:

Theorem 6.2.3 (Ito) *Let (G, Ω) be a 2-transitive group of degree n. Let $H \leq G$ be such that (H, Ω) is not transitive, $|G:H| = n$ and $F_\Omega(H) = \emptyset$. If (H, Ω) has an orbit Λ such that (H, Λ) is not faithful, then $PSL(m+1, q) \leq G \leq P\Gamma L(m+1, q)$ for some natural number $m \geq 2$ and some prime power q.*

Proof Just as in the proof of theorem 6.2.2 we put: $\mathfrak{B} = \{X \subseteq \Omega | X = \Lambda^\sigma\}$ and $\mathbf{D} = (\Omega, \mathfrak{B})$. Then $\mathbf{D}$ is a symmetric design such that $G \leq \mathrm{Aut}\,\mathbf{D}$ and H coincides with $G_{\langle\Lambda\rangle}$ which is the stabilizer of one point Λ of the permutation group $(G, \mathfrak{B})$. Putting $K = H_\Lambda$, we have $1 \neq K \trianglelefteq H$ and $F_\Omega(K) \neq \emptyset$ by assumption, hence it suffices to prove that K contains a nilpotent characteristic subgroup that is not equal to 1 (by theorem 6.2.2).

We first prove that $\sigma \in K$ $(\sigma \neq 1)$ has at most one fixed point in $\Omega - \Lambda$. (For, let $a \in \Omega - \Lambda$ be such that $a^\sigma = a$. Let Λ' be a block of $\mathbf{D}$ containing a. Since Λ' is the only block containing both a and $\Lambda \cap \Lambda'$, we have $\Lambda'^\sigma = \Lambda'$, i.e. σ leaves all blocks containing a fixed. Therefore σ leaves all lines through a fixed, hence σ is a central automorphism of $\mathbf{D}$ with centre a. The assertion follows now from theorem 1.5.12.) So, if K does not operate semi-regularly on $\Omega - \Lambda$, then K operates faithfully as a Frobenius group on some orbit, and the Frobenius kernel ($\neq 1$) of that Frobenius group is a nilpotent characteristic subgroup of K by theorem 2.11.5. So we may assume that K operates semi-regularly on $\Omega - \Lambda$.

Let us put $|\Lambda| = h$ and $|K| = l$. Since Ω is divided into two orbits under H by theorem 6.1.1, $\mathfrak{B}$ is also divided into two orbits under H by theorem 1.5.8, hence $(H, \mathfrak{B} - \{\Lambda\})$ is transitive. $(K, \mathfrak{B} - \{\Lambda\})$ is $\frac{1}{2}$-transitive since $K \trianglelefteq H$, therefore all orbits of $(K, \mathfrak{B} - \{\Lambda\})$ have the same length $l_1 \neq 1$ and $l_1 | l$. Put $l = l_1 l_2$. Since K leaves all points of Λ fixed, for an element $a \in \Lambda$, $\{X \in \mathfrak{B} | X \ni a, X \neq \Lambda\}$ is an invariant domain of $(K, \mathfrak{B} - \{\Lambda\})$ and hence $l_1 | h - 1$. (Notice that $|\{X \in \mathfrak{B} | X \ni a, X \neq \Lambda\}| = h - 1$ because the number of points of a block of **D** equals the number of blocks containing a given point, since **D** is symmetric.) Since the number of orbits of (K, Ω) and the number of orbits of $(K, \mathfrak{B})$ are the same by theorem 1.5.8 we have:

$$h + \frac{n-h}{l} = 1 + \frac{n-1}{l_1}. \tag{6.2.1}$$

Suppose $l = l_1$, then $l = l_1 = 1$, contradiction, hence $l \neq l_1$. Let p be a prime divisor of l, and write: $l = p^r l'$, $l_1 = p^{r_1} l'_1$, $l'_2 = p^{r_2} l'_2$, where $(l', p) = 1$, $l' = l'_1 l'_2$ and $r = r_1 + r_2$. Let $K(p)$ denote a Sylow p-subgroup of K, let Δ be an orbit of $(K, \mathfrak{B} - \{\Lambda\})$ and let s denote the number of orbits of $(K(p), \Delta)$. Since $K \trianglelefteq H$, the number s is independent from the choice of Δ. Since the number of orbits of $(K(p), \Omega)$ equals the number of those of $(K(p), \mathfrak{B})$ for the same reasons given above, we have

$$h + \frac{n-h}{p^r} = 1 + \frac{n-1}{l_1} s. \tag{6.2.2}$$

Eliminating l_1 from (6.2.1) and (6.2.2) gives:

$$(h-1)(s-1) = \frac{(n-h)(l'-s)}{l}. \tag{6.2.3}$$

Rewriting (6.2.1) we get

$$h + \frac{n-h}{l} = 1 + \frac{n-1}{l_1} = 1 + \frac{l_2(n-h)}{l} + \frac{h-1}{l_1},$$

hence

$$\frac{(n-h)(l_2-1)}{l} = \frac{(h-1)(l_1-1)}{l_1}. \tag{6.2.4}$$

From (6.2.3) and (6.2.4) we get:

$$l_1(l_2-1)(s-1) = (l_1-1)(l'-s) \tag{6.2.5}$$

hence $s \equiv 0 (\mathrm{mod}\ l'_1)$. Let $\Delta = \Delta_1 + \ldots + \Delta_s$ be the orbit decomposition of Δ under $K(p)$. Choosing $\Lambda_i \in \Delta_i$, we have: $|\Delta| = l_1 = |K : K_{\langle\Lambda_i\rangle}|$ and $|\Delta_i| = |K(p) : K(p)_{\langle\Lambda_i\rangle}|$. Since

$$\begin{aligned}|K : K_{\langle\Lambda_i\rangle}||K_{\langle\Lambda_i\rangle} : K(p)_{\langle\Lambda_i\rangle}| &= |K : K(p)_{\langle\Lambda_i\rangle}| \\ &= |K : K(p)||K(p) : K(p)_{\langle\Lambda_i\rangle}|,\end{aligned}$$

we have $p^{r_1} \mid |\Delta_i|$. Putting $|\Delta_i| = p^{r_1} n_i$, we have

$$l_1 = |\Delta| = \sum_{i=1}^{s} |\Delta_i| = p^{r_1} \sum_{i=1}^{s} n_i \quad \forall n_i \geq 1.$$

Therefore, $l'_1 = \sum_{i=1}^{s} n_i \geq s$, hence $n_1 = \ldots = n_s = 1$ and $s = l'_1$. Combining this with (6.2.5), we get

$$p^{r_1}(l_2-1)(l'_1-1) = (l_1-1)(l'_2-1). \tag{6.2.6}$$

Since $l \neq l_1$, we have $l_2 - 1 \neq 0$. Suppose $l'_2 - 1 \neq 0$, then $l'_1 \neq 1$ since $l_1 > 1$, hence we get from (6.2.6):

$$p^{r_1}\frac{p^{r_2}l'_2-1}{l'_2-1} = \frac{p^{r_1}l'_1-1}{l'_1-1}. \tag{6.2.7}$$

From:

$$p^{r_2} \leq \frac{p^{r_2}l'_2-1}{l'_2-1} \quad \text{and} \quad \frac{p^{r_1}l'_1-1}{l'_1-1} \leq 2p^{r_1}-1$$

and (6.2.7) we conclude $p^{r_1}p^{r_2} \leq 2p^{r_1} - 1$. Therefore, $r_2 = 0$, hence $r_1 = 0$ by (6.2.7) and so $r = r_1 + r_2 = 0$. Contradiction, therefore $l'_2 - 1 = 0$. Since $l_2 - 1 \neq 0$ we conclude from (6.2.6) $l'_1 = 1$, hence $l' = l'_1 l'_2 = 1$, i.e. $K = K(p)$ and K itself is nilpotent. ■

Epilogue

I would like to supplement the main text by including a list of the books and articles that I used while writing this book and the papers where the main results of this book first appeared.

First, references [1]–[9] served as general references.

[1] E. Artin: *Geometric Algebra* (Interscience, 1957).
[2] P. Dembowski: *Finite Geometry* (Springer, 1968).
[3] M. Hall: *The Theory of Groups* (Macmillan, (1957).
[4] B. Huppert: *Endliche Gruppen* (Springer, 1967).
[5] D. Passman: *Permutation Groups* (Benjamin, 1968).
[6] H. Wielandt: *Finite Permutation Groups* (Academic Press, 1964).
[7] H. Wielandt: *Permutation Groups through Invariant Relations and Invariant Functions* (Lecture Notes, Ohio State University, 1969).
[8] H. Wielandt: *Verlagerung und Permutationsgruppen* (Vorlesung an der Universität Tübingen, 1970/71).
[9] H. Nagao: *Groups and Designs* (Iwanami Shoten, 1974, in Japanese).

The following three books are quite different in content from this one, but I mention them as typical accounts concerned with the modern development of group theory:

[10] D. Gorenstein: *Finite Groups* (Harper & Row, 1968).
[11] M. Suzuki: *Group Theory*, Vols. I and II (Iwanami Shoten, 1979, in Japanese).
[12] N. Ito: *Theory of Finite Groups* (Kyōritsu Shuppan, 1970, in Japanese)

[10] and [11], are indispensable for an understanding of modern research in finite simple groups, whilst [12] is about the classification

problem of Zassenhaus groups, which could be called the starting point of modern development of finite group theory.

The definition of determinants by J. Dieudonne in §1.4 follows the method of E. Artin [1], pp. 151–8.

I used [5] as a general reference for the material in chapters 2 and 3. Theorem 2.11.5 used to be called 'Frobenius' conjecture' and was unproved for a long time. Its proof appeared in

[13] J. Thompson: Finite groups with fixed-point-free automorphisms of prime order, *Proc. Nat. Acad. Sci.*, **45** (1959) 578–81,

and this result has deeply influenced the subsequent development of the theory of finite groups. It also plays an essential role, in this book.

Theorem 3.6.1 of §3.6 is a classic and famous result of E. Galois. With regard to this theorem, I want to make the following additional remarks:

Let $f(x) = x^n + a_1 x^{n-1} + \ldots + a_n$ be a polynomial of degree n over $\mathbb{Q}$, and let $\theta_1, \ldots, \theta_n$ the roots of the equation $f(x) = 0$ in $\mathbb{C}$. The smallest subfield K (of $\mathbf{C}$) containing $\theta_1, \ldots, \theta_n$ is called the minimal splitting field of $f(x)$. If σ is an isomorphism form K into $\mathbf{C}$, then θ^σ is a root of $f(x) = 0$ for all roots θ, and σ becomes an automorphism of K inducing a permutation of $\theta_1, \ldots, \theta_n$. The automorphism group of K (i.e. the group consisting of all automorphisms of K) is called the Galois group of the polynomial $f(x)$ and indicated by $G_{f(x)}$. $G_{f(x)}$ can be regarded as a permutation group (of degree n) on $\{\theta_1, \ldots, \theta_n\}$. The roots of an equation of the first or second degree can be expressed in the coefficients of the equation using the four basic operations and square roots. That is:

$$\text{the root of } x + a = 0 \text{ is given by } x = -a$$

and the roots of

$$x^2 + ax + b = 0$$

are given by

$$x = \frac{-a \pm \sqrt{(a^2 - 4b)}}{2}.$$

If the roots of the equation $f(x)=0$ of degree n can be expressed in the coefficients of $f(x)$ using the four basic operations and roots, then the equation $f(x)=0$ is called algebraically solvable. It has been known for a long time that equations of degree 3 and 4 are algebraically solvable, but this is not necessarily so for equations of degree 5 (and higher). This famous result was discovered by N.H. Abel. Galois found necessary and sufficient conditions for an equation $f(x)=0$ of degree n to be algebraically solvable. His result is:

Theorem *The equation* $f(x)=0$ *is algebraically solvable* $\Leftrightarrow G_{f(x)}$ *is a solvable group.*

The symmetric groups S_n are solvable for $n \leq 4$, but not solvable for $n \geq 5$ (corollary 3.5.11). Galois determined the structure of $G_{f(x)}$ for the case of a polynomial $f(x)$ of prime degree such that the equation $f(x)=0$ is algebraically solvable. His result is contained in theorem 3.6.1. These results, dating from the early 19th century are called classics of modern mathematics.

If the degree of the polynomial $f(x)$ is not a prime, we do not have a simple result such as theorem 3.6.1 describing the permutation group. In general, it is a fundamental problem to determine all permutation groups (G, Ω) of a given degree n, but little is known about this. For example, let us restrict ourselves to the case of permutation groups of prime degree. We know the structure of these groups if they are solvable by theorem 3.6.1. Then what kind of groups are non-solvable permutation groups of prime degree? The following is a list of all known such groups:

(1) S_p, A_p $(p \geq 5)$.

(2) The 11th degree permutation representation of $PSL(2, 11)$ (theorem 4.7.13).

(3) Two permutation groups M_{11} and M_{23} of degree 11 and 23 respectively, called Mathieu groups.

(4) The projective transformation groups G regarded as permutation groups on a projective space such that $(q^n-1)/(q-1)$ is a prime, where q is power of a prime and $P\Gamma L(n, q) \geq G \geq PSL(n, q)$.

It has been conjectured that all the possibilities are contained in this list. But almost nothing is known and this is a very interesting problem. For some special cases, N. Ito, P. Neumann and others have published some results:

[14] N. Ito: Transitive permutation groups of degree $p = 2q + 1$, p and q being prime numbers, *Bull. A.M.S.*, **69**(1963), 165–92.

[15] P. Neumann: Transitive permutation groups of prime degree, *Bull. London Math. Soc.*, **4** (1972), 337–9.

For a detailed account of this problem and for references, see

[16] N. Ito: Sosūji no chikangun ni tsuite, Sūgaku, Volume 15, (1963–4), pp. 129–140 (in Japanese).

The primitive permutation groups that we study in §3.7 are a topic of current research. Most of the basic results are collected and given in [6], [7] and [8]. Theorems 3.7.9 and 3.7.16 can be found in

[17] C.C. Sims: Graphs and finite permutation group, *Math. Z.*, **95**(1967), 76–86.

In this article it is shown that the case $m = 6$ in theorem 3.7.16 does not occur. Furthermore in

[18] W.J. Wong: Determination of a class of primitive permutation groups, *Math. Z.*, **99**(1967), 235–46,

all permutation groups satisfying the conditions of theorem 3.7.16 are determined. All results from theorem 3.7.25 to 3.7.28 are contained in

[19] P.J. Cameron: Permutation groups with multiply transitive suborbits, *Proc. London Math. Soc.*, (3) **25** (1972), 427–40,

[20] Knapp: On the point stabilizer in a primitive permutation group, *Math. Z.*, **133** (1973), 137–68,

with much more fruitful results. We did not touch upon the recently discovered so-called sporadic simple groups and their connection with primitive permutation groups. In this area much research has been done, for which we refer the reader to

[21] T. Tsuzuku, T. Kondō, H. Kimura and E. Bannai: *Yūgen-*

gunron no genjō, Kagaku (Iwanami Shoten, 1974) pp. 522–35 (in Japanese).

The main objective of chapter 5 is to give a proof of theorem 5.1.34. This theorem is the most celebrated result in this field and is usually called the theorem of Ostrom–Wagner. It was first proved in the form of theorem 5.1.33 in

[22] T.G. Ostrom and A. Wagner: On projective and affine planes with transitive collineation groups, *Math. Z.*, **71** (1959), 186–99.

Later it was extended to the form of theorem 5.1.34 in

[23] A. Wagner: On finite affine line transitive plane, *Math. Z.*, **89** (1965), 1–11.

Theorem 5.2.4 is a famous result that has been known for a long time. Theorem 5.3.3 appeared in

[24] P. Dembowski and A. Wagner: Some characterization of finite projective spaces, *Arch. Math.*, **11** (1960), 465–9.

The many results proved in chapters 1–5 are preparation for chapter 6, the most important chapter in the book.

Many interesting groups can be represented as automorphism groups of geometric structures. It is a basic problem to determine which properties of a group reflect the characteristics of the geometric structure. There are several interesting results in this direction, of which theorems 6.2.1 and 6.2.3 are beautiful examples. They appear as the main results of

[25] O'Nan: A characterization of $L_n(q)$ as a permutation group, *Math. Z.*, **127** (1972), 301–14;

and

[26] N. Ito: On a class of doubly, but not triply transitive groups, *Arch. Math.*, **13** (1967), 564–70,

respectively. In this book we prove theorem 6.2.3 by first reducing it to theorem 6.2.1. These theorems are characterizations of $PSL(n, q)$ for $n \geq 3$. The corresponding problem for $n = 2$ has a long history as the classification problem of Zassenhaus groups. This problem was completely solved in

[27] W. Feit: On a class of doubly transitive permutation groups, *Illinois J. Math.*, **4** (1960), 170–86,

[28] M. Suzuki: On a class of doubly transitive groups, *Ann. of Math.*, **75** (1962), 105–45,

[29] N. Ito: On a class of doubly transitive permutation groups, *Illinois J. Math.*, **6** (1962), 341–52.

The discovery of the Suzuki simple groups during work on this problem marked a turning point in the history of finite groups – especially in the history of the classification problem of simple groups. For details concerning the classification problem of Zassenhaus groups we refer the reader to [12]. In connection with theorem 6.2.1 we mention

[30] O'Nan: Normal structure of the one-point stabilizer of a doubly transitive permutation group, Part I *Trans. Amer. Math. Soc.*, **214** (1975), 1–42; Part II, *Trans. Amer. Math. Soc.*, **214** (1975), 43–74.

for a detailed study of the structure of doubly transitive groups.

Even if we restrict ourselves to the simple groups that are known at present, the number that can be represented as doubly transitive groups is extremely limited. Therefore, it would be worthwhile to have results corresponding to theorem 6.2.1 (or theorem 6.2.3) for permutation groups that are not doubly transitive. Let us consider the projective symplectic group $PS_p(2n, K)$ as the simplest example of a simple group that cannot be represented as a doubly transitive permutation group. The group $PS_p(2n, K)$ is defined as follows. Let K be a commutative field, let A be a non-singular square matrix of degree $2n$ with entries in K satisfying ${}^tA = -A$, and put

$$S_p(2\mathrm{n}, K) = \{X \in GL(2n, K) \mid {}^tX A X = A\}.$$

It is easy to check that $S_p(2n, K)$ is a subgroup of $GL(2n, K)$ called the symplectic group. The quotient group of $S_p(2n, K)$ with regard to its centre (in this case $\pm E$) is known to be a simple group (with a few exceptions), called the projective symplectic group and denoted by $PS_p(2n, K)$. The symplectic group $S_p(2n, K)$ (as a subgroup of $GL(2n, K)$) operates on the projective space $P(2n-1, K)$ and the permutation groups $(PS_p(2n, K), P(2n-1, K))$ is a permutation group of rank 3 that is not doubly transitive. In the famous article

[31] D.G. Higman: Finite permutation groups of rank 3, *Math. Z.*, **86** (1964), 145–56,

which could be regarded as the starting point for research on permutation groups of rank 3, Higman obtained the following result for the problem to characterize $PS_p(2n, K)$ as a permutation group of rank 3:

Theorem *Let (G, Ω) be a permutation group of rank 3 and let the lengths of the orbits of G_a $(a \in \Omega)$ be 1, $q(a+1)$ and q^3, where q is an integer greater than or equal to 2. Let Δ denote the orbit of (G_a, Ω) of length $q(q+1)$. If one of the following conditions*

(i) $|G_{\{a\} \cup \Delta}| \geq q$,
(ii) $|G_{\{a\} \cup \Delta}| \neq 1$ *and* $q = p^r$ *where* p *is a prime and* r *is a power of* 2,

is satisfied, then

$$\text{Aut } PS_p(4, q) \geq G \geq PS_p(4, q).$$

For attempts to extend this theorem to higher dimensions and to other groups of rank 3 – unitary groups, orthogonal groups – the reader is referred to

[32] T. Tsuzuku: On a problem of D.G. Higman, *Hokkaido Math. J.*, **4** (1975), 300–2,
[33] A. Yanushka: A characterization of the symplectic groups $PS_p(2m, q)$ as rank 3 permutation groups, *Pacific J. Math.*, **59** (1975), 611–22,
[34] W.M. Kantor: Rank 3 characterizations of classical geometries, *J. of Algebra*, **36** (1975), 309–13.

For an understanding of Kantor's work, the following article is essential.

[35] F. Buekenhout-E. Shult: On the fundations of polar geometry, *Geometriae Dedicate*, **3** (1974), 155–70.

Supplement to Epilogue

About five years has passed since the epilogue was written. I would now like to add some more references and comments.

First I supply the following reference.

[36] Lecture Notes Prepared in Connection with the Summer Institute on Finite Group Theory, held at the University of California, Santa Cruz, California, June 25–July 20, 1979.

At the present time it seems that most mathematicians working on finite groups have begun to believe that all finite simple groups will be known to us in the near future. This greatly influences other topics in finite groups, and some people have started to check up which problems have been solved and which will still be left open when all finite simple groups have been found.

In the epilogue I listed all non-solvable transitive permutation groups of prime degree known then and stated that it had been conjectured that there would be no other non-solvable transitive permutation groups of prime degree, and that this problem was quite interesting although little was known about it. In the reference [36] supplied above, O'Nan discusses this problem among others in the following lecture:

[37] O'Nan: Some open problems in primitive permutation groups,

and remarked that all doubly transitive permutation groups not only non-solvable transitive permutation groups of prime degree are known immediately by the classification of all finite simple groups, using the following results.

[38] C. Curtis, W. Kantor, and G. Seitz,: The 2-transitive permutation representations of the finite Chevalley groups, *Trans. A.M.S.* **218** (1976), 1–59.

[39] C. Hering: Transitive linear groups and linear groups which contain irreducible subgroups of prime order, *Geometrica Dedicate*, **2** (1973), 425–59.

[40] C. Hering: Transitive linear groups and linear groups which contain irreducible subgroups of prime order II (preprint).

So this problem is less interesting now, though a purely permutational proof of this problem is still to be found. W. Feit also referred to this problem.

[41] W. Feit: Modular characters and the classification of finite simple groups (in [36]).

The study of finite groups related to finite geometries, which is the main subject of this book, and which is treated only for a few special topics and methods in this book, will still be alive as a subject to be studied after the completion of the classification of all finite simple groups. This subject is spreading along many different paths, not only into the field of classical finite geometries but also into combinatorics and graph theory. The reader will find several lectures about this subject in [36].

Index